Bonobo Cognition and Behaviour

Bonobo Cognition and Behaviour

Edited by

Brian Hare and Shinya Yamamoto

BRILL

Leiden • Boston
2015

Library of Congress Control Number: 2015956161

ISBN: 978 90 04 30416 1

E_ISBN: 978 90 04 30417 8

Koninklijke Brill NV incorporates the imprints Brill, Brill Hes & De Graaf, Brill Nijhoff, Brill Rodopi and Hotei Publishing.

Printed by Printforce, the Netherlands

Printed in The Netherlands

CONTENTS

The page numbers in the above Table of Contents and in the Index refer to the bracketed page numbers in this volume. The other page numbers are the page numbers in Behaviour 152/3–4. When citing a chapter from this book, refer to Behaviour 152 (2015) and the page numbers without brackets.

[When citing this chapter, refer to Behaviour 152 (2015) 247–258]

Editorial

Moving bonobos off the scientifically endangered list

Brian Hare [a,b,*] **and Shinya Yamamoto** [c,d]

[a] Department of Evolutionary Anthropology, Duke University, 104 Biological Sciences Building, Box 90383, Durham, NC 27708-9976, USA
[b] Center for Cognitive Neuroscience, Levine Science Research Center, Duke University, Box 90999, Durham, NC 27708, USA
[c] Graduate School of Intercultural Studies, Kobe University, 1-2-1 Tsurukabuto, Nada-ku, 657-8501 Kobe, Japan
[d] Wildlife Research Center, Kyoto University, Yoshida-honmachi, Sakyo-ku, Kyoto 606-8501, Japan
[*]Corresponding author's e-mail address: b.hare@duke.edu

Accepted 14 January 2015; published online 2 February 2015

Abstract

This Special Issue of *Behaviour* includes twelve novel empirical papers focusing on the behaviour and cognition of both captive and wild bonobos (*Pan paniscus*). As our species less known closest relative, the bonobo has gone from being little studied to increasingly popular as a species of focus over the past decade. We suggest that bonobos are ready to come off the scientific endangered list as a result. This Special Issue is exhibit A in showing that a renaissance in bonobo research is well underway. In this Editorial we review a number of traits in which bonobos and chimpanzees are more similar to humans than they are each other. We show how this means that bonobos provide an extremely powerful test of ideas about human uniqueness as well as being crucial to determining the evolutionary processes by which cognitive traits evolve in apes. This introduction places the twelve empirical contributions within the Special Issue in the larger evolutionary context to which they contribute. Overall this Special Issue demonstrates how anyone interested in understanding humans or chimpanzees must also know bonobos.

Editorial

Almost twenty years ago Frans de Waal and Frans Lanting married art and science in publishing the book *Bonobos*: *The forgotten ape* (de Waal & Lanting, 1997). The book tells the story of our closest relative that has been largely ignored by both the public and Western science alike. Compared to our other close relative the chimpanzee, bonobos were only recently recognized as a species, have been little studied, and are little known to the general public as a result. This can distort our view of ape and human evolution, but worse makes this endangered species particularly vulnerable to extinction since a much smaller group of scientists are involved in protecting them. The book remains a call to action for behavioural scientists in particular to give bonobos the equal attention they deserve.

Perhaps *The forgotten ape's* biggest contribution is to make it clear that the lack of attention is largely an accident of history and not that chimpanzees inherently have more to teach us. Nowhere is the scientific value of bonobos more obvious then when trying to answer questions about how ape and human cognition evolves. Bonobos and chimpanzees are each more similar to humans than they are to each other across a number of traits that also need to be explained relative to our own species' evolution (Table 1). This means that understanding how bonobos and chimpanzees diverged from one another can allow for inference about cognitive evolution in similar traits in our own species (Hare, 2007, 2009, 2011). These significant phenotypic differences are particularly exciting given how genetically similar the two species are. Comparisons between bonobos and chimpanzees raise the spectre of identifying the genetic basis and evolutionary origin of traits that otherwise would be too technically challenging to tackle given the relative gulf between human and chimpanzee (Prufer et al., 2012). Moreover, a careful comparison of traits likes those in Table 1 show the danger of only considering chimpanzees when determining what behavioural or cognitive traits in humans are unique. Focusing exclusively on human comparisons to chimpanzees would lead us to erroneously conclude that humans are unique among apes for non-conceptive sex, a reliance on mothers in adulthood, for showing adult play, sharing with strangers or having female alliances. For each of these traits bonobos are more similar to humans than to chimpanzees — meaning these traits are either shared between bonobos and humans through common descent or convergent evolution. Either outcome will be extremely important in understanding the evolution of these traits in humans.

With the publication of this Special Issue of *Behaviour* we are signalling that the bonobo is ready to be moved off the scientific endangered list. *Pan paniscus* has gone from forgotten to popular as a species for scientific study. With relative stability within bonobo habitat in the Democratic Republic of Congo over the past decade, field researchers are observing wild bonobos

Table 1.
Behaviours in bonobos and chimpanzees more similar to humans than each other.

	Bonobo	Chimpanzee	Human foragers	References
Extractive foraging	Only captivity	Frequent	Frequent	Hohmann & Fruth (2003a); Gruber et al. (2010); Furuichi et al., this issue; Hopkins et al., this issue
Non-conceptive sexual behaviour	Frequent	Absent	Frequent	Kano (1992); Hohmann & Fruth (2000); Hashimoto & Furuichi (2006); Hare et al. (2007); Hare & Woods (2011); Ryu et al., this issue; Clay & de Waal, this issue
Lethal aggression between groups	Absent	Present	Present	Wrangham (1999); Wilson et al. (2014)
Mother's importance to adult offspring	High	Low	High	de Lathouwers & Van Elsacker (2006); Surbeck et al. (2011); Schubert et al. (2013)
Infanticide/female coercion	Absent	Present	Present	Hohmann & Fruth (2002); Surbeck et al. (2011)
Levels of adult play	High	Low	High	Palagi & Paoli (2007); Wobber et al. (2010)
Cooperative hunting	Absent	Present	Present	Ihobe (1992); Watts & Mitani (2002); Surbeck & Hohmann (2008)
Sharing between strangers	Present	Absent	Present	Tan & Hare (2013); Tan et al., this issue; Yamamoto et al., data not shown
Male–male alliances	Absent	Frequent	Frequent	Kano (1992); Wrangham (1999)
Female gregariousness	High	Low	High	Furuichi (2011); Stevens et al., this issue

Even with the central role bonobos can play in testing hypothesis regarding ape and human evolution, bonobo research lags far behind work with chimpanzees. Searching ISI Web of Science and Google Scholar for 'bonobo' and 'chimpanzee' reveals that the bonobo makes up only 3% and 9% of the total citations indexed for both species collectively. This is far from the 50% that phylogeny alone would predict.

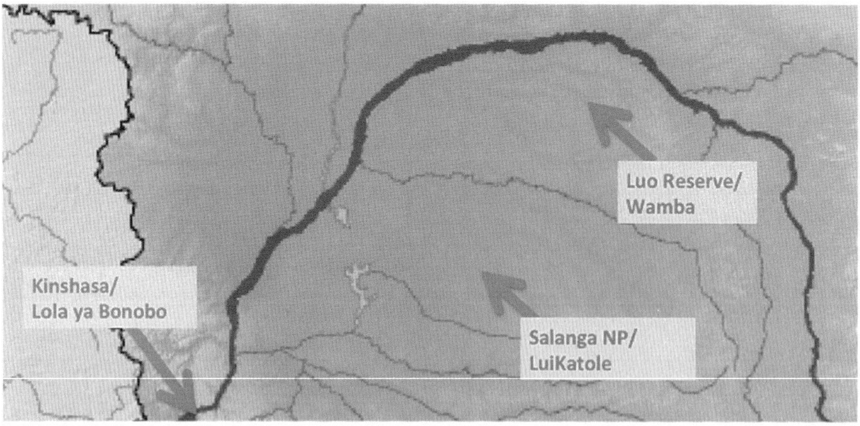

Figure 1. Bonobos are endangered and are only endemic to tropical forest South of the Congo River in the Democratic Republic of Congo. Their suspected historical range (approx. 500 000 km^2) is nearly the size of France and bigger than California (see http://www.iucnredlist.org/details/full/15932/0). The map shows the Congo River and the location of the most productive research sites over the last decade for bonobo behaviour and cognition. Wamba in the Luo Reserve represents the first and oldest study of wild bonobos established in 1973 by Takayoshi Kano (Kano, 1992), Lui Katole in Salonga National Park has likely been the most productive study site on wild bonobos after over a decade of support from the Max Planck Society (Hohmann & Fruth, 2003b), and Lola ya Bonobo Sanctuary outside the capital of Kinshasa has allowed for dozens of cognitive and developmental studies focusing on the sanctuaries' rehabilitated orphan bonobos rescued from the bushmeat trade (Wobber & Hare, 2011). Map from wikicommons: File:CongoLualaba_watershed_topo.png. This figure is published in colour in the online edition of this journal, which can be accessed via http://booksandjournals.brillonline.com/content/journals/1568539x.

more than ever and Lola ya Bonobo, the bushmeat orphanage in Kinshasa, has hosted scores of researchers from over a dozen institutions (Figure 1). Moreover, researchers have increasingly published on bonobo behaviour from zoo populations — particularly bonobo colonies in Europe. Captive bonobo researchers have also finally begun to break the shackles of small sample size from which most bonobo research has long suffered. Lola ya Bonobo has allowed for a series of large-scale experimental comparisons of behaviour and cognition between the two *Panins* where at least 20–30 individuals of each species have participated (e.g., Hare et al., 2007; Herrmann et al., 2010, 2011; Wobber et al., 2010a, b, 2014; Maclean & Hare, 2012, 2013; Rosati & Hare, 2012, 2013). Researchers in European zoos have also been able to collaborate and boost sample sizes. A number of recent studies have sampled several zoo populations and allow for powerful analyses (Stevens et al., 2007; Jaeggi et al., 2010; Behringer et al., 2014a, b). These experiments and large-scale studies have corroborated many earlier observational studies (e.g., de Waal, 1987; Kuroda, 1989; Kano, 1992), revealed many new phenomenon, and suggested at least one novel hypothesis for bonobo evolution (Hare et al., 2012). Thus, this Special Issue is exhibit A in demonstrating that a renaissance in bonobo research is well underway.

The issue includes a dozen new empirical papers from the wild and captivity illustrating why anyone wanting to understand humans or chimpanzees must also know bonobos. Bonobos are not only equal to chimpanzees as our relatives, but they are also unique (see Table 1). The majority of papers in this issue show that whether you are interested in the evolution of culture and tool use, social relationships and sharing or foraging ecology and cognition, bonobos have a major contribution to make.

The bonobo female is the focus of four papers that further show exactly how unusual bonobo female relationships are. Stevens and colleagues examine relationship value in bonobos by observing social interactions in one of the world's largest zoo colonies of bonobos. They report that unlike chimpanzees who show the strongest relationship values between males, bonobos show the strongest values between females. Ryu and colleagues present evidence that bonobo swellings may act to attract females as much as males. In comparison to female chimpanzees whose swellings are only attractive to males and act as honest indicators of ovulation, bonobos are more like humans in disguising their ovulation. Ryu and colleagues present new evidence that the extended swelling cycle in bonobos is as crucial to cementing relationships between females as they are in attracting males. Clay and de Waal revisit the question of the function of bonobo sexual contact. They find evidence that bonobo sexual contact is in large part about conflict resolution. Bonobos — and in particular female bonobos — often make social sexual contact during reconciliation or consolation events following a conflict. Unlike chimpanzees, bonobos are able to use genital contact and social sexual behaviour to maintain relatively low intensity aggression. Finally, Yamamoto reports that the bonobos at the Wamba field site frequently shared pieces of large fruit that are seasonally available. Unlike chimpanzees where most sharing occurs between males, in bonobos the majority of sharing occurred between adult females. In addition, this paper suggests the existence of 'courtesy' food sharing characterized by begging for social bond rather than food itself, since the recipients begged for fruit, which could be obtained by themselves without any cooperation or specialized skills. This is proposed to enhance female–female bonding in bonobos. All four papers provide further evidence that the behaviour and psychology of bonobo females is radically different from that observed in chimpanzees.

Foraging behaviour and cognition of bonobos is the focus of three papers that each show important ways that bonobos spatial cognition differs remarkably from chimpanzees. Beaune and colleagues present years of field data from the Lui Katole site that shows the pattern of seed dispersal in bonobos. They find that in violation of optimal foraging theory bonobos homogenously disperse a variety of fruits that vary in both the quality and quantity at the patch of origin. Plants are thought to evolve strategies to manipulate the dispersal patterns of their dispersers, but this is not the case in bonobos apparently. Instead, bonobos uniformly disperse seeds a minimum of 100 m and typically as much as a kilometre away from their origin. This suggests that the quality and quantity of food a tree provides does not affect the foraging decisions that bonobos make on a daily basis. Two studies on foraging cognition might help explain how cognitive abilities may free bonobos from being manipulated by plant evolution similar to other species. Rosati reports a set of experiments that measure the spatial memory strategies that bonobos rely on when searching for food. Bonobos are clearly not completely egocentric and show evidence of using an allocentric — or more flexible landmark based strategy to remember things. But what is most interesting is what is not observed — a developmental shift seen in chimpanzees. Further, Wobber and Herrmann test for the possible link between testosterone and cognition found in other species. While they find the familiar pattern of male testosterone and spatial cognition being linked in chimpanzees they find no such link in male bonobos. This adds to the growing evidence that the hormonal profile of bonobos — and in particular male bonobos — seems to differ significantly from chimpanzees. Bonobos are the ultimate seed dispersers in the Congo Basin, have the ability to use landmarks when finding things in space but violate optimal foraging theory, and unlike chimpanzees develop these abilities differently and are unaffected by testosterone. Bonobos seem to break all the rules when it comes to foraging.

Two papers are relevant to solving the puzzle of why bonobos are expert extractive foragers in captivity but have never been seen using tools to obtain food in the wild. Furuichi and colleagues provide powerful new data from Wamba to document that while bonobos do have a rich tool use tradition in Wamba they do not use tools from extractive foraging as seen in chimpanzees at a comparable site in Congo-Brazzaville. More interesting is the fact that ecological differences between the two sites do not seem to explain the prevalence of extractive foraging in the chimpanzees and the lack of this type of tool use in bonobos. Hopkins and colleagues conducted experimental comparisons of handedness in bonobos and chimpanzees. While chimpanzees have been repeatedly shown to show right hand bias, bonobos do not show the same lateralized pattern as chimpanzees. Hopkins and colleagues suggest that the chimpanzee pattern may be a result of their increased reliance on tools while the bonobo pattern may be more representative of the basal state in our last common ancestor. The two papers raise as many questions as they answer and make bonobo tool use one of the biggest puzzles within cognitive ethology.

What is particularly exciting is that for the first time captive and wild bonobos are regularly being directly compared to chimpanzees. This work is not only highlighting ways in which bonobos are different, but also surprising ways they are the same. Finding similarity is as important as finding differences since it puts any differences into a larger phenotypic context. Three papers in this issue also reveal surprising similarities. Maclean and Hare test the hypothesis that bonobos will be more skilled at reading human cooperative — communicative intentions than chimpanzees. Despite their prediction both species were equally skilled in reading cooperative — communicative gestures and equally unskillful at reading the same gestures in a more competitive context. The results suggest that both bonobos and chimpanzees differ from human infants in their use of human gestures in a similar way. Tan and colleagues tested whether bonobos would show a prosocial tendency in what has become a 'standard' paradigm to test for proactive sharing in nonhumans. Despite recent evidence suggesting that bonobos might show a strong tendency to share in this same paradigm where chimpanzees do not, the authors found no evidence for proactive sharing. However, they interpret the results to suggest that there are major limitations to the paradigm used and that it should be abandoned in favor of other assessments of prosociality that have been successfully validated across species. Finally, Schroepfer-Walker and Hare experimentally measure the effect of grooming and play on the social preferences of bonobos and chimpanzees. While it has long been assumed that non-food social currencies such as grooming and play carry social value, this assumption has never been explicitly tested. The authors find that both bonobos and chimpanzees shift their preferences toward a human who recently groomed or played with them even though no food was exchanged. Both species can shift social preference based on the social currency of grooming or play alone. Interestingly against the authors predictions the two species did not differ in how their preferences shifted with males of both species showing a robust effect and females showing a small or non-significant shift.

Taken together this new body of work clearly shows that progress is being made understanding in what way bonobos and chimpanzees share traits and where they do not — as well as pointing to very exciting directions for future research. Given the quantity and quality of research being produced, we predict a very exciting future for bonobo research. This means the future is also bright for a better understanding of our species' evolution as well as understanding primate behaviour more generally. Despite growing momentum there are still major pieces of infrastructure that are missing for sustainable growth. The first and most obvious is the simple fact that we are unaware of a dedicated bonobo field researcher that has received a tenure-tracked assistant professorship at a major research university in the past decade or more. Over the same period a host of researchers focused primarily on chimpanzee behaviour have deservedly taken up positions at top universities. Things should become more balanced soon as a new crop of young bonobo researchers — many of whom published in this issue — are reaching the stage where they will successfully compete for future positions. Likewise, as more high impact research is produced departments will want to lead in this exciting area of research. Another gap that must be filled is the shortage of Congolese scientists that focus primarily on bonobo behaviour or conservation (i.e., only one paper in the current Special Issue includes a Congolese collaborator). It will be these scientists that teach the Congolese public, politicians and students about the value of the only ape that is 100% Congolese. Finally, primatologists across the board desperately need to pivot to Asia. China, in particular, is having increasing influence on all areas where endangered primates live — the Congo Basin being no exception. Students need to be recruited from China to work with apes in Africa, conferences on conservation need to be held there and Chinese academics and zoological societies must be engaged to build a vibrant conservation community that can respectfully respond to future challenges. We are hopeful that progress will be made in these areas rapidly and we hope that this Special Issue plays a small role in moving bonobo research into the forefront where it belongs. The bonobo is forgotten no more!

References

Beaune, D., Bretagnolle, F., Bollache, L., Hohmann, G. & Fruth, B. (2015). Can fruiting plants control animal behavior and seed dispersal distance? — Behaviour 152: 359-374.

Behringer, V., Deschner, T., Deimel, C., Stevens, J. & Hohmann, G. (2014a). Age related changes in urinary testosterone levels suggest differences in puberty onset and divergent life history strategies in bonobos and chimpanzees. — Horm. Behav. 66: 525-533.

Behringer, V., Deschner, T., Murtagh, R., Stevens, J. & Hohmann, G. (2014b). Age-related changes in Thyroid hormone levels of bonobos and chimpanzees indicate heterochrony in development. — J. Hum. Evol. 66: 83-88.

Clay, Z. & de Waal, F.B.M. (2015). Sex and strife: post conflict sexual contacts in bonobos. — Behaviour 152: 313-334.

de Lathouwers, M. & Van Elsacker, L. (2006). Comparing infant and juvenile behavior in bonobos (*Pan paniscus*) and chimpanzees (*Pan troglodytes*): a preliminary study. — Primates 47: 287-293.

de Waal, F.B.M. (1987). Tension regulation and non-reproductive functions of sex in captive bonobos. — Nat. Geogr. Res. 3: 318-335.

de Waal, F.B.M. & Lanting, F. (1997). Bonobo: the forgotten ape. — California University Press, Berkeley, CA.

Furuichi, T. (2011). Female contributions to the peaceful nature of bonobo society. — Evol. Anthropol. 20: 131-142.

Furuichi, T., Sanz, C., Koops, K., Sakamaki, T., Ryu, H., Tokuyama, N. & Morgan, D. (2015). Why do wild bonobos not use tools like chimpanzees do? — Behaviour 152: 425-460.

Gruber, T., Clay, Z. & Zuberbühler, K. (2010). A comparison of bonobo and chimpanzee tool use: evidence for a female bias in the *Pan* lineage. — Anim. Behav. 80: 1023-1033.

Hare, B. (2007). From nonhuman to human mind: what changed and why. — Curr. Dir. Psychol. Sci. 16: 60-64.

Hare, B. (2009). What is the effect of affect on bonobo and chimpanzee problem solving? — In: The neurobiology of the umwelt: how living beings perceive the world (Berthoz, A. & Christen, Y., eds). Springer, Berlin, p. 89-102.

Hare, B. (2011). From hominoid to hominid mind: what changed and why? — Ann. Rev. Anthropol. 40: 293-309.

Hare, B. & Woods, V. (2011). Bonobo but not chimpanzee infants use socio-sexual contact with peers. — Primates 52: 111-116.

Hare, B., Melis, A., Woods, V., Hastings, S. & Wrangham, R. (2007). Tolerance allows bonobos to outperform chimpanzees in a cooperative task. — Curr. Biol. 17: 619-623.

Hare, B., Wobber, T. & Wrangham, R. (2012). The self-domestication hypothesis: bonobo psychology evolved due to selection against male aggression. — Anim. Behav. 83: 573-585.

Hashimoto, C. & Furuichi, T. (2006). Comparison of behavioral sequence of copulation between chimpanzees and bonobos. — Primates 47: 51-55.

Herrmann, E., Hare, B., Call, J. & Tomasello, M. (2010). Differences in the cognitive skills of bonobos and chimpanzees. — PLoS One 5: e12438.

Herrmann, E., Hare, B., Cisseski, J. & Tomasello, M. (2011). The origins of human temperament: children avoid novelty more than other apes. — Dev. Sci. 14: 1393-1405.

Herrmann, E., Keupp, S., Hare, B., Vaish, A. & Tomasello, M. (2013). Direct and indirect reputation formation in great apes and human children. — J. Comp. Psychol. 127: 63-75.

Hohmann, G. & Fruth, B. (2000). Use and function of genital contacts among female bonobos. — Anim. Behav. 60: 107-120.

Hohmann, G. & Fruth, B. (2002). Dynamics in social organization of bonobos (*Pan paniscus*). — In: Behavioural diversity in chimpanzee (Boesch, C., Marchant, L. & Hohmann, G., eds). Cambridge University Press, Cambridge, p. 138-150.

Hohmann, G. & Fruth, B. (2003a). Culture in bonobos? Between-species and within-species variation in behavior. — Curr. Anthropol. 44: 563-609.

Hohmann, G. & Fruth, B. (2003b). Lui Kotal — a new site for field research on bonobos in Salonga National Park. — Pan African News 10: 25-27, available online at http://mahale.main.jp/PAN/10_2/10(2)_05.html.

Hopkins, W., Schaeffer, J., Russell, J., Bogart, S., Meguerditchian, A. & Coulon, O. (2015). A comparative assessment of handedeness and its potential neuroanatomical correlates in chimpanzees and bonobos. — Behaviour 152: 461-492.

Ihobe, H. (1992). Male–male relationships among wild bonobos at Wamba, Republic of Zaire. — Primates 33: 163-179.

Jaeggi, A., Stevens, J. & van Schaik, C. (2010). Tolerant food sharing and reciprocity is precluded by despotism among bonobos but not chimpanzees. — Am. J. Phys. Anthropol. 143: 41-51.

Kano, T. (1992). The last ape: pygmy chimpanzee behavior and ecology. — Stanford University Press, Stanford, CA.

Kuroda, S. (1989). Developmental retardation and behavioral characteristics of pygmy chimpanzees. — In: Understanding chimpanzees (Heltne, P. & Marquardt, L., eds). Harvard University Press, Cambridge, MA, p. 184-193.

Maclean, E. & Hare, B. (2012). Bonobos and chimpanzees infer the target of an actor's attention. — Anim. Behav. 83: 345-353.

Maclean, E. & Hare, B. (2013). Spontaneous triadic play in bonobos and chimpanzees. — J. Comp. Psychol. 127: 245-255.

Maclean, E. & Hare, B. (2015). Bonobos and chimpanzees read helpful gestures better than prohibitive gestures. — Behaviour 152: 493-520.

Palagi, E. & Paoli, T. (2007). Play in adult bonobos (*Pan paniscus*): modality and potential meaning. — Am. J. Phys. Anthropol. 134: 219-225.

Prufer, K., Munch, K., Hellmann, I., Akagi, K., Miller, J.R., Walenz, B., Koren, S., Sutton, G., Kodira, C., Winer, R., Knight, J.R., Mullikin, J.C., Meader, S.J., Ponting, C.P., Lunter, G., Higashino, S., Hobolth, A., Dutheil, J., Karakoç, E., Alkan, C., Sajjadian, S., Catacchio, C.R., Ventura, M., Marques-Bonet, T., Eichler, E.E., André, C., Atencia, R., Mugisha, L.,

Junhold, J., Patterson, N., Siebauer, M., Good, J.M., Fischer, A., Ptak, S.E., Lachmann, M., Symer, D.E., Mailund, T., Schierup, M.H., Andrés, A.M., Kelso, J. & Pääbo, S. (2012). The bonobo genome compared to the chimpanzee and human genome. — Nature 486: 527-531.

Rosati, A. (2015). Context influences spatial frames of reference in bonobos (*Pan paniscus*). — Behaviour 152: 375-406.

Rosati, A.G. & Hare, B. (2012). Decision making across social contexts: competition increases preferences for risk in chimpanzees and bonobos. — Anim. Behav. 84: 869-879.

Rosati, A. & Hare, B. (2013). Chimpanzees and bonobos exhibit emotional responses to decision outcomes. — PLoS One 8: e63058.

Ryu, H., Hill, D. & Furuichi, T. (2015). Prolonged maximal sexual swelling in wild bonobos facilitates affiliative interactions between females. — Behaviour 152: 285-311.

Schroepfer-Walker, K., Wobber, T. & Hare, B. (2015). Experimental evidence that grooming and play are social currency in bonobos and chimpanzees. — Behaviour 152: 545-562.

Schubert, G., Vigilant, L., Boesch, C., Klenke, R., Langergraber, K., Mundry, R., Surbeck, M. & Hohmann, G. (2013). Co-residence between males and their mothers and grandmothers is more frequent in bonobos than chimpanzees. — PLoS One 8: e83870.

Stevens, J.M.G., Vervaecke, H., de Vries, H. & Van Elsacker, L. (2007). Sex differences in the steepness of dominance hierarchies in captive bonobo groups. — Int. J. Primatol. 28: 1417-1430.

Stevens, J., De Groot, E. & Staes, N. (2015). Relationship quality in captive bonobo groups. — Behaviour 152: 259-283.

Surbeck, M. & Hohmann, G. (2008). Primate hunting by bonobos at LuiKotale, Salonga National Park. — Curr. Biol. 18: R906-R907.

Surbeck, M., Mundry, R. & Hohmann, G. (2011). Mothers matter! Maternal support, dominance status and mating success in male bonobos (*Pan paniscus*). — Proc. Roy. Soc. Lond. B: Biol. Sci. 278: 590-598.

Tan, J. & Hare, B. (2013). Bonobos share with strangers. — PLoS One 8: e51922.

Tan, J., Kwetuenda, S. & Hare, B. (2015). Preference or Paradigm? Bonobos do not share in "the" prosocial choice task. — Behaviour 152: 521-544.

Watts, D. & Mitani, J. (2002). Hunting behaviour of chimpanzees at Ngogo, Kibale National Park, Uganda. — In: Behavioral diversity in chimpanzees and bonobos (Boesch, C., Hohmann, G. & Marchant, L., eds). Cambridge University Press, Cambridge, p. 244-257.

Wilson, M., Boesch, C., Fruth, B., Furuichi, T., Gilby, I.C., Hashimoto, C., Hobaiter, C.L., Hohmann, G., Itoh, N., Koops, K., Lloyd, J.N., Matsuzawa, T., Mitani, J.C., Mjungu, D.C., Morgan, D., Muller, M.N., Mundry, R., Nakamura, M., Pruetz, J., Pusey, A.E., Riedel, J., Sanz, C., Schel, A.M., Simmons, N., Waller, M., Watts, D.P., White, F., Wittig, R.M., Zuberbühler, K. & Wrangham, R.W. (2014). Lethal aggression better explained by adaptive strategies than by human impact. — Nature 513: 414-417.

Wobber, T. & Hare, B. (2011). Psychological health of orphan bonobos and chimpanzees in African sanctuaries. — PLoS One 6: e17147.

Wobber, V. & Herrmann, E. (2015). The influence of testosterone of cognitive performance in bonobos and chimpanzees. — Behaviour 152: 407-423.
Wobber, V., Wrangham, R. & Hare, B. (2010a). Evidence for delayed development of social behavior and cognition in bonobos relative to chimpanzees. — Curr. Biol. 20: 226-230.
Wobber, V., Hare, B., Maboto, J., Lipson, S., Wrangham, R. & Ellison, P. (2010b). Differential reactivity of steroid hormones in chimpanzees and bonobos when anticipating food competition. — Proc. Natl. Acad. Sci. 107: 12457-12462.
Wobber, T., Herrmann, E., Hare, B., Wrangham, R. & Tomasello, M. (2014). The evolution of cognitive development in *Pan* and *Homo*. — Dev. Psychobiol. 5: 547-573.
Wrangham, R. (1999). Evolution of coalitionary killing. — Yb. Phys. Anthropol. 42: 1-30.
Yamamoto, S. (2015). Non-reciprocal but peaceful fruit sharing in the wild bonobos of Wamba. — Behaviour 152: 335-357.

Bonobo (*Pan paniscus*). Photo taken by Shinya Yamamoto at Wamba, Democratic Republic of the Congo.

[When citing this chapter, refer to Behaviour 152 (2015) 259–283]

Relationship quality in captive bonobo groups

Jeroen M.G. Stevens [a,*], **Evelien de Groot** [a] **and Nicky Staes** [a,b]

[a] Centre for Research and Conservation, Royal Zoological Society of Antwerp, Koningin Astridplein 20–26, 2018 Antwerp, Belgium

[b] Department of Biology, University of Antwerp, Groenenborgerlaan 171, 2020 Antwerp, Belgium

[*]Corresponding author's e-mail address: jeroen.stevens@kmda.org

Accepted 4 January 2015; published online 28 January 2015

Abstract

We use Principal Component Analyses (PCA) to describe components of social relationship quality in bonobos. We find a three component structure, with the first two components, labelled Value and Compatibility, closely matching the theoretical constructs as well as components reported for chimpanzees and other primates. The third component differed but was abandoned based on Parallel Analysis. Among bonobos, female–female dyads have higher Value and Compatibility. Relationships between males are characterised by low Value and Compatibility. Dyads that had been housed together for a longer time and maternally related ones also have more valuable relationships, while individuals close in rank have low compatibility. The results confirm the strong bonds among female bonobos, but for the first time can describe how they differ qualitatively from close bonds reported for captive chimpanzee females. We suggest future studies should also include Parallel Analysis to more accurately describe the number of components in relationship quality.

Keywords

social organisation, *Pan paniscus*, relationship quality.

1. Introduction

Living in social groups can bring about fitness benefits (Kappeler & van Schaik, 2002). Within social groups not all relationships are equal, and individuals of many species are known to have diversified social relationships, resulting in complex social networks. Social relationships can be beneficial for individual fitness, and recent studies have shown that strong friendships between females can result in higher longevity, fecundity and offspring survival (Silk et al., 2003, 2009, 2010; Silk, 2007; Cameron et al., 2009; Frère et al., 2010). Similarly, males can increase their fitness by forming bonds or friendships with other males (Feh, 1999; Schülke et al., 2010; Wiszniewski et al., 2012; Gilby et al., 2013). Finally, there is accumulating evidence that male–female bonds can also increase the fitness of males in baboons, macaques and chimpanzees (Moscovice et al., 2010; Massen et al., 2012; Langergraber et al., 2013).

While the influence of high quality social relationships on individual fitness is becoming increasingly clear, describing and measuring the quality of these social bonds is still under much debate. Hinde (1976) and Kummer (1978) describe social relationships as an emergent property of repeated social interactions between two individuals and can be seen as investments, where individuals are expected to invest more in valuable partners. Following these concepts, Cords & Aureli (2000) proposed that relationship quality could be described by three components, labelled Value, Compatibility and Security. Value refers to the direct benefits gained by the relationship, such as food sharing or agonistic support. Compatibility is measured by tolerance and affiliation, based on a history of shared history of social exchanges. Security means that the behaviour of partners is predictable and consistent over time. Subsequently, several studies have sought to specify these relationship components and relate them to other behaviours, conciliatory tendencies for example. However chosen operational measures of each component often differ between studies (reviewed in Cords & Aureli, 2000 and Fraser et al., 2008). Traditionally, behavioural ecologists have looked at the frequency of one or a few behaviours, such as proximity, grooming or support, to measure relationship value (e.g., Lehman & Boesch, 2009; Langergraber et al., 2012; Massen et al., 2012); alternatively, they inferred closeness of social relationships indirectly, for example by assuming that kin have higher quality relationships than non-kin, or that members of the dispersing sex have lower relationship quality than dyads that are philopatric. The latter assumption may be problematic (see below). More recently, principal component analysis (PCA) has been used to describe relationship quality in captive chimpanzees (Fraser et al., 2008). The benefit of Principal Component Analyses is that behavioural variables can be objectively reduced to a few behavioural dimensions, which are comprehensive and conceptually more coherent and can then be related to different variables. In a study on zoo housed chimpanzees, Fraser et al. (2008) found three components of relationship quality from nine behavioural variables, which matched to the three theoretical constructs that had been proposed. Once these dimensions of relationship quality have been assessed, it is then possible to look at variation across different dyads. Fraser et al. (2008) found that value was higher among related dyads, between chimpanzees that had been housed together longer or where age differences were smaller. Compatibility was higher among female–female than among male–male or male–female dyads, and was also higher among male–female compared to male–male dyads. Furthermore, this component was higher for related chimpanzees, and among individuals that had been housed together longer. Finally, security again was higher in related dyads, among male–female dyads and dyads of similar age, but was lower in dyads that had been housed together longer. Thus, this first study could clearly identify components of relationship quality and show that certain preconceptions about the influence of kin and sex-combination were not present for this captive group of chimpanzees. Subsequent studies have used the same approach and a three-component model has been found in a different group of chimpanzees (Koski et al., 2012), in Japanese macaques (Majolo et al., 2010), Barbary macaques (McFarland & Majolo, 2011) as well as in ravens (Fraser & Bugnyar, 2010). In wild spider monkeys the PCA resulted in only two components, showing the need for a wider taxonomic angle (Rebecchini et al., 2011). Moreover, it has been noted that while these models use dyadic values, relationships within dyads can be asymmetric (Majolo et al., 2010). All these studies seem to indicate that multiple components can be described, but more studies are needed to see how consistent they are across taxa.

While two studies have investigated relationship quality in captive chimpanzees, much less is known about relationship quality in the chimpanzees' sibling species, the bonobo (*Pan paniscus*). Quantifying the relationship values can shed light on the ongoing debate about how different these two species really are, and can provide more insight on the influence of kinship in social bonding. Traditionally, differences between these two sibling species have been highlighted. Both species live in societies that are comparable in size, and in both species females migrate to neighbouring communities when they reach adolescence (Goodall, 1986; Kano, 1992; Furuichi et al., 1998; Stumpf, 2007). However, chimpanzee society is described as male bonded and females are relatively more solitary (Goodall, 1986; Stumpf, 2007) while bonobo society is female-centred with strong relationships among females, between females and their adult sons, and perhaps even between females and unrelated males (Badrian & Badrian, 1984; White, 1988; Furuichi, 1989, 2011; Parish, 1996; Hohmann & Fruth, 2002; Stevens et al., 2006). These close bonds between female bonobos have received much attention, as they challenge assumptions about socio-ecological models and kin selection that predict closer bonds between related individuals, and even about our own evolutionary past (Parish & de Waal, 2000).

However, the chimpanzee-bonobo dichotomy in terms of social bonding has been questioned on two levels. Firstly, the idea of chimpanzees as being only male bonded has been challenged. Chimpanzee males do indeed form meaningful relationships with other males (Mitani, 2009; Gilby et al., 2013), but so do female chimpanzees in the wild and in captivity (Baker & Smuts, 1994; Langergraber et al., 2009; Lehman & Boesch, 2009). Furthermore, kinship did not predict bonding in wild male and female chimpanzees (Langergraber et al., 2007, 2009). Recently, long-term male–female relationships have been shown to influence male reproductive success in chimpanzees in East-Africa (Langergraber et al., 2013). Secondly, the strong female bonding of bonobos has been questioned. Previous research in captivity has shown very strong bonds between female bonobos (Parish, 1996), but often these females were closely related and in groups containing only one or two males that were always unrelated to the females. Therefore, the existence of female bonding in bonobos has been questioned and it has been proposed as a side effect of life in captivity (Stanford, 1998; Franz, 1999; Hohmann et al., 1999). It has been suggested that relationships between male and female bonobos are more important, even when controlling for kinship effects (Hohmann et al., 1999; Hohmann & Fruth, 2002). Our own subsequent research on a larger dataset of several groups of zoo housed bonobos, in which females were generally unrelated and mother–son dyads were present, showed that grooming, proximity and proximity maintenance was not stronger among females than among females and unrelated males or among males. We found that only coalitionary support was consistently significantly more common among females than among males or among females and unrelated males (Stevens et al., 2006). However, the significance of these behavioural variables and their interdependence, in terms of the three proposed components of relationship quality, was not examined. Jaeggi et al. (2010) compared relationships in chimpanzees and bonobos, using a PCA on three behavioural variables (grooming frequency, proximity and support given). In the chimpanzees this resulted in one component of relationship quality, which had high loadings on grooming given, support and proximity. In the bonobos however, the data split into two components; the first was labelled 'affiliative relationship' and included a high loading of grooming and proximity and a negative loading of support. The second component was labelled 'political relationship' and had a high loading of support, minor loading of proximity and negative loading of grooming. This suggests that relationship quality in bonobos may differ from chimpanzees. However, the analyses of Jaeggi et al. (2010) were based on one group of six bonobos, studied for 125 h, and were based on a limited set of behavioural variables. Therefore, a study that is based on a larger sample size and is more comparable to the chimpanzee studies (Fraser et al., 2008; Koski et al., 2012) is needed.

Here we use a larger set of data from captive bonobo groups, not including the data from Jaeggi et al. (2012) group, and include eight behavioural variables, resembling those used in the chimpanzee studies (Fraser et al., 2008; Koski et al., 2012). Our first aim is to describe components of Relationship Quality among captive bonobos, using Principal Component Analyses. We predict that Relationship Quality can be described in similar components to those found in chimpanzees. Second, we test which components are stronger in female–female dyads to further clarify the nature of female bonding in bonobos. Third, we look at other variables that have been found to affect the components of Relationship Quality in other species: kinship, tenure of the relationship, age difference and rank difference.

2. Material and methods

2.1. *Study groups*

We studied six groups of bonobos housed in four European Zoos (Wuppertal, Apenheul, Twycross, Planckendael, see Table 1). Data were only collected from individuals above seven years old, which is the earliest age of reproduction in captivity for male and female bonobos (Pereboom et al., 2011). In each group, the bonobos were housed together for most of the day and night and were only separated briefly during maintenance or cleaning work. Observations ran from the morning (usually around 8 h 30 min) until dusk, when bonobos started building nests for the night (usually around 5.30 p.m., depending on season). The group in Planckendael was studied three times (in 1999, 2001 and 2011), however the group composition differed substantially between each period, and only a few dyads were present in more than one

Table 1.
Composition and observation time for each of the five study groups.

Group and year of study	N of males > 7 years	N of females > 7 years	N of immatures < 7 years	Observation time (h)
Planckendael 1999	3	4	4	203
Planckendael 2001	3	3	2	503
Planckendael 2011	3	3	2	295
Wuppertal 1999	4	2	1	193
Apenheul 2001	3	5	3	493
Twycross 2001	3	3	2	490

Immatures were not included in any of the analyses.

group setting (overlap between period Planckendael 1999 and Planckendael 2001: 6 out of 15 dyads; overlap period Planckendael 2001 and Planckendael 2010: 1 out of 15 dyads; there was no further overlap in dyads between any of the other groups studied). When a dyad occurred in more than one study period, we chose the data for the dyad based on the period with the longest observation time. This resulted in a total sample of 102 different dyads composed of 17 females and 16 males. Each individual was represented in 6.18 ± 0.23 dyads (mean ± SE). Data on Planckendael in 2011 were collected by EDG, all other data were collected by JMGS, following a standardised ethogram for social behaviour in bonobos.

2.2. *Data collection and analyses*

We used behaviour sampling to score social behaviours among individuals that were older than seven years of age. All occurrences of aggression, grooming, coalitionary support and peering were recorded. In each grooming bout, the participation of each partner was scored once. Subsequent switches between the active and passive role where not counted as new bouts (Vervaecke et al., 2000a) and frequency of grooming bouts within a dyad was calculated. Support was defined as all instances where an individual A intervenes with an aggression within 30 s in an agonistic interaction between two other individuals B and C to aid in attack or in defence (de Waal, 1978). We only studied triadic interventions because in polyadic interactions, where more than three individuals were involved, it was often unclear to whom the support was directed. When more individuals joined in support, these interactions were not considered. All aggressions from individual C towards B were considered as opportunities for support, regardless of the reaction of the victim B. We then corrected for the opportunity for individual A to provide support to B in a conflict with C, by dividing the total number of supports between A and B by the total number of conflicts that individuals A or B had with other group members, excluding the conflicts they had with each other, and then multiplied this number by 100 (Vervaecke et al., 2000b; Stevens et al., 2006). Every support pro for an individual implied a contra support against a victim (de Waal, 1978). Here too, we corrected for the number of conflicts. Thus the target of a coalition is considered as the receiver of 'counter-intervention' (or contrasupport, sensu Vervaecke et al., 2000b). Proximity was scored using scan sampling, noting every fifteen minutes which individuals were within arm's reach (or less than one meter) from each other. For analyses, seven data points were chosen at random per observation day, to avoid interdependency of these data (Martin & Bateson, 1993; Stevens et al., 2006). To calculate a proportional value the total number of randomised samples was then divided by the total number of samples taken per group.

For each dyad we used eight behavioural variables to include in the first Principal Component Analysis (PCA1; Table 2). We chose behavioural vari-

Table 2.
Behavioural variables entered in the Principal Component Analysis.

Behavioural variable	Definition	Mean ± SE per dyad
Grooming frequency	Number of grooming bouts exchanged within a dyad (i.e. the sum of all bouts from A to B and from B to A)	0.28 ± 0.03
Grooming symmetry	Symmetry of grooming within a dyad (see text)	0.26 ± 0.02
Proximity	Proportion of scans spent within arm's reach	0.12 ± 0.07
Aggression frequency	Frequency of all aggressive interactions within a dyad	0.17 ± 0.02
Aggression symmetry	Symmetry of aggression within a dyad (see text)	0.20 ± 0.18
Support	Index of agonistic support (frequency of support/opportunity to support)	0.48 ± 0.08
Counter-intervention	Index of counter-intervention (frequency of counter-intervention/ opportunity to intervene)	0.46 ± 0.07
Peering	Frequency of peering	0.12 ± 0.01

ables that closely matched the studies on chimpanzees by Fraser et al. (2008) and Koski et al. (2012). To allow intergroup comparison with observation times that differed between groups, we transformed all behaviours either to frequencies by dividing by the number of group observation hours, or to proportions (see below). For frequencies of grooming and aggression between individuals A and B, we added the total number of aggressions and grooming bouts from A to B and from B to A, and divided by the total group observation time. Following Fraser et al. (2008), we calculated symmetry in grooming between individuals A and B using the following formula: A grooms B/(A grooms B + B grooms A). For each dyad, the lowest of the two values obtained reversing A's and B's roles was chosen to represent the degree of symmetry, so values ranged from 0 to 0.5, with higher values indicating more symmetrical exchanges. Symmetry in aggression was calculated in the same way. Food sharing was not consistently scored in all of the study groups; therefore, successful begging could not be included as a variable. Instead, peering behaviour (i.e., "the actor stares at the receiver's face from very close distance, up to a few centimetres" Kano, 1992, p. 200) was used as a proxy for tolerance. The function of peering in bonobos is poorly understood, but according to some it is a begging gesture (Kano, 1980), while others have stated it is "a solicitation for initiating affiliative interactions" (Idani, 1995) or 'a request for social tolerance' (Stevens et al., 2005a). Consistency of affiliation could not be measured in our study, since observation periods lasted between one to three months. Unlike the more conspicuous behaviours, approaches and their response could not reliably be scored using the all occurrence sampling and were therefore not included in the analyses.

We analysed the variables with a first principal component analysis (PCA1) with varimax rotation and Kaiser normalization. To improve normality, we transformed the data using square root transformation on all frequency data (grooming, aggression, support, counter-intervention, peering). Kaiser's measure of sampling adequacy was 0.65, which is on the low side, but acceptable (Budaev, 2010). To determine the number of components, we first followed Kaiser's rule and accepted all components which had an eigenvalue greater than 1 (Kaiser, 1960), in combination with visual inspection of the scree plot (Cattell, 1966). Coefficients of correlation greater than 0.5 or less than −0.5 were considered to be high loadings. Although Kaiser's rules and/or investigation of scree plots have also been used as criteria in all previous studies investigating relationship quality to determine the number of components (Fraser et al., 2008; Fraser & Bugnyar, 2010; Majolo et al., 2010; McFarland & Majolo, 2011; Rebecchini et al., 2011; Koski et al., 2012), this procedure may lead to overestimating the number of factors, and the use of Parallel Analysis has been shown to be more reliable to determine the number of extracted components (Horn, 1965; Franklin et al., 1995; O'Connor, 2000). Parallel Analysis is a "sample based adaptation of the population based Kaiser's rule" (Zwick & Velicer, 1986) and allows the researchers to determine the significance of components (Franklin et al., 1995). Only when eigenvalues from PCA are larger than the corresponding eigenvalues from Parallel Analysis, should the components be considered as true components. We therefore used SAS to conduct Parallel Analysis to confirm the number of extracted factors (O'Connor, 2000). When the Parallel Analysis indicated only two out of the three original components should be contained in the final analysis, we performed a second Principal Component Analysis extracting just two components to determine the variable loadings per component and individual component scores (PCA2).

Next, we used Linear Mixed Models to investigate the influence of kinship (separate factor for maternal and paternal kin), tenure (in years), sex combination (female–female; male–female; male–male), rank difference, and age difference (in years) on the extracted scores from PCA2. Kinship information was derived from the studbook; within each group all paternities were known (Pereboom et al., 2011). We only included kin with relatedness coefficients larger than 0.125 and treated it as a binary variable (kin or non-kin). For the factor 'maternal kin', this included one mother–daughter pair, six mother–son pairs and three maternal half-sibling pairs; for paternal kinship this included five father–son pairs and three paternal half-sibling pairs. Since females usually are transferred between zoos when they reach adolescence, our study groups contained no father–daughter pairs. There were also no full siblings. Relationship tenure was calculated from the studbook and included the number of years two individuals had spent together (with an error margin of 0.5 years; minimum 0.5, maximum 22 years). Rank differences were taken from Stevens et al. (2007) for five of the groups and calculated in the same way for Planckendael 2010 by constructing an ordinal dominance hierarchy based on the analyses of 'fleeing upon aggression' (Vervaecke et al., 2000a), and calculating net differences between individuals' ordinal ranks using MATMAN software (de Vries et al., 1993). Age difference was calculated by taking the differences of (estimated) years of birth (taken from

Pereboom et al., 2011) for the two members of a dyad. Because of the relatively low number of related individuals in our groups (which were mostly mother–son dyads), we could not include interaction factors for sex combination, kin or age in the linear models. To account for interdependence of the data, we included the identities of the two bonobos forming each dyad, as well as the factor 'group' as random variables in the statistical model. We used F-tests and backward selection to construct a final model that only retained variables with significant effects. When a significant effect was found for categorical variables, we performed post-hoc pairwise comparisons, using Tukey tests for multiple comparisons of means. PCA, Parallel Analysis, and linear mixed models were conducted using SAS software (version 9.0) with the critical α set at 0.05.

3. Results

3.1. Components of relationship quality

Using Kaiser's criterion, three components were extracted from PCA1, explaining 31.03%, 17.57% and 13.64% of overall variance respectively, totalling 62.24%. The first component had high positive loadings for proximity, grooming frequency, support and peering. These variables are very close to the definition of relationship value by other studies (Cords & Aureli, 2000; Fraser et al., 2008) and thus could be labelled 'Value'. The second component included high values for aggression frequency and high values for aggression symmetry (indicating asymmetric aggression) variables that suggest incompatibility (Fraser et al., 2008). In the further analyses, we reversed the signs for the scores obtained for each dyad for that component so that the values represented the degree of compatibility as opposed to the incompatibility of the dyad (cf. Fraser et al., 2008). The third component included a high positive loading for counter-intervention and a negative loading for grooming symmetry. Since grooming symmetry loaded high and positive on Fraser et al. (2008)'s security component, our component seems to measure insecurity. However, subsequent Parallel Analysis showed that only the two first components had higher eigenvalues than randomly generated values (Figure 1). Therefore, the third factor should not be considered further, and was dropped from further analyses. The factor loadings for the two-component model, resulting from PCA2, are shown in Table 3. In this

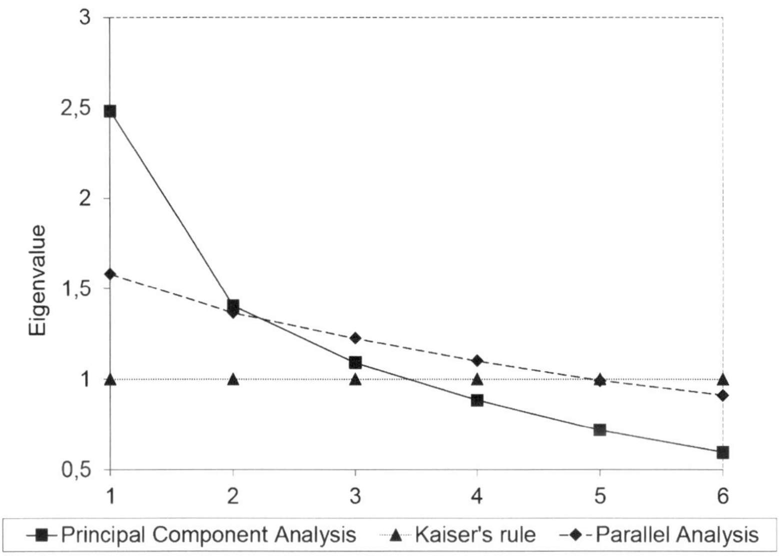

Figure 1. Comparison of eigenvalues of the components extracted in the Principal Component Analysis with Kaiser's criterion and with eigenvalues exctracted based on Parallel Analysis. Based on Kaiser's rule, three components would be retained. Parallel Analysis shows that the third component is not valid.

model, the first component explained 34.80% of total variance, and contained high positive loadings for proximity, grooming, support, peering and grooming reciprocity, and therefore corresponds to the theoretical component 'Value'. The second component explained 19.25% of total variance and contained high positive loadings for aggression frequency and aggression reciprocity and we therefore labelled it 'Compatibility'. The behavioural variable counter-intervention did not load high on either of the two components and therefore was abandoned in the final model.

3.2. Factors affecting relationship quality

The first component, labelled Value, was significantly influenced by the following factors: sex combination, age difference, maternal kinship and tenure (Table 4). Post-hoc comparisons showed that female–female dyads had significantly higher value than male–female ($p_{adj} < 0.0001$) and male–male dyads ($p_{adj} < 0.0001$), and male–female dyads had significantly higher values than male–male dyads ($p_{adj} = 0.0013$) (Figure 2). Maternal kin had significantly higher value than non-kin, but there was no effect of paternal

Table 3.
Varimax rotated component matrix of the final Principal Component Analyses (PCA2) run on the dyadic scores for 8 variables.

Variable	Value	(in)Compatibility
Proximity	0.88*	0.09
Groom frequency	0.72*	0.28
Support	0.66*	−0.28
Peering	0.61*	−0.07
Groom symmetry	0.55*	−0.17
Aggression symmetry	0.09	0.76*
Aggression frequency	0.00	0.67*
Counter-intervention	−0.25	0.44
% of variation explained	34.80	19.25
Eigenvalue	2.44	1.35

An asterisk indicates high loadings (>0.5 or <−0.5).

Table 4.
Effects of the variables on the components Value and Compatibility.

Component	$B \pm$ SE	t value	p
Value			
Intercept	−0.38 ± 0.38	−1.00	0.32
Sex combination (MM vs FF)	1.55 ± 0.25	6.21	<0.0001
Sex combination (MM vs MF)	0.84 ± 0.20	4.31	<0.0001
Maternal kin	−0.86 ± 0.26	−3.32	0.001
Tenure	0.054 ± 0.017	3.16	0.003
Age difference	0.017 ± 0.013	1.31	0.19
Rank difference	0.002 ± 0.051	0.04	0.97
Paternal kin	−0.27 ± 0.39	−0.70	0.49
Compatibility			
Intercept	0.50 ± 0.29	1.70	0.10
Sex combination (MM vs FF)	0.79 ± 0.27	2.90	0.005
Sex combination (MM vs MF)	−0.058 ± 0.22	−0.26	0.80
Rank difference	0.11 ± 0.06	1.95	0.05
Maternal kin	0.47 ± 0.29	1.60	0.11
Paternal kin	0.60 ± 0.42	1.44	0.15
Tenure	0.002 ± 0.018	0.10	0.91
Age difference	0.004 ± 0.016	0.24	0.80

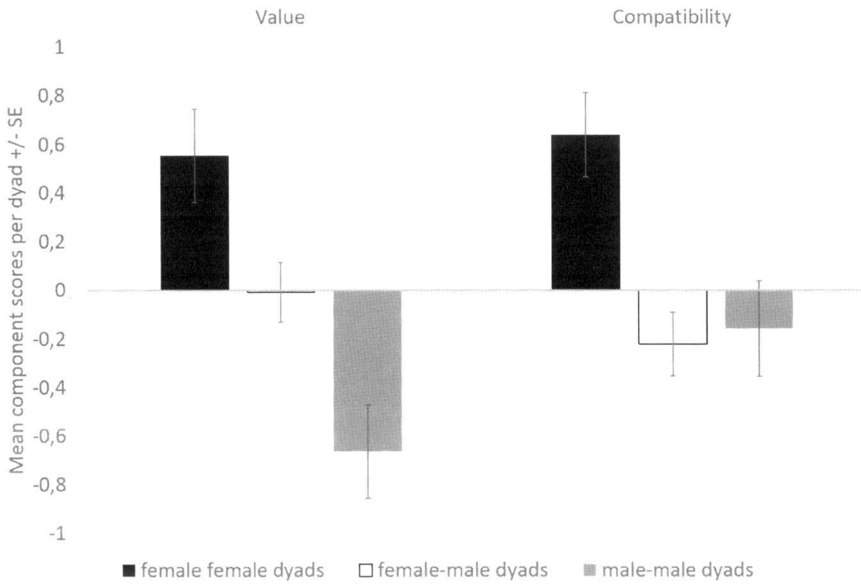

Figure 2. Mean relationship quality (PCA component scores) of female–female, female–male and male–male dyads.

kin. Dyads that had longer tenure had higher value, and value increased with age difference.

Compatibility was influenced by sex combination and rank difference. Female–female dyads had higher compatibility than male–female dyads ($p_{adj} = 0.0001$) and male–male dyads ($p_{adj} = 0.06$). There was no significant difference in compatibility between male–male and male–female dyads ($p_{adj} = 0.45$). Compatibility increased with increasing rank distance, indicating that dyads which were close in the hierarchy were less compatible.

4. Discussion

Using traditional methods, we found a three-component model for relationship quality in bonobos, similar to that reported for chimpanzees (Fraser et al., 2008; Koski et al., 2012), although some behavioural variables could not be included in this study. This contradicts an earlier attempt to use PCA to measure relationship quality in bonobos, using less behavioural variables and including only one study group (Jaeggi et al., 2010), but provides further evidence that, even when using different behavioural variables, the three component structure as proposed by Cords & Aureli (2000) can emerge in various species even when social organisation differs (Majolo et al., 2010; Fraser & Bugnyar, 2011; McFarland & Majolo, 2011). However, when we used Parallel Analysis to avoid overestimation of the number of factors (Horn, 1965), we only found the first two components of relationship quality. Perhaps the second method can be seen as more conservative and more reliable. Since other studies have not used Parallel Analysis, a direct comparison becomes more difficult. On the other hand, the results of the two PCAs were very similar, with only grooming reciprocity changing from component and counter-intervention did not load on either of the components in the second PCA.

Irrespective of our methods, in bonobos the first component 'Value' is very similar to the one reported for chimpanzees (Fraser et al., 2008; Koski et al., 2012), macaques (Majolo et al., 2010; McFarland & Majolo, 2011) and ravens (Fraser & Bugnyar, 2010), suggesting that this factor is consistent across taxa. Our second component, 'Compatibility' included aggression frequency and aggression symmetry. Most studies report aggression frequency in the compatibility factor, but aggression symmetry has not been included in any of the previous studies. Other variables reported in compatibility components include: tolerance to approaches (Fraser et al., 2008; Fraser & Bugnyar, 2010; Majolo et al., 2010; McFarland & Majolo, 2011), proximity during feeding (Majolo et al., 2010), support (McFarland & Majolo, 2011), and counter-intervention (Fraser et al., 2008; Fraser & Bugnyar, 2010; Koski et al., 2012). In our study 'tolerance to approaches' and 'proximity during feeding' could not be included for practical reasons; 'support' grouped with the first component, as it does in most of the other studies; and 'counter-intervention' only had weaker loadings on the compatibility factor in either of the PCAs, if we follow Budaev's advice that minimum loadings should be 0.5 or higher (but see Koski et al., 2012 for the use of minimal loadings of 0.4 or higher). In general, our compatibility component matches closely to that reported in previous studies. In our traditional analysis, using Kaiser's criterion, the third component of relationship quality contained 'grooming symmetry' and 'counter-intervention' and we labelled it, tentatively, as security. However, since the Parallel Analysis indicated this last component was not statistically valid, we suggest the third factor should not be retained, and suggest a more cautious interpretation of this third factor.

In general, the full three-component model, as proposed by Cords & Aureli (2000), has only been found in chimpanzees (Fraser et al., 2008) and ravens (Fraser & Bugnyar, 2011). Notably the third component, 'security', defined as "the predictability of the partner's response to social interactions" (Cords & Aureli, 2000), seems to be least consistent across studies, and is not always found or differs greatly between studies. This may be partly due to different behavioural variables entered in the model. In one chimpanzee study, the security component contained a measure of consistency of affiliation and grooming symmetry (Fraser et al., 2008). Grooming symmetry was the only variable in a component labelled 'security' in a study on Barbary macaques (McFarland & Majolo, 2011), but in Japanese macaques the component labelled as '(in)security' comprised aggression and a measure for grooming variability (Majolo et al., 2010). Furthermore, few studies (including this one) could actually include measures of temporal consistency within a dyad. Finally, these consistency measures may also depend on the length of the observation period. In general, this component is less consistent across studies and labelling seems to be more subjective than in the first two components. As mentioned, the inclusion of the Parallel Analysis suggests that the third factor may also not be statistically valid in our study, and therefore, we recommend that future studies also apply this procedure when studying relationship quality and especially before making generalisations about the three-component structure of relationship quality.

Our review above shows that finding ubiquitous components of relationship quality across species can be challenging. Deciding which behavioural variables are entered in the model will depend on practicalities and observational conditions, but the question remains whether there are universal indicators of relationship quality and whether these can or should be complemented with species-specific indicators of relationship quality. The six studies so far all have some behavioural variables in common but differ in many others. Measures of grooming/preening frequency, proximity and aggression frequency are common variables in all studies to date, and consistently group together on the first two components. Other variables, such as food transfers in chimpanzees (Fraser et al., 2008), or embraces in spider monkeys (Rebecchini et al., 2011), are more species specific and may never or only rarely be observed in the other species. For example in spider monkeys, support in was not observed in sufficient number of dyads and was not included in the analyses (Rebecchini et al., 2011). Finally, there are variables that do not consistently cluster onto the same components, such as agonistic support, which in Barbary macaques was found to load on the factor labelled Compatibility (McFarland & Majolo, 2011), but loaded on the Value component in Japanese macaques (Majolo et al., 2010), ravens (Fraser & Bugnyar, 2012), chimpanzees (Fraser et al., 2008; Koski et al., 2012) and bonobos (this study). It has been suggested that studying relationships at a dyadic level may not always be appropriate, since relationships are rarely symmetrical and may have different implications for a subordinate member of a dyad compared to a dominant member of the dyad (Majolo et al., 2010; McFarland & Majolo, 2011). Previous studies in bonobos have shown that certain behaviours, such as grooming, can be fairly asymmetric and can be influenced by dominance status of the receiver (Vervaecke et al., 2000b, c; Stevens et al., 2005b; Jaeggi et al., 2010). In the current study we attempted to include symmetry measures for grooming and for aggression in the dyadic analyses, but found that they always grouped with the frequency variables in the Principal Component Analysis. A future study looking at relationship quality on individual rather than dyadic levels, such as has been done in macaques (Majolo et al., 2010; McFarland & Majolo, 2011), may also be useful in bonobos, so that the asymmetric nature of relationships can be further investigated. Studying relationship value at a dyadic level has allowed us to investigate in more detail the effects of kinship, sex combination, relationship tenure and similarities in rank and age in a more objective way than was previously possible.

4.1. Kinship

Kinship had no significant effect on compatibility, but bonobos that were maternally related were found to have higher relationship value than unrelated dyads. This is in line with expectations of inclusive fitness benefits between related individuals (Silk, 2002, 2007) and this higher value was also found in related dyads in chimpanzees (Fraser et al., 2008; Koski et al., 2012), ravens (Fraser & Bugnyar, 2010) and macaques (Majolo et al., 2010). Due to small sample size it was not possible to test mother–son and maternal kin relationships separately, and both were included in our factor 'kin'. In general, maternal (half-) siblings in bonobos do not seem to form close bonds (Kano, 1992), so the main effect of kin should be attributed to mother–son dyads. In bonobos, mother–son bonds have been shown to be important, as mothers provide agonistic support to their sons even into adulthood (Stevens et al., 2006; Furuichi, 2011) and can influence the mating success of their adult sons (Surbeck et al., 2011) In contrast to other studies in macaques (Widdig, 2007; Schülke et al., 2013), but similar to studies on wild chimpanzees (Langergraber et al., 2007; Mitani, 2009), we did not find strong effects of paternal kin relationships on either component of relationship quality.

4.2. Sex combination

In captive bonobo groups, female–female dyads were characterised by high value and compatibility. Unlike in previous studies on social bonding in female bonobos (Parish, 1996), all but one female–female dyads in our study were unrelated, demonstrating that high value of female–female value and compatibility in captive groups need not be a side effect of relatedness and underscoring that kinship is not always required for maintenance of high value relationships (Langergraber et al., 2007; Mitani, 2009; Seyfarth & Cheney, 2012). These results were not unexpected, since we previously reported higher frequency of support among unrelated female bonobos, even though grooming was never more common among female–female dyads and proximity was not always higher in these dyads (Stevens et al., 2006). By using the composite measure value, we can show that the component value is significantly higher in female–female dyads. In matrilocal macaques, female–female dyads also have higher relationship value (Majolo et al., 2010; McFarland & Majolo, 2011), but studies on chimpanzees found no effect of sex combination on relationship value. This makes bonobos unique in being a female-dispersal species, where higher value is found between unrelated females than between mostly related males. Like in our results on bonobos, the chimpanzee studies showed higher compatibility among female–female dyads (Fraser et al., 2008; Koski et al., 2012) and security was highest in male–female dyads and lowest in male–male dyads (Fraser et al., 2008).

While previous studies have suggested that unrelated female chimpanzees can also form close bonds (Baker & Smuts, 1994; Langergraber et al., 2009; Lehmann & Boesch, 2009), our study, and comparisons with relationship quality data of captive chimpanzees (Fraser et al., 2008; Koski et al., 2012), suggests that these bonds may differ between species in the aspect of relationship value, which is high in female–female dyads among bonobos but not chimpanzees. In bonobos it still remains to be studied what the effects of high value and high compatibility female bonds are on individual fitness (cf., Silk, 2007). Female–female coalitions may help individual females to obtain and maintain a higher rank, but fitness effects of dominance have not been shown in female bonobos. In chimpanzees, such rank effects have been documented in the wild, and high-ranking females have higher reproductive success (Pusey et al., 1997). Data on the direct link between relationship quality and reproductive success are still missing for chimpanzees and bonobos. In baboons, high-ranking females also tend to mature at an earlier age, grow faster, produce healthier infants, have shorter interbirth intervals, live longer and have higher lifetime fitness than lower ranking females (Silk et al., 2003, 2009, 2010; Cheney et al., 2004), but also direct relations between females with strong social relationships and higher reproductive success are reported (Silk et al., 2003, 2009; Silk, 2007). Such fitness effects may be hard to detect in captive primates, and wild studies on the effects of female bonding and dominance on fitness outcomes will be more appropriate.

For male–female dyads, value was significantly lower than female–female dyads but still higher compared to male–male dyads. This can be partly attributed to the mother–son dyads that were included in this category, but it has also been suggested for wild bonobos that males may gain fitness benefits from maintaining long-term friendships with unrelated females (Hohmann & Fruth, 2003; Surbeck et al., 2012). Compatibility in male–female dyads was lower than female–female dyads but did not significantly differ from male–male dyads. Indeed aggression is most common between males and from females to males, but much more rare between females or from males to females (Hohmann & Fruth, 2003). Male–male dyads had lower value than male–female or female–female dyads, and lower compatibility than female–female, but not compared to male–female dyads. Similar patterns of behaviour are observed in the wild, where males rarely groom or support each other (Ihobe, 1992; Kano, 1992), but do show high grooming rates with females (Surbeck et al., 2012) and male–male aggression is most common (Kano, 1992; Hohmann & Fruth, 2003; Surbeck et al., 2012). This opens intriguing questions why male–male relationships in bonobos seem to be more different from those in chimpanzees. In chimpanzees, participation in male coalitions has been shown to influence male mating success, providing a direct link between social bonding and fitness (Gilby et al., 2013). For bonobos, coalitions between males are uncommon (Kano, 1992; Stevens et al., 2005b) and seem to have no fitness benefits. Future studies could focus more on the nature of male–male relationships in wild and captive conditions.

4.3. *Rank difference*

Rank distance can have a large influence on affiliative behaviour (de Waal, 1991). Previous studies in bonobos have shown that individuals with similar rank groomed more reciprocally but not more frequently (Vervaecke et al., 2000c). In our study, there was no effect of rank difference and relationship Value. However, relationship Compatibility was influenced by rank difference, indicating that aggression was more frequent and more reciprocal in dyads that had smaller rank differences. Since we did not discriminate between mild and severe forms of aggression, it is possible that individuals close in rank have frequent aggressive encounters that are relatively mild; the threshold to reciprocate aggression may be lower in closely ranked individuals, leading to more reciprocal aggression. Rank distance was not investigated in the studies on relationship quality of chimpanzees (Fraser et al., 2008; Koski et al., 2012), and proved to have no effect on components of relationship quality in Japanese macaques (Majolo et al., 2010) or Barbary macaques (McFarland & Majolo, 2011). The fact that we included aggression symmetry in our analyses, and that this variable grouped with aggression frequency, may also explain why we found an effect. Previous studies in wild and zoo-housed bonobos have shown that group hierarchies in captive bonobos are fairly steep, and male hierarchies especially are steep and asymmetric (Stevens et al., 2007; Surbeck et al., 2010). This underscores the asymmetric nature of social relationships as suggested by Majolo et al. (2010).

4.4. *Relationship tenure*

We found that dyads with longer relationship tenure had higher value. This effect was also found in chimpanzees, where security also decreased with increasing tenure, which the authors could not explain (Fraser et al., 2008). Compatibility (indicating frequency and symmetry of aggression) was not influenced by relationship tenure in our study, similar to what has been found in chimpanzees (Fraser et al., 2008). The fact that value increased with relationship tenure indicates that it may take some time before high value relationships are established, even though the response of bonobos towards young (female) strangers has been shown to be positive (Idani, 1991; Tan & Hare, 2013). The reciprocal interchange of grooming for support may act as a functional mechanism to strengthen bonds over time, especially between female bonobos (Vervaecke et al., 2000a, b, c; Stevens et al., 2005b). Further studies should compare these interchange mechanisms between chimpanzees and bonobos.

4.5. Age difference

We found no significant effects of age difference on either component of relationship quality.

Age effects on relationship quality are rarely found. Fraser et al. (2008) found higher value in dyads of more similar age and explained this using the similarity principle proposed by de Waal & Luttrel (1986), which states that individuals of similar age might have similar needs and possibilities in a social group. There could, however, be an interaction between age and relationship length or tenure. Fraser et al. (2008) studied one group of chimpanzees, in which no adult chimpanzees were introduced and individuals in the study were often born in the group and grew up together. However, in our sampled group of bonobos, dyads with longer tenure do not necessarily have more similar ages, as there has been a high frequency of individual transfers between groups due to the breeding program, so that some individuals with very similar ages were only recently introduced to one another.

Additionally, not only age differences but the actual age of the individuals in the dyads may be influencing the results. Relationships between adolescent bonobos could differ from relationships between adult bonobos in aspects of value and compatibility.

As this is the first large-scale study on relationship quality in bonobos, repeatability of this model should be tested by including more captive and wild populations. Nonetheless, thus far it seems that the first two components, value and compatibility, are very similar to those reported for chimpanzees. However, at least in captivity, the two species differ in that female–female dyads have high relationship value in bonobos, but not in chimpanzees, while in both species relationship compatibility is higher in female–female dyads. Relationships between unrelated bonobo female are characterised by high relationship value and compatibility, which may be the key to their success in maintaining long term bonds, which seem to strengthen over time, perhaps by mechanisms of reciprocal altruism.

Acknowledgements

We are very grateful to Brian Hare and Shinya Yamamoto for inviting us to contribute to the IPS Congress 2013 Symposium and this accompanying Special Issue on 'Bonobos: The newly discovered ape'. We thank the curators and keepers of Planckendael Wild Animal Park; Wuppertal Zoo; Apenheul Primate Park and Twycross Zoo for their kind assistance during data collection in these institutions. We thank two anonymous reviewers for their constructive comments on an earlier version of the manuscript. Sumir Keenan did careful language editing. The CRC is structurally supported by the Flemish Government.

References

Badrian, A. & Badrian, N. (1984). Social organization of *Pan paniscus* in the Lomako Forest, Zaire. — In: The pygmy chimpanzee: evolutionary ecology and behaviour (Susman, R.L., ed.). Plenum Press, New York, NY, p. 325-346.

Baker, K.C. & Smuts, B.B. (1994). Social relationships of female chimpanzees: diversity between captive social groups. — In: Chimpanzee cultures (Wrangham, R.W., McGrew, W.C., de Waal, F.B.M. & Heltne, P.G., eds). Harvard University Press, Cambridge, MA, p. 227-242.

Budaev, S.V. (2010). Using principal components and factor analysis in animal behaviour research: caveats and guidelines. — Ethology 116: 472-480.

Cattell, R.B. (1966). The scree test for the number of factors. — Multivar. Behav. Res. 1: 245-276.

Cheney, D.L., Seyfarth, R.M., Fischer, J., Beehner, J.C., Bergman, T.J., Johnson, S.E., Kitchen, D.M., Palombit, R.A., Rendall, D. & Silk, J.B. (2004). Factors affecting reproduction and mortality among baboons in the Okavango Delta, Botswana. — Int. J. Primatol. 25: 401-428.

Cords, M. & Aureli, F. (2000). Reconciliation and relationship qualities. — In: Natural conflict resolution (Aureli, F. & de Waal, F.B.M., eds). University of California Press, Berkeley, CA, p. 177-198.

de Vries, H., Netto, W.J. & Hanegraaf, P.L.H. (1993). Matman: a program for the analysis of sociometric matrices and behavioural transition matrices. — Behaviour 125: 157-175.

de Waal, F.B.M. (1978). Join-aggression and protective-aggression among captive *Macaca fascicularis*. — In: Recent advances in primatology (Chivers, D.J. & Herbert, J., eds). Academic Press, New York, NY, p. 577-579.

de Waal, F.B.M. (1991). Rank distance as a central feature of rhesus monkey social organization: a sociometric analysis. — Anim. Behav. 41: 383-395.

de Waal, F.B.M. & Luttrell, L.M. (1986). The similarity principle underlying social bonding among female rhesus monkeys. — Folia Primatol. 46: 215-234.

Feh, C. (1999). Alliances and reproductive success in Camargue stallions. — Anim. Behav. 57: 705-713.

Franklin, S.B., Gibson, D.J., Robertson, P.A., Pohlmann, J.T. & Fralish, J.S. (1995). Parallel analysis: a method for determining significant principal components. — J. Vegetat. Sci. 6: 99-106.

Franz, C. (1999). Allogrooming behavior and grooming site preferences in captive bonobos (*Pan paniscus*): association with female dominance. — Int. J. Primatol. 20: 525-546.

Fraser, O.N. & Bugnyar, T. (2010). The quality of social relationships in ravens. — Anim. Behav. 79: 927-933.

Fraser, O.N., Schino, G. & Aureli, F. (2008). Components of relationship quality in chimpanzees. — Ethology 114: 834-843.

Frère, C.H., Krützen, M., Mann, J., Connor, R.C., Bejder, L. & Sherwin, W.B. (2010). Social and genetic interactions drive fitness variation in a free-living dolphin population. — Proc. Natl. Acad. Sci. USA 107: 19949-19954.

Furuichi, T. (1989). Social interactions and the life history of female *Pan paniscus* in Wamba, Zaire. — Int. J. Primatol. 10: 173-197.

Furuichi, T. (2011). Female contributions to the peaceful nature of bonobo society. — Evol. Anthropol. 20: 131-142.

Furuichi, T., Idani, G., Ihobe, H., Kuroda, S., Kitamura, K., Mori, A., Enomoto, T., Okayasu, N., Hashimoto, C. & Kano, T. (1998). Population dynamics of wild bonobos (*Pan paniscus*) at Wamba. — Int. J. Primatol. 19: 1029-1043.

Gilby, I.C., Brent, L.J.N., Wroblewski, E.E., Rudicell, R.S., Hahn, B.H., Goodall, J. & Pusey, A.E. (2013). Fitness benefits of coalitionary aggression in male chimpanzees. — Behav. Ecol. Sociobiol. 67: 373-381.

Goodall, J. (1986). The chimpanzees of Gombe: patterns of behavior. — The Belknap Press, Cambridge, MA.

Hinde, R.A. (1976). Interactions, relationships and social structure. — Man 11: 1-17.

Hohmann, G. & Fruth, B. (2002). Dynamics in social organization of bonobos (*Pan paniscus*). — In: Behavioural diversity in chimpanzees and bonobos (Boesch, C., Hohmann, G. & Marchant, L.F., eds). Cambridge University Press, Cambridge, p. 138-150.

Hohmann, G. & Fruth, B. (2003). Intra- and inter-sexual aggression by bonobos in the context of mating. — Behaviour 140: 1389-1413.

Hohmann, G., Gerloff, U. & Fruth, B. (1999). Social bonds and genetic tests: kinship, association and affiliation in a community of bonobos (*Pan paniscus*). — Behaviour 136: 1219-1235.

Horn, J.L. (1965). A rationale and test for the number of factors in factor analysis. — Psychometrika 30: 179-185.

Idani, G. (1995). Function of peering behaviour among bonobos (*Pan paniscus*) at Wamba, Zaire. — Primates 36: 377-383.

Ihobe, H. (1992). Male–male relationships among wild bonobos (*Pan paniscus*) at Wamba, Republic of Zaire. — Primates 33: 163-179.

Jaeggi, A.V., Stevens, J.M.G. & van Schaik, C.P. (2010). Tolerant food sharing and reciprocity is precluded by despotism among bonobos but not chimpanzees. — Am. J. Phys. Anthropol. 143: 41-51.

Kaiser, H.F. (1960). The application of electronic computers to factor analysis. — Ed. Psychol. Meas. 20: 141-151.

Kano, T. (1980). Social behavior of wild pygmy chimpanzees (*Pan paniscus*) of Wamba: a preliminary report. — J. Hum. Evol. 9: 243-260.

Kano, T. (1992). The last ape: pygmy chimpanzee behavior and ecology. — Stanford University Press, Stanford, CA.

Kappeler, P.M. & van Schaik, C.P.M. (2002). Evolution of primate social systems. — Int. J. Primatol. 23: 707-740.

Koski, S.E., de Vries, H., van de Kraats, A. & Sterck, E.H.M. (2012). Stability and change of social relationship quality in captive chimpanzees. — Int. J. Primatol. 33: 905-921.

Kummer, H. (1978). On the value of social relationships to non-human primates: a heuristic scheme. — Soc. Sci. Inform. 17: 697-705.

Langergraber, K.E., Mitani, J. & Vigilant, L. (2007). The limited impact of kinship on cooperation in wild chimpanzees. — Proc. Natl. Acad. Sci. USA 104: 7786-7790.

Langergraber, K.E., Mitani, J. & Vigilant, L. (2009). Kinship and social bonds in female chimpanzees (*Pan troglodytes*). — Am. J. Primatol. 71: 840-851.

Langergraber, K.E., Prüferc, K., Rowney, C., Boesch, C., Crockford, C., Fawcett, K., Inoue, E., Inoue-Muruyama, M., Mitani, J.C., Muller, M.N., Robbins, M.M., Schubert, G., Stoinski, T.S., Viola, B., Watts, D., Wittig, R.M., Wrangham, R.W., Zuberbühler, K., Pääbo, S. & Vigilant, L. (2012). Generation times in wild chimpanzees and gorillas suggest earlier divergence times in great ape and human evolution. — Proc. Natl. Acad. Sci. USA 109: 15716-15721.

Langergraber, K.E., Mitani, J.C., Watts, D.P. & Vigilant, L. (2013). Male–female socio-spatial relationships and reproduction in wild chimpanzees. — Behav. Ecol. Sociobiol. 67: 861-873.

Lehmann, J. & Boesch, C. (2009). Sociality of the dispersing sex: the nature of social bonds in West African female chimpanzees, *Pan troglodytes*. — Anim. Behav. 77: 377-387.

Majolo, B., Ventura, R. & Schino, G. (2010). Asymmetry and dimensions of relationship quality in the Japanese macaque. — Int. J. Primatol. 31: 736-750.

Martin, P. & Bateson, P. (1993). Measuring behaviour: an introductory guide, 2nd edn. — Cambridge University Press, Cambridge.

Massen, J.J., Overduin-de Vries, A.M., de Vos-Rouweler, A.J., Spruijt, B.M., Doxiadis, G.G. & Sterck, E.H. (2012). Male mating tactics in captive rhesus macaques (*Macaca mulatta*): the influence of dominance, markets, and relationship quality. — Int. J. Primatol. 33: 73-92.

McFarland, R. & Majolo, B. (2011). Exploring the components, asymmetry and distribution of relationship quality in wild barbary macaques (*Macaca sylvanus*). — Plos One 6: e28826.

Mitani, J.C. (2009). Male chimpanzees form enduring and equitable social bonds. — Anim. Behav. 77: 633-640.

Moscovice, L.R., Fiore, A.D., Crockford, C., Kitchen, D.M., Wittig, R., Seyfarth, R.M. & Cheney, D.L. (2010). Hedging their bets? Male and female chacma baboons form friendships based on likelihood of paternity. — Anim. Behav. 79: 1007-1015.

O'Connor, B.P. (2000). SPSS and SAS programs for determining the number of components using parallel analysis and velicer's MAP test. — Behav. Res. Methods Instrum. Comput. 32: 396-402.

Parish, A.R. (1996). Female relationships in bonobos (*Pan paniscus*). — Hum. Nat. 7: 61-96.

Parish, A.R. & de Waal, F.B.M. (2000). The other "closest living relative": how bonobos (*Pan paniscus*) challenge traditional assumptions about females, dominance, intra- and intersexual interactions, and hominid evolution. — Ann. NY Acad. Sci. 907: 97-113.

Pereboom, Z., Stevens, J.M.G. & Van der Eerden, I. (2011). International studbook for bonobo, *Pan paniscus* Schwarz 1929. — Royal Zoological Society of Antwerp, Antwerp.

Pusey, A., Williams, J. & Goodall, J. (1997). The influence of dominance rank on reproductive success of female chimpanzees. — Science 277: 828-831.

Rebecchini, L., Schaffner, C. & Aureli, F. (2011). Risk is a component of social relationships in spider monkeys. — Ethology 117: 691-699.

Schülke, O., Bhagavatula, J., Vigilant, L. & Ostner, J. (2010). Social bonds enhance reproductive success in male macaques. — Curr. Biol. 20: 2207-2210.

Schülke, O., Wenzel, S. & Ostner, J. (2013). Paternal relatedness predicts the strength of social bonds among female rhesus macaques. — Plos One 8: e59189.

Seyfarth, R.M. & Cheney, D.L. (2012). The evolutionary origins of friendship. — Annu. Rev. Psychol. 63: 153-177.

Silk, J.B. (2007). Social components of fitness in primate groups. — Science 317: 1347-1351.

Silk, J.B., Alberts, S.C. & Altman, J. (2003). Social bonds of female baboons enhance infant survival. — Science 302: 1231-1234.

Silk, J.B., Beehner, J.C., Berglan, T.J., Crockford, C., Engh, A.L., Moscovice, L.R., Wittig, R.M., Seyfarth, R.M. & Cheney, D.L. (2009). The benefits of social capital: close social bonds among female baboons enhance offspring survival. — Proc. Roy. Soc. Lond. B: Biol. Sci. 276: 3099-3104.

Silk, J.B., Beehner, J.C., Berglan, T.J., Crockford, C., Engh, A.L., Moscovice, L.R., Wittig, R.M., Seyfarth, R.M. & Cheney, D.L. (2010). Strong and consistent social bonds enhance the longevity of female baboons. — Curr. Biol. 20: 1359-1361.

Stanford, C.B. (1998). The social behavior of chimpanzees and bonobos. — Curr. Anthropol. 39: 399-420.

Stevens, J.M.G., Vervaecke, H., de Vries, H. & Van Elsacker, L. (2005a). Peering is not a formal indicator of subordination in bonobos (*Pan paniscus*). — Am. J. Prim. 65: 255-267.

Stevens, J.M.G., Vervaecke, H., de Vries, H. & Van Elsacker, L. (2005b). The influence of the steepness of dominance hierarchies on reciprocity and interchange in captive groups of bonobos (*Pan paniscus*). — Behaviour 142: 941-960.

Stevens, J.M.G., Vervaecke, H., de Vries, H. & Van Elsacker, L. (2006). Social structures in *Pan paniscus*: testing the female bonding hypothesis. — Primates 47: 210-217.

Stevens, J.M.G., Vervaecke, H., de Vries, H. & Van Elsacker, L. (2007). Sex differences in the steepness of dominance hierarchies in captive bonobo groups. — Int. J. Primatol. 28: 1417-1430.

Stumpf, R. (2007). Chimpanzees and bonobos: diversity within and between species. — In: Primates in perspective (Campbell, C.J., Fuentes, A., MacKinnon, K.C., Panger, M. & Bearder, S.K., eds). Oxford University Press, Oxford, p. 321-344.

Surbeck, M., Mundry, R. & Hohmann, G. (2011). Mothers matter! Maternal support, dominance status and mating success in male bonobos (*Pan paniscus*). — Proc. Roy. Soc. Lond. B: Biol. Sci. 278: 590-598.

Surbeck, M., Deschner, T., Schubert, G., Weltring, A. & Hohmann, G. (2012). Mate competition, testosterone and intersexual relationships in bonobos, *Pan paniscus*. — Anim. Behav. 85: 659-669.

Tan, J. & Hare, B. (2013). Bonobos share with strangers. — Plos One 8: e51922.

Vervaecke, H., de Vries, H. & Van Elsacker, L. (2000a). Dominance and its behavioral measures in a captive group of bonobos (*Pan paniscus*). — Int. J. Primatol. 21: 47-68.

Vervaecke, H., de Vries, H. & Van Elsacker, L. (2000b). Function and distribution of coalitions in captive bonobos (*Pan paniscus*). — Primates 41: 249-265.

Vervaecke, H., de Vries, H. & Van Elsacker, L. (2000c). The pivotal role of rank in grooming and suport behavior in a captive group of bonobos (*Pan paniscus*). — Behaviour 137: 1463-1485.

White, F.J. (1988). Party composition and dynamics in *Pan paniscus*. — Int. J. Primatol. 9: 179-193.

Widdig, A. (2007). Paternal kin discrimination: the evidence and likely mechanisms. — Biol. Rev. 82: 319-334.

Wiszniewski, J., Corrigan, S., Beheregaray, L.B. & Möller, L.M. (2012). Male reproductive success increases with alliance size in Indo-Pacific bottlenose dolphins (*Tursiops aduncus*). — J. Anim. Ecol. 81: 423-431.

Zwick, W.R. & Velicer, W.F. (1986). Comparison of five rules for determining the number of components to retain. — Psychol. Bull. 99: 432-442.

[When citing this chapter, refer to Behaviour 152 (2015) 285–311]

Prolonged maximal sexual swelling in wild bonobos facilitates affiliative interactions between females

Heungjin Ryu, David A. Hill and Takeshi Furuichi *

Primate Research Institute, Kyoto University, Inuyama, Aichi, Japan

*Corresponding author's e-mail address: furuichi@pri.kyoto-u.ac.jp

Accepted 6 May 2014; published online 12 June 2014

Abstract

Perineal sexual skin swelling in relation to menstrual cycle occurs in a variety of primate taxa. However, sexual swelling with exaggerated size and colour is found only in some Old World monkeys and the two *Pan* species. Although several hypotheses have been proposed (e.g., reliable indicator hypothesis and graded signal hypothesis), it seems unlikely that a single explanation can account for the significance of the sexual swelling in all of these species. Bonobos (*Pan paniscus*) provide an excellent opportunity for studying sexual swelling since they have the most prolonged maximal swelling periods among primates. In this study we propose a new hypothesis that sexual swelling in female bonobos increases their attractiveness to other females and thereby facilitates affiliative social interaction with them. We found that free-ranging female bonobos with maximal sexual swelling engaged in affiliative social interactions with other females, including genito-genital rubbing, staying in close proximity and grooming, more frequently than females without maximal swelling. These tendencies suggest that females with maximal swelling were attractive to other females. The results also suggest that the benefits of maximal swelling might vary among females depending on their life-history stage. In particular, young females may get more benefits from prolonged maximal swelling through increased grooming reciprocity and staying in close proximity to other females. Thus our study supported the hypothesis that one function of prolonged maximal swelling in bonobos is to increase attractiveness to other females, thereby enhancing affiliative relationships between females in a male-philopatric social system.

Keywords

sexual swelling, female bonobos, life-history, social interactions.

1. Introduction

The term sexual swelling is used to describe swelling of the skin of the perineum in relation to the menstrual cycle of females. It has been described in a variety of primate species including prosimians (e.g., ring-tailed lemur,

Lemur catta (Jolly, 1966), and Horsfield's tarsier, *Tarsius bancanus* (Wright et al., 1986)), Old World monkeys (e.g., Japanese macaques, *Macaca fuscata* (Tokuda, 1961) and baboons, *Papio hamadryas* (Pocock, 1906)), and ape species (e.g., bonobos, *Pan paniscus* (Savage-Rumbaugh & Wilkerson, 1978; Kuroda, 1980) and chimpanzees, *Pan troglodytes* (Zuckerman & Fulton, 1934)).

Sexual swelling of very conspicuous size and colour, which has been described as 'exaggerated sexual swelling' (Nunn, 1999), is found only in several Old World monkey species and the two *Pan* species. In the rest of this paper, the term 'sexual swelling' is used synonymously with Nunn's (1999) definition of the exaggerated sexual swelling. It has also been suggested that this conspicuous sexual swelling has evolved independently in Old World anthropoids at least three times (Dixson, 1983; Nunn, 1999). In many species, sexual swelling reaches its maximal size near ovulation (e.g., *Macaca nigra* (Dixson, 1977); *Papio anubis* and *Papio cynocephalus* (Hendrickx & Kraemer, 1969); and *Pan troglodytes* (Dahl et al., 1991)) and most mating activity is concentrated in the maximal swelling periods (Saayman, 1970; Furuichi, 1987; Bercovitch, 1988; Wallis, 1992; Young et al., 2013). Sexual swelling is thought to incur some cost for females, such as possible hindrance of metabolism from water retention (Nunn, 1999), increased chance of injuries (Hausfater, 1975) and increased body weight (Bielert & Busse, 1983), which can increase the energetic cost of locomotion (Nunn, 1999). Therefore, it is assumed that there must also be advantages to females of having sexual swelling that outweigh the costs (Nunn, 1999; Zinner et al., 2004; Alberts & Fitzpatrick, 2012).

Several hypotheses have been proposed to explain the function and evolution of sexual swelling in the context of female reproductive strategies. Some researchers have suggested that sexual swelling invokes frequent promiscuous mating and inter-male competition that are beneficial for females (the best-male hypothesis (Clutton-Brock & Harvey, 1976); many-male hypothesis (Hrdy & Whitten, 1987)). Others have suggested that the sexual swelling encourages dominant males to mate-guard and recognize paternity by accurately indicating the timing of ovulation, and thus to provide support for the females and/or offspring (obvious-ovulation hypothesis (Hamilton, 1984); male-services hypothesis (reviewed in Nunn, 1999); and paternal care hypothesis (Alberts & Fitzpatrick, 2012)). Other explanations include a comprehensive hypothesis which puts together the functions of sexual swelling derived from several hypotheses (graded signal hypothesis; Nunn, 1999) and the reliable quality indicator hypothesis (Pagel, 1994) which assumes that sexual swelling may indicate reproductive value of females. Furthermore, some groups of researchers have suggested variations of a social passport hypothesis. For example, that sexual swelling facilitates between-group transfer of adolescent females by showing resident males their oestrous status (Anderson & Bielert, 1994), increases support from males in female–female competitive situations especially during immigration (Boesch & Boesch-Achermann, 2000), or helps female–female interactions (Paoli et al., 2006). However, it seems unlikely that a single benefit inferred from a single hypothesis could account for all aspects of the phenomenon in all taxa. Although sexual swelling is found only in species with multi-male societies (Clutton-Brock & Harvey, 1976), social systems vary among those species. It seems likely, therefore, that benefits of sexual swelling for females also differ between species and should be considered in the context of the social system in which they occur.

Bonobos (*Pan paniscus*) provide a unique opportunity for examining the function of sexual swelling as they have very prolonged periods of maximal swelling (Thompson-Handler et al., 1984; Furuichi, 1987; Dahl et al., 1991; Kano, 1992) and also females that achieve high social status in a male philopatric society (Furuichi, 1997, 2011; Surbeck et al., 2011). In many primate species sexual swelling reaches its maximal size near ovulation (Wildt et al., 1977; Dahl et al., 1991; Emery & Whitten, 2003; Young et al., 2013) and affects male behaviours related to mating, such as consortship (Tokuda, 1961; Hall & de Vore, 1965; Hill, 1987) or coercive mate guarding (Smuts & Smuts, 1993; Muller et al., 2007). However, bonobos seem to deviate from this general trend, since the timing of their ovulation and onset of maximal swelling are highly variable (Furuichi, 1987; Heistermann et al., 1996; van Schaik et al., 2000; Reichert et al., 2002, but also see Surbeck et al., 2011) even compared with those of chimpanzees (Deschner et al., 2003). The high level of variation of timing of ovulation and onset of maximal swelling in this species seem to reflect effects of very prolonged periods of maximal swelling (Furuichi, 1987) of female bonobos. Therefore the key to understanding sexual swelling in bonobos may be found in the prolonged period of maximal swelling, which sometimes continues for more than 20 days (Furuichi, 1987; Ryu, unpublished data).

Most previous research on the prolonged period of maximal swelling in bonobos has focused on its role in controlling the behaviour of males, by making females attractive to males even outside the oestrous periods (Furuichi, 1987, 2011; Kano, 1992; de Waal, 1995). This manipulative function of sexual swelling has been noted with the fact that there is no known case of coercive mate guarding and infanticide by males in bonobos (Furuichi & Hashimoto, 2004; Paoli, 2009). It is, therefore, recognized as a successful female counter-strategy toward male mating strategy. However, without considering that female–female and female–male bonding play an important social role and that sexual swelling seems to have an important role in social bonding (Parish, 1994; Furuichi, 2011) in this male philopatric species, it will be impossible for us to reach a comprehensive understanding of the role and evolution of sexual swelling. Therefore it is important to take, not only mating-related function, but also function for social bonding into account for better understanding. This will provide a new direction and another important dimension for future studies. Some researchers have already mentioned other possible functions of this prolonged sexual swelling, such as maximal sexual swelling for genito-genital rubbing (hereafter g-g rubbing; Kuroda, 1980) rather than copulations (Takahata et al., 1996), or providing protection against male coercion by attracting other females rather than males (Paoli et al., 2006; Paoli, 2009), but these have not been considered thoroughly, or examined quantitatively. Considering the finding that g-g rubbing increases for females with maximal swelling (Kano, 1992; Hohmann & Fruth, 2000, but also see Takahata et al., 1996), it is reasonable to assume that sexual swelling influences female–female social relationships in bonobos (Kano, 1992; Takahata et al., 1996; Paoli et al., 2006). In bonobos, new immigrant females, which are likely to have no stable social relationship with others in the group, are eager to interact with senior females (Furuichi, 1989; Idani, 1991), and such new immigrant females show very long-lasting maximal swelling (Furuichi, 1987). Therefore, if females can gain benefits from having maximal swellings in terms of their social interactions with other females, those benefits are likely to vary in relation to their life-history stage.

Based on these characteristics, we hypothesized that one function of prolonged sexual swelling in bonobos is to increase attractiveness to other females and facilitate social interaction with them (Swelling for female attraction). This hypothesis leads to the following predictions:

1. Females with maximal swelling will be invited and/or engage in g-g rubbing more frequently than females without maximal swelling.
2. Females with maximal swelling will have more females in close proximity to them than females without maximal swelling.
3. Females with maximal swelling will spend more time in grooming interactions with other females than females without maximal swelling.
4. Young females with maximal swelling will receive more reciprocated grooming in interactions with other females.

To test this hypothesis and its predictions, we first investigated frequency of g-g rubbing as well as copulation in relation to age and swelling status. Second, we compared the number of females found within 1 m or 5 m of the focal animal, and those found in the same party. Finally, we investigated the frequency and reciprocity of grooming interactions in relation to age and swelling status of females.

2. Materials and methods

2.1. *Study site and subjects*

Data were collected at the long-term bonobo field site at Wamba in the northern sector of the Luo Scientific Reserve, D.R. Congo (Kano, 1992; Furuichi et al., 2012). The research camp is situated in the center of the northern section of the reserve (0°11′07.6″N, 22°37′57.5″E; WGS84). One fully habituated unit-group (or community) of free-ranging wild bonobos, called E1 group, was followed. From 1974 to 1996 researchers conducted field research using provisioning. Field research was stopped in 1996 by the civil war. Since researchers resumed field research in 2003, no artificial food has been provided. E1 group ranges over primary, old and young secondary, and swamp forests and sometimes agricultural fields (Mulavwa et al., 2010). January and February have less rain (Mulavwa et al., 2008), although there is no clear division of dry season and rainy season.

The E1 group consisted of 28 to 31 individuals, including 7 adult and 3 adolescent males and 9 adult and 2 adolescent females during the study periods. Nine adult females, whose estimated ages ranged from 14 to 41 years old in 2012, were the targets of focal sampling (Altmann, 1974). All of them had immigrated into the E1 group and had successfully given birth at least once after their immigration. Among the 9 adult females, 2 females who were more than 38 years old (Table 1) were classified as old, and 4 females from 21 to 29 years old were classified as middle-aged. Three females of less than 20 years old were classified as young. All of the young adult females had just one record of successful parturition in the E1 group after their immigration and had an infant during the study periods. One of 2 adolescent females (Nc) was the daughter of the oldest female (No) and the other (Zn) immigrated in October 2011, and they were not included in focal sampling. Hs and Sl were already members of the E1 group at the time that field study was resumed in 2003, therefore their immigration was earlier than August 2003. Yk and Jk joined the E1 group in 2004 with infants, which was considered a possible example of group fusion (for more details, see Hashimoto et al., 2008).

2.2. *Assessment of sexual skin swelling*

Sexual swelling was scored on a daily basis in relation to firmness and size of each individual (Furuichi, 1987). Based on these records, the sexual swelling status of females was assigned to one of three categories: non-swelling (sw1), intermediate swelling (sw2) and maximal swelling (sw3). Firmness of sexual swelling was the key feature used to distinguish sw3 from sw2 since the size of swelling between 2 and 3, especially of young females, was not very different. Moreover because of great individual variation in swelling size, it has been suggested that firmness is a more reliable measure of sexual swelling in bonobo than size (Furuichi, 1987). To avoid observation errors and individual bias in the assessment of sexual skin swelling, at least one researcher and two research assistants discussed the swelling status of each female during field observations and during daily meetings in the evening. Sexual swelling of one female (Sl) did not reach sw3 during the study periods. This might reflect its pregnancy and postnatal infertility since it gave birth in December 2012 and was nursing its infant until the end of the second study period.

2.3. *Behavioural observation*

Data were collected by the first author from September 2011 to January 2012, and June to September 2012. During the study period, bonobos were followed for 1004.6 h on 134 days (7.57 ± 2.46 h per day). Bonobos were followed from their night beds, located before 06:00 h, until they made new

Table 1.
Composition of females of E1 group and focal following hours (2012).

Name	Age	Age class	Immigration	Latest delivery	Focal following hour (party)					
					Swelling 1		Swelling 2		Swelling 3	
No	41	Old	1983.11	2009.05	11.5	(218)	6.4	(148)	9.6	(307)
Ki	38	Old	1984.12	2009.07	4.3	(94)	15.3	(298)	7.8	(290)
Hs	29	Middle	>2003.08	2009.08	6.9	(105)	12.9	(232)	7.1	(203)
Yk	29	Middle	2004.04	2009.10	7.1	(109)	10.2	(229)	9.8	(279)
Jk	24	Middle	2004.04	2012.01	16.7	(285)	7.4	(262)	2.9	(84)
Sl	21	Middle	>2003.08	2011.12	9.5	(211)	17.1	(365)	–	
Nv	17	Young	2007.08	2008.09	1.6	(47)	9.9	(199)	15.0	(353)
Ot	15	Young	2008.06	2011.02	1.7	(37)	17.5	(432)	7.3	(152)
Fk	14	Young	2008.06	2011.02	15.8	(326)	7.1	(264)	3.5	(121)

The age of each female was estimated based on immigration date and several morphological cues. All of the latest deliveries survived so all adult females had infants during the study periods. Each number in the following hour column refers total focal sampling hours for each individual in relation to swelling status (sw1, non-swelling; sw2, intermediate swelling; sw3, maximal swelling). Numbers in parentheses show the total numbers of the 1-h party segment in which each female was observed.

night beds, usually later than 17:00 h, by one or two researchers and at least two research assistants (one mainly followed the main party and the other usually helped researchers). Bonobos live in fission–fusion social systems (Kuroda, 1979; Kano, 1982; White, 1988), in which the composition of the membership of a party changes continuously (Aureli et al., 2008). When they split into several parties during daily ranging, the largest party was followed as far as possible. Party composition was recorded every single hour during party following (the 1-h party method; Hashimoto et al., 2001).

While bonobos were followed, one of the females in the party was randomly selected for continuous focal sampling. A random order for focal sampling for each day was generated before the beginning of daily observation. After bonobos were located in the morning at their bed site, HR and one research assistant looked for the first animal in the random sampling order for at least 30 min. If we could not find the first one, we moved to the next one in the order, and so on. Some females had priority for focal animal selection because of a shortage of focal sampling data for them which was mainly due to the fission–fusion social system (sometimes some females were not observed for more than a week). One focal session was continued for 20 min and if a focal animal could not be located for more than 5 min within a focal session, the focal sampling was terminated. To keep a degree of independence of focal sampling of the same individual, focal follows for the same individual were separated by at least 100 min. Within a focal session, all activities (e.g., feeding, moving, resting, grooming, agonistic interactions) involving the focal animal were recorded. All rare events, such as agonistic interactions (e.g., bite, hit, push and chase), copulations (Hashimoto, 1997) and g-g rubbing (Kuroda, 1980), which were observed during focal animals and parties following, were recorded. At the beginning of a focal session and every 5 min within the session (in total 5 scan points in a focal session), the names of neighbouring individuals within 5 m of the focal animal and their behaviours were recorded. The total focal following time for all individuals was 241.99 h (744 focal sessions; 26.89 $\pm$ 0.39 h per individual) during 1004.3 h of party following on 134 days (7.57 $\pm$ 2.46 h). The average length of a focal session was 19.5 $\pm$ 1.1 min and each focal session consisted of 4.83 $\pm$ 0.42 scans (3591 scans in 744 focal sessions). More details for observed hours of each individual in relation to swelling status are shown in Table 1.

2.4. *Data analysis*

All analyses were done using R 2.15.2 (R Core Team, 2012) in R-studio 0.97.248 (RStudio, 2012) with the following packages: lme4 (Douglas et al., 2012), glmmADMB (Sakug et al., 2012), nlme (Pinheiro et al., 2012), lmtest (Zeileis & Hothorn, 2002), plyr (Wickham, 2002), fitdistrplus (Delignette-Muller et al., 2012), ggplot2 (Wickham, 2009). We used generalized linear mixed models (GLMM) or linear mixed models (LMM) to control for random effects of individual differences in the models. All models included swelling status (3 levels; non-swelling, intermediate swelling, maximal swelling) and age classes (3 levels; old, middle, young) as predictor variables. Interactions between swelling status and age classes were also examined in the models by visual inspection rather than by numerical inspection with data explorations, to avoid too many variables in the model.

2.5. *Analysis of copulation and genito-genital rubbing (g-g rubbing)*

To compare the frequency of copulation and g-g rubbing of females in relation to sexual swelling status, we used a generalized linear mixed model with negative binomial distribution (GLMM; Zuur et al., 2009). To examine the effects of swelling status and age on frequency of copulation and g-g rubbing (response variables), we included swelling status (3 levels: sw1–3) and age class (3 levels: young, middle, old) as predictor variables and ID of individuals as random variables in the model. Repeated copulations within the same dyad within 5 min (without any interruptions by others or copulation with others between repeated copulation) were considered to be one event. The number of copulations and g-g rubbing interactions on each day for each female were response variables. Most of copulations and g-g rubbings which occurred within an observed party were detected by researchers and research assistants. However, because of the fission–fusion social system, daily observed time for each individual was different from each other. To control for the difference of observation time among individuals, we included the number of 1-h parties in which each female was observed in a day as an offset term in the model.

2.6. *Analysis of neighbouring female pattern*

To examine whether a female with maximal swelling attracted more females than females with other swelling status (non-swelling or intermediate swelling), the number of females in proximity to the focal females was analysed using GLMM with negative binomial distribution. Two separate GLMMs were conducted, including the number of females within 1 m and 5 m respectively in response variables, with swelling status and age classes as predictor variables. Because one focal session consisted of several scan points (4.83 ± 0.42 scan points), each scan point could be related to others to some degree. To control for this spatio-temporal autocorrelation, each scan in a focal session was nested in a focal session and the IDs of focal individuals were included as random variables in the model. Because it was also possible that spatio-temporal distribution of females in a 1-h party could have some effect on the neighbouring females, the number of females in a 1-h party (response variable) in relation to swelling status and age class (predictor variable) were also examined by running another GLMM.

2.7. Analysis of grooming interactions

Because bonobos almost exclusively engaged in dyadic grooming interactions and very rarely in mutual grooming interactions (Sakamaki, 2013), we analysed only dyadic grooming interactions without mutual grooming (mutual grooming was observed only once in one dyad for 40 s in this study). To examine the effect of sexual swelling and age on the grooming interactions of females (which might reflect sociality of females), first, we examined occurrence of grooming in each focal session (response variable 0 or 1) in relation to swelling status and age class (predictor variables) by running a GLMM. Second, we compared the grooming reciprocity index (response variable) only in the focal sessions in which grooming interaction was observed using a linear mixed model (LMM), to examine the effect of sexual swelling and age class (predictor variables) on grooming reciprocity. In these two-step analyses, individual differences were controlled by including them as a random effect of the model. The grooming reciprocity index was calculated as follows:

$$\mathrm{R(A)} = \mathrm{Ge(B)} - \mathrm{Gr(A)}/\mathrm{Ge(B)} + \mathrm{Gr(A)}$$

R(A): Grooming reciprocity index of A, while A was a focal animal (grooming with B)

Gr(A): $\sum$(A groomed B, while A was a focal animal)

Ge(B): $\sum$(A groomed by B, while A was a focal animal)

The value of the grooming reciprocity index: R(A) ranges from −1 (which means that A groomed B, but A was not groomed by B at all during the focal session) to 1 (which means that A was groomed by B, although A did not groom B at all).

3. Results

3.1. Copulation in relation to swelling status and age

In total 275 copulations were observed involving 9 adult females and 10 adult and adolescent males. These comprised 208 copulation events, when copulations that were repeated within 5 min were counted as one event. Females with maximal sexual swelling had more copulations than females with any other sexual swelling status (Figure 1a; for more details of GLMM, see Table 2). Young females also had more copulations than middle-aged or old females (Figure 1a; for more details of GLMM, see Table 2). The two old females did not copulate outside of their maximal swelling periods.

3.2. G-g rubbing and solicitation of g-g rubbing in relation to swelling status and age

184 cases of g-g rubbing were observed involving the 9 adult females. Females with maximal sexual swelling engaged in g-g rubbing more frequently than females with non-maximal swelling (Figure 1b; for more details of GLMM, see Table 2). There was no relationship between frequency of g-g rubbing and age. G-g rubbing typically in feeding contexts, such as immediately after entering a fruiting tree, or after encountering a food item (89.8%; 141 out of 157 g-g rubbing interactions in which context could be clearly confirmed). Because solicitation or initiation of g-g rubbing was confirmed in only 54 cases out of 184 (29.3%), we did not conduct statistical analysis for solicitation of the interaction. However, females with maximal swelling solicited other females with maximal swelling most frequently; in 25 of 54 g-g rubbings (44.64%). The second most frequently observed solicitation was between females with intermediate sexual swelling (16.07%), and the third most frequently observed one was females with intermediate sexual swelling soliciting females with maximal swelling (14.29%). Solicitation between other combinations of females was less than 10% for each.

3.3. Females in the proximity and in the same party

Focal females had other females within 1 m (including females in contact with the focal females) in 17.3% of scans (621 of 3591), and the mean

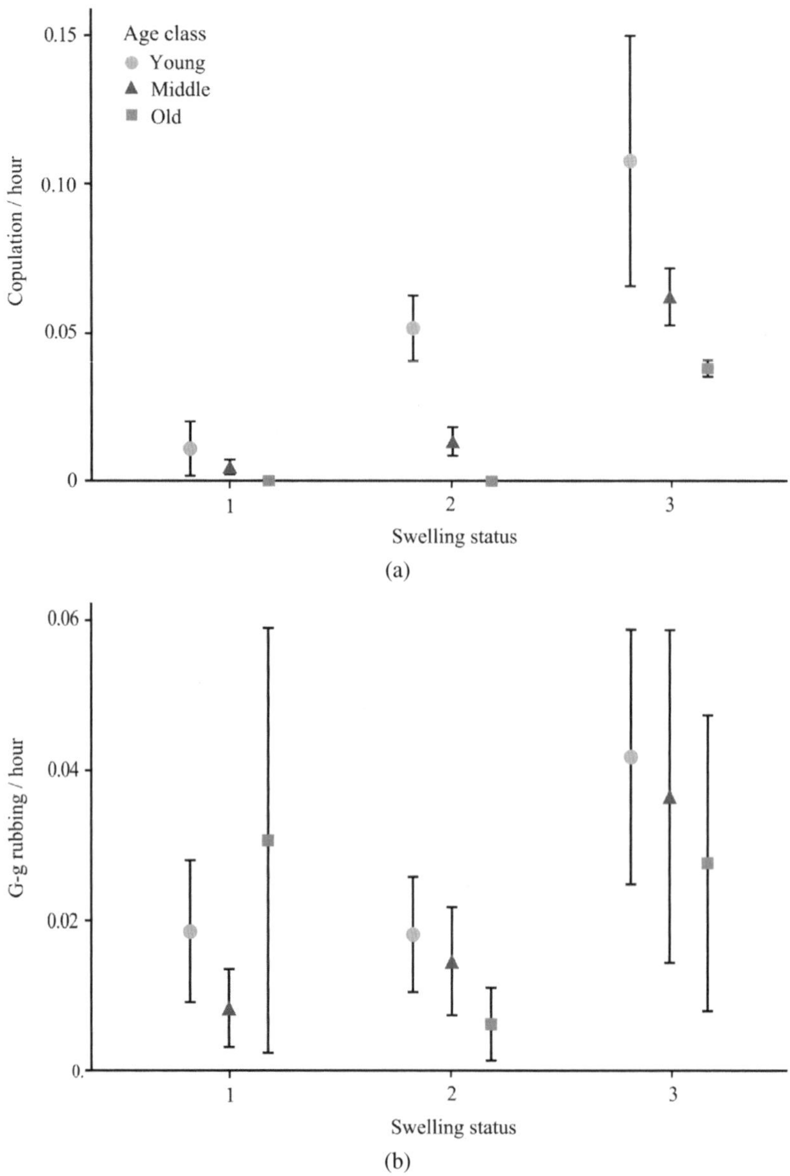

Figure 1. Mean number of copulations and g-g rubbing interactions per hour with standard error. (a) Copulations in relation to swelling status and age. (b) G-g rubbing interactions in relation to swelling status and age. This figure is published in colour in the online edition of this journal, which can be accessed via http://booksandjournals.brillonline.com/content/journals/1568539x.

Table 2.
Negative binomial GLMM results of copulation and genito-genital rubbing in relation to swelling status and age.

Predictor variable	Copulation				g-g rubbing			
	Estimate	SE	Z	$p(>\|Z\|)$	Estimate	SE	Z	$p(>\|Z\|)$
Intercept	−5.380	0.416	−12.94	<0.001	−5.782	0.540	−10.71	<0.001
Swelling 1 vs. 3	−4.311	0.536	−8.04	<0.001*	−1.629	0.387	−4.21	<0.001*
Swelling 2 vs. 3	−2.357	0.383	−6.15	<0.001*	−1.403	0.334	−4.20	<0.001*
Middle-aged vs. Young	−1.159	0.492	−2.35	0.019*	−0.185	0.677	−0.27	0.79
Old vs. Young	−3.259	0.625	−5.21	<0.001*	−0.830	0.807	−1.03	0.30

number of females within 1 m was 0.21 ± 0.51 (Figure 2a). Focal females had other females within 5 m in 40.8% of scans (1465 of 3591), and the mean number of females within 5 m was 0.63 ± 0.93 (Figure 2b). Focal females had 3.39 ± 2.23 female companions within the same 1-h party (Figure 2c). Females with maximal sexual swelling had more females within 1 m than those of other swelling status (GLMM; Table 3). Females with maximal swelling did not have females within 5 m compared with non-swelling females, although there was a significant difference compared with females with intermediate swelling (GLMM; Table 3). Females with maximal swelling, joined larger parties which consisted of more females than non-swelling females (Table 3), but there was no significant difference compared with females with intermediate swelling (Table 3).

3.4. Grooming frequency

Grooming interactions involving the focal animal were observed in 14.9% of focal sessions (111 focal sessions of 744). Females spent 8.69% of all focal animal following time (20.99 focal following hours of 241.99) in grooming interaction (sw1: 7.71%; 5.79 h, sw2: 6.93%; 7.17 h, sw3: 12.74%; 8.04 h, respectively). Comparison of the number of focal sessions (only in which grooming interaction was observed) in relation to swelling status showed that females with maximal swelling engaged in grooming interactions more often than females of other swelling status (non-swelling: 12.6% (29 focal sessions of 230), intermediate swelling: 12.8% (41 focal sessions of 320), maximal swelling: 21.1% (41 focal sessions of 194); Figure 3a, GLMM;

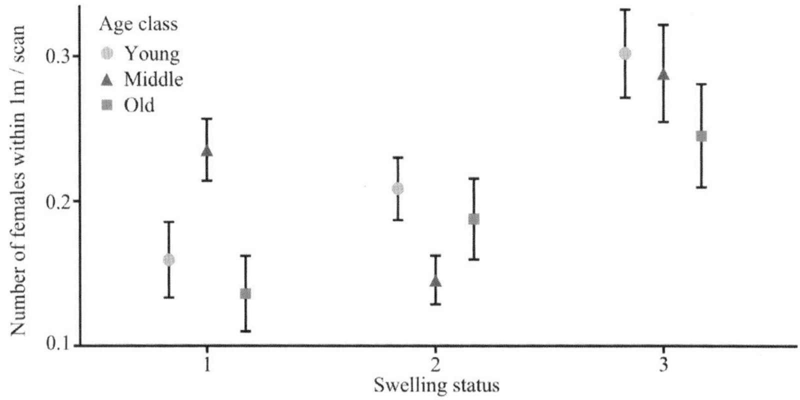

(a)

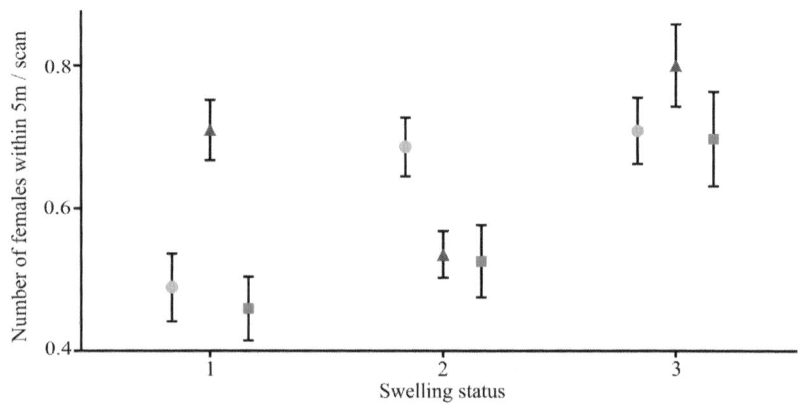

(b)

Figure 2. Mean number of females within 1 m, 5 m and party. (a) Number of females within 1 m in relation to swelling status and age. (b) Number of females within 5 m in relation to swelling status and age. (c) Number of females in 1-h party in relation to swelling status and age. This figure is published in colour in the online edition of this journal, which can be accessed via http://booksandjournals.brillonline.com/content/journals/1568539x.

Table 4). Young females engaged in grooming interactions more often than old females, but not more than middle-aged females (GLMM; Table 4).

3.5. Grooming reciprocity

The grooming reciprocity index was calculated independently for 111 focal sessions. Although old and middle-aged females seemed to have a higher overall reciprocity index (old: 0.275, middle: 0.078, young: −0.332) which might imply that they received more grooming than young females, this

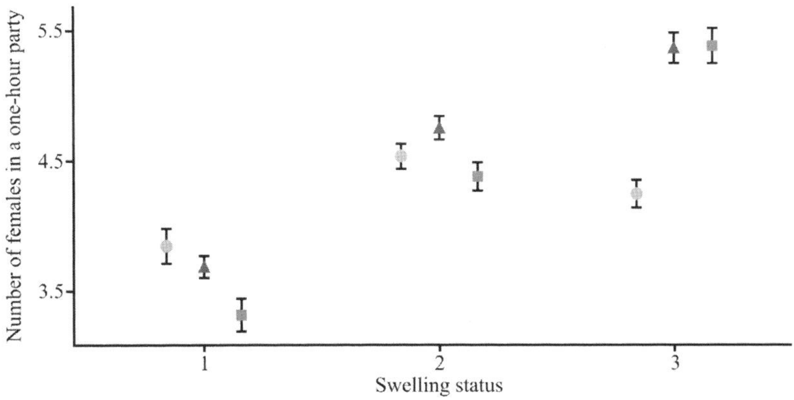

(c)

Figure 2. (Continued.)

difference was not statistically significant (Figure 3b, LMM; Table 4). Females with maximal swelling had significant higher reciprocity index scores than those of intermediate swelling status (LMM; Table 4), but no difference was found compared with non-swelling status (LMM; Table 4). Although increase of reciprocity index of young females according to swelling sta-

Table 3.
Negative binomial GLMM results of number of females within 1 m, within 5 m and within 1-h party in relation to swelling status and age.

	GLMM	Intercept	Swelling 1 vs. 3	Swelling 2 vs. 3	Middle-aged vs. Young	Old vs. Young
Within 1 m	Estimate	−2.517	−0.613	−0.788	−0.069	−0.349
	SE	0.240	0.257	0.239	0.224	0.267
	Z	−10.480	−2.390	−3.300	−0.310	−1.310
	$p(>\|Z\|)$	<0.001	0.017*	0.001*	0.756	0.192
Within 5 m	Estimate	−0.933	−0.174	−0.307	−0.030	−0.196
	SE	0.180	0.166	0.142	0.205	0.244
	Z	−5.180	−1.050	−2.160	−0.150	−0.800
	$p(>\|Z\|)$	<0.001	0.295	0.031*	0.883	0.422
Within party	Estimate	1.471	−0.350	−0.091	−0.082	−0.038
	SE	0.046	0.053	0.049	0.046	0.054
	Z	31.930	−6.560	−1.860	1.790	0.710
	$p(>\|Z\|)$	<0.001	<0.001*	0.063	0.073	0.473

* Significant difference ($p < 0.05$).

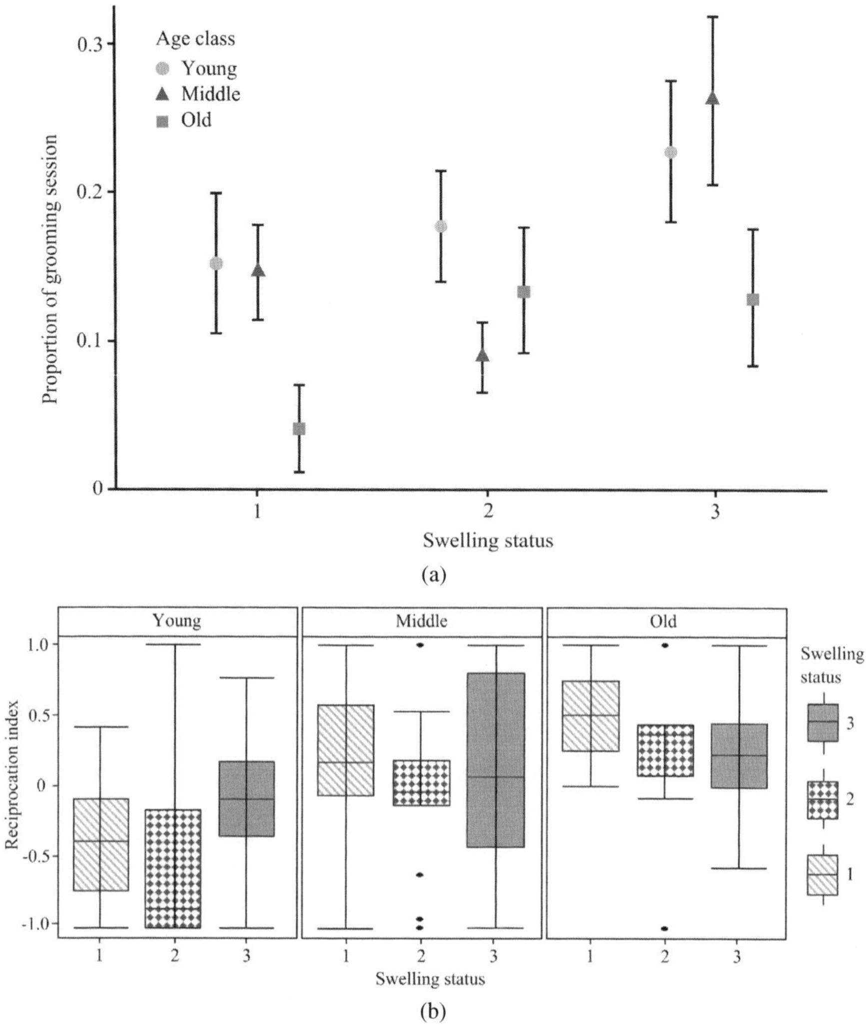

Figure 3. Grooming frequency and reciprocity. (a) Proportion of focal sessions in which grooming interaction was observed. (b) Grooming reciprocity index. Each box indicates upper quartile and lower quartile and dots indicate outliers. This figure is published in colour in the online edition of this journal, which can be accessed via http://booksandjournals.brillonline.com/content/journals/1568539x.

tus appeared to be different from the tendencies found in the other two age classes (Figure 3b), interaction between the two predictor variables (age class and swelling) was not statistically significant. However, further analysis with linear mixed model (LMM) on each age category revealed that there was no

Table 4.
Binomial GLMM results of comparison of the number of focal sessions which included grooming interactions and linear mixed model (LMM) result of comparison of grooming reciprocity index in relation to swelling and age.

Predictor variable	Focal session with grooming				Grooming reciprocity index			
	Estimate	SE	Z	$p(>\|Z\|)$	Estimate	SE	T	$p(>\|Z\|)$
Intercept	−1.076	0.211	−5.10	<0.001	−0.205	0.174	−1.182	0.240
Swelling 1 vs. 3	−0.610	0.271	−2.26	0.024*	−0.072	0.147	−0.487	0.628
Swelling 2 vs. 3	−0.612	0.246	−2.49	0.013*	−0.259	0.124	−2.077	0.040*
Middle-aged vs. Young	−0.242	0.232	−1.04	0.296	0.340	0.221	1.541	0.174
Old vs. Young	−0.668	0.300	−2.23	0.026*	0.630	0.266	2.372	0.055

* Significant difference ($p < 0.05$).

difference in grooming reciprocity index in middle-aged and old females in relation to sexual swelling, but there was a significant difference among swelling phases in young females. The pairwise test with LMM showed that the reciprocity index in young females was significantly higher in maximal swelling than in intermediate swelling ($p = 0.013$, $t = -2.599$, SE $= 0.184$), though the difference between maximal swelling and non-swelling was a non-significant tendency ($p = 0.097$, $t = -1.695$, SE $= 0.098$).

4. Discussion

Female bonobos are highly social (Furuichi, 1987; White, 1988; Kano, 1992; Hohmann et al., 1999) and more central in their society than female chimpanzees (Furuichi, 1989; Kano, 1992; Parish, 1994; Furuichi, 2011; Surbeck et al., 2011). Previous studies have suggested that prolonged sexual swelling of this species may contribute to the more female-centred society of bonobos by diminishing the competitive element of male mating strategy (Furuichi, 2011; Surbeck et al., 2011). Although this study also supported the idea that sexual swelling attracts males as a sexual signal, the four predictions concerning female–female interactions were also supported. Thus our results support the hypothesis that the prolonged sexual swellings of female bonobos also facilitate female–female interactions by making females with maximal swelling more attractive to other females. Therefore, this study suggests that prolonged sexual swelling might be a physiological adaptation in this species to maintain high level of affiliative social bonding among females and may contribute to high social status of females in the society (de Waal, 1995; Parish et al., 2000; Furuichi, 2011). In addition, the function of sexual swelling in bonobos and chimpanzees may differ in that sexual swelling in chimpanzees seems to function mainly to attract males (Boesch & Boesch-Achermann, 2000; Hashimoto et al., 2001). It would be very useful to examine the extent of variation between populations or sub-species in chimpanzees and bonobos for a better understanding of the function of sexual swelling in *Pan* species.

4.1. Increased socio-sexual behaviour during the maximal swelling period of females

Our data showed that the frequency of copulation was clearly related to the swelling status in all age classes, which is consistent with previous reports from wild populations (Furuichi, 1987; Surbeck et al., 2011). Although some captive studies have reported that females copulated with males throughout swelling cycles, with the exception of a few days (Savage-Rumbaugh & Wilkerson, 1978; de Waal, 1995), this study showed that this tendency was prominent for young females only. Young females (less than 5 years of tenure) had higher copulation frequencies than other females even in non-maximal swelling periods. By contrast, two old females never copulated with males outside of maximal swelling status. This result may indicate that the need for socio-sexual activity is greater for young females than old females, because they have relatively short tenure in the group and therefore, may be more vulnerable to competition even with males. On the other hand, it could be that they need male support in agonistic interactions with other females, which is predicted by the social passport hypothesis (Boesch & Boesch-Achermann, 2000; Deschner & Boesch, 2007).

Considering the fact that g-g rubbing was most frequent in female dyads with maximal swelling and this interaction occurs most frequently in feeding contexts (89.8% in this study, 79.5% in Hohmann & Fruth, 2000), females with maximal swelling can take advantage of their condition in feeding contexts. In particular, young females, whose rank is usually low (Furuichi, 1997; Ryu, unpublished data) and are, therefore, at a disadvantage in feeding competition, can gain more benefits from keeping maximal swelling since genito-genital rubbing requires body contact and therefore decreases distance and may also reduce tension between two participants (Furuichi, 1989;

Hohmann & Fruth, 2000). This may also explain the benefits and the function of very long-lasting maximal swelling of new immigrant females (Furuichi, 1987, 1992) although they were not the subjects of this study.

4.2. Increased social interaction and attractiveness of females with maximal swelling

Females with maximal swelling tended to have more females in proximity, and this tendency was more significant in 1 m proximity, which corresponds to the 'personal distance' defined in the proxemics as established by Hall (1966). Females with maximal swelling also had more females in the same 1-h party, although the difference between intermediate swelling and maximal swelling was not significant. These results suggest that females with maximal swelling attended bigger parties, or that they attracted other females to attend parties, although it was not possible to distinguish between these two possibilities in this study.

It has been reported that time spent in proximity is highly correlated with frequency of affiliative social interaction in primates (rhesus macaques (*Macaca mulatta*) (Hill, 1987); baboon (*Papio cynocephalus*) (Silk et al., 2003); ring-tailed lemurs (*Lemur catta*) (Sbeglia et al., 2010)) and, thus, staying in very close proximity, such as within 1 m, probably facilitates social interaction or at least indicates mutual or one-directional attraction among females. Therefore, together with grooming interaction and g-g rubbing data including solicitation of interaction, the increased number of females within 1 m of females with maximal swelling implies that females with maximal swelling were attractive for other females for social interaction.

This study also suggested that sexual swelling played a positive role in grooming interactions. The high frequency of grooming interactions of young females is also compatible with previous reports that young females had higher needs of social bonding with other females than older females (Furuichi, 1989; Idani, 1991). This implies that the benefits and reason to keep maximal swelling might vary in relation to female age. Although young females showed lower grooming reciprocity indices in general, it shifted to almost the same as old and middle-aged females when they had maximal swelling.

Newly-immigrated young female bonobos show very prolonged periods of maximal swelling (Furuichi, 1987, 1992) which were longer than adult females which are already settled in the group. In chimpanzees, it has been suggested that maximal swelling of new immigrant females is to attract males to facilitate the immigration process (Boesch & Boesch-Achermann, 2000). However this study provides an additional scenario in which maximal swelling of new immigrant females in bonobos is also used to facilitate female–female social interaction as well as male attraction. Increasing the periods of maximal swelling may help new immigrant females to integrate into the new group more easily. If females with maximal swelling are more socially active and attract other females for social interactions with them, prolonged periods of maximal swelling are more beneficial for immigrant young females whose needs of social bonding with more senior females seem to be greater (Furuichi, 1989; Idani, 1991).

Grooming in primates has been studied intensively and has been shown to have important social functions in maintaining and strengthening social relationships (Seyfarth, 1977; McKenna, 1978; Hill, 1987; Spruijt et al., 1992), regulating stress and tension (Spruijt et al., 1992), and may also function as currency exchanged for tolerance or agonistic support (Henzi & Barrett, 1999). Therefore, increased reciprocity in grooming interactions of young females with maximal swelling may result in direct benefits such as reducing stress, ecto-parasite load and facilitating social bonding. It may also help young females to strengthen social bonding with other females in the long-term. Such benefit seems to be greater for bonobos, in which senior females tend to have high social status and play central social roles (Parish, 1994, 1996; Furuichi, 2011).

Overall, this study suggests that one of the keys to understanding the function of sexual swelling in bonobos (especially for young females) is increased attractiveness of females for social interactions during prolonged maximal swelling periods. Female bonobos show prolonged periods of maximal swelling throughout their lifetimes (ca. 27% of time inter-birth interval within the average 4.8 years birth interval; Furuichi & Hashimoto, 2002). However, it is not reasonable to assume that these prolonged periods of maximal swelling evolved only for male attraction, since female bonobos showed lower copulation rates than female chimpanzees (Furuichi & Hashimoto, 2002) during maximal swelling periods and also females can be receptive even in non-maximal swelling cycles, as was found in young females in this study.

Therefore we suggest, in conclusion, that attraction of other females has become an additional function of sexual swelling in bonobos, which originally evolved as a sexual signal to males. In other words, prolonged periods of maximal swelling in bonobos may have evolved partly to facilitate social interaction among females and may contribute to a female-centred social structure by facilitating female–female bonding in a male-philopatric society (White, 1988; Parish, 1994; Furuichi, 2011; Surbeck et al., 2011). Together with the results of this study and previous suggestions that sexual swellings have disappeared and re-emerged at least three times during speciation (Nunn, 1999), the function of sexual swelling may differ among species. Therefore, we suggest that it is useful to consider the demography and socio-ecological factors of each species for a more comprehensive understanding of the function and evolution of sexual swelling in Old World monkeys and apes.

During the long life-span of bonobos, they encounter demographical and socio-environmental changes which have to be dealt with in order to live in the group and minimize fitness loss. Therefore it seems to be more advantageous for females if they have some degree of flexibility in the length of maximal swelling periods in relation to their life-history or socio-environmental or demographical changes since sexual swelling also incurs some costs. Although it has yet to be investigated thoroughly and quantitatively, there seems to be some degree of flexibility and individual difference in occurrence of maximal swelling (Furuichi, 1987; Furuichi, 1992). Long-term field study will clarify this speculation about the flexibility of keeping maximal swelling and will provide valuable perspectives on understanding female sexuality and its evolution, not only in this species, but also in humans.

Acknowledgements

We thank Dr. MacIntosh for his advice on statistical analysis. We also thank our research assistants at Wamba, especially Mr. Bafutsa, Mr. Bafaluka, Mr. Eteny, Mr. Batsindelia, Mr. Isoluka and Mr. Bafike, for their great help. Finally we thank the members of Department of Ecology and Social Behavior and CICASP at the Primate Research Institute of Kyoto University, in particular Dr. Bercovitch, Dr. Huffman, Dr. Sakamaki, Ms. Tokuyama, Dr. Tsuji and Sojung Han for comments and help. This study was supported by the Environment Research and Technology Development Fund (D-1007 to T.F.), Japan Society for the Promotion of Science (JSPS) Grants-in-aid for Scientific Research (22255007 to T.F.), the JSPS Asia-Africa Science Platform Program (2009–2011, 2012–2014 to T.F.), JSPS ITP-HOPE Project (ITP-23-012 to H.R.) and JSPS AS-HOPE Project (AS-24-007 to H.R.). We thank the

Ministry of Scientific and Technological Research of the D.R. Congo for research permissions (No MIN.RS/SG/013/2011, MIN.ESURS/SG/010/2012) and the Research Centre of Ecology and Forestry for helping us and maintaining Luo Scientific Reserve.

References

Alberts, S.C. & Fitzpatrick, C.L. (2012). Paternal care and the evolution of exaggerated sexual swellings in primates. — Behav. Ecol. 23: 699-706.

Altmann, J. (1974). Observational study of behavior: sampling methods. — Behaviour 49: 227-267.

Anderson, C.M. & Bielert, C.F. (1994). Adolescent exaggeration in female catarrhine primates. — Primates 35: 283-300.

Aureli, F., Schaffner, C.M., Boesch, C., Bearder, S.K., Call, J., Chapman, C.A., Connor, R., Fiore, A.D., Dunbar, R.I.M., Henzi, S.P., Holekamp, K., Korstjens, A.H., Layton, R., Lee, P., Lehmann, J., Manson, J.H., Ramos-Fernandez, G., Strier, K.B. & van Schaik, C.P. (2008). Fission–fusion dynamics: new research frameworks. — Curr. Anthropol. 49: 627-654.

Bercovitch, F.B. (1988). Coalitions, cooperation and reproductive tactics among adult male baboons. — Anim. Behav. 36: 1198-1209.

Bielert, C. & Busse, C. (1983). Influences of ovarian hormones on the food intake and feeding of captive and wild female chacma baboons (*Papio ursinus*). — Physiol. Behav. 30: 103-111.

Boesch, C. & Boesch-Achermann, H. (2000). The chimpanzees of the Taï Forest: behavioural ecology and evolution. — Oxford University Press, New York, NY.

Clutton-Brock, T.H. & Harvey, P.H. (1976). Evolutionary rules and primate societies. — In: Growing points in ethology (Bateson, P.P.G. & Hinde, R.A., eds). Cambridge University Press, Cambridge, p. 195-237.

Dahl, J.F., Nadler, R.D. & Collins, D.C. (1991). Monitoring the ovarian cycles of *Pan troglodytes* and *P. paniscus*: a comparative approach. — Am. J. Primatol. 24: 195-209.

de Waal, F.B.M. (1995). Bonobo sex and society. — Sci. Am. 272: 82-88.

Delignette-Muller, M.L., Pouillot, R., Denis, J.-B. & Dutang, C. (2012). fitdistrplus: help to fit of a parametric distribution to non-censored or censored data. — R package version 1.0-0.

Deschner, T. & Boesch, C. (2007). Can the patterns of sexual swelling cycles in female Taï chimpanzees be explained by the cost-of-sexual-attraction hypothesis? — Int. J. Primatol. 28: 389-406.

Deschner, T., Heistermann, M., Hodges, K. & Boesch, C. (2003). Timing and probability of ovulation in relation to sex skin swelling in wild West African chimpanzees, *Pan troglodytes verus*. — Anim. Behav. 66: 551-560.

Dixson, A.F. (1977). Observations on the displays, menstrual cycles and sexual behaviour of the "Black ape" of Celebes (*Macaca nigra*). — J. Zool. 182: 63-84.

Dixson, A.F. (1983). Observations on the evolution and behavioral significance of "sexual skin" in female primates. — Adv. Stud. Behav. 13: 63-106.

Douglas, B., Martin, M. & Ben, B. (2012). lme4: linear mixed-effects models using S4 classes. — RStudio, Boston, MA.

Emery, M.A. & Whitten, P.L. (2003). Size of sexual swellings reflects ovarian function in chimpanzees (*Pan troglodytes*). — Behav. Ecol. Sociobiol. 54: 340-351.

Furuichi, T. (1987). Sexual swelling, receptivity, and grouping of wild pygmy chimpanzee females at Wamba, Zaire. — Primates 28: 309-318.

Furuichi, T. (1989). Social interactions and the life history of female *Pan paniscus* in Wamba, Zaire. — Int. J. Primatol. 10: 173-197.

Furuichi, T. (1992). The prolonged estrus of females and factors influencing mating in a wild group of bonobos (*Pan paniscus*) in Wamba, Zaire. — In: Topics in primatology 2. Behavior, ecology, and conservation (Itoigawa, N., Sugiyama, Y., Sackett, G.P. & Thompson, R.K.R., eds). University of Tokyo Press, Tokyo, p. 179-190.

Furuichi, T. (1997). Agonistic interactions and matrifocal dominance rank of wild bonobos (*Pan paniscus*) at Wamba. — Int. J. Primatol. 18: 855-875.

Furuichi, T. (2011). Female contributions to the peaceful nature of bonobo society. — Evol. Anthropol. Issues News Rev. 20: 131-142.

Furuichi, T. & Hashimoto, C. (2002). Why female bonobos have a lower copulation rate during oestrus than chimpanzees. — In: Behavioural diversity in chimpanzees and bonobos (Boesch, C., Hohmann, G. & Marchant, L.F., eds). Cambridge University Press, Cambridge, p. 156-167.

Furuichi, T. & Hashimoto, C. (2004). Sex differences in copulation attempts in wild bonobos at Wamba. — Primates 45: 59-62.

Furuichi, T., Idani, G., Ihobe, H., Hashimoto, C., Tashiro, Y., Sakamaki, T., Mulavwa, M.N., Yangozene, K. & Kuroda, S. (2012). Long-term studies on wild bonobos at Wamba, Luo Scientific Reserve, D. R. Congo: towards the understanding of female life history in a male-philopatric species. — In: Long-term field studies of primates (Kappeler, P.M. & Watts, D.P., eds). Springer, Berlin, p. 413-433.

Hall, E.T. (1966). The hidden dimension. — Doubleday, New York, NY.

Hall, K.R.L. & de Vore, I. (1965). Baboon social behavior. — In: Primate behavior: field studies of monkeys and apes (De Vore, I., ed.). Holt, Rinehart and Winston, New York, NY, p. 55-110.

Hamilton, W.J.I. (1984). Significance of paternal investment by primates to the evolution of adult male–female associations. — In: Primate paternalism (Taub, D.M., ed.). Van Nostrand Reinhold, New York, NY, p. 309-335.

Hashimoto, C. (1997). Context and development of sexual behavior of wild bonobos (*Pan paniscus*) at Wamba, Zaire. — Int. J. Primatol. 18: 1-21.

Hashimoto, C., Furuichi, T. & Tashiro, Y. (2001). What factors affect the size of chimpanzee parties in the Kalinzu Forest, Uganda? Examination of fruit abundance and number of estrous females. — Int. J. Primatol. 22: 947-959.

Hashimoto, C., Tashiro, Y., Hibino, E., Mulavwa, M., Yangozene, K., Furuichi, T., Idani, G. & Takenaka, O. (2008). Longitudinal structure of a unit-group of bonobos: male philopatry and possible fusion of unit-groups. — In: The bonobos: behavior, ecology, and conservation (Furuichi, T. & Thompson, J., eds). Springer, New York, NY, p. 107-119.

Hausfater, G. (1975). Dominance and reproduction in baboons (*Papio cynocephalus*): a quantitative analysis. — In: Contributions to primatology. Karger, Basel, p. 1-150.

Heistermann, M., Möhle, U., Vervaecke, H., Van Elsacker, L. & Hodges, J.K. (1996). Application of urinary and fecal steroid measurements for monitoring ovarian function and pregnancy in the bonobo (*Pan paniscus*) and evaluation of perineal swelling patterns in relation to endocrine events. — Biol. Reprod. 55: 844-853.

Hendrickx, A.G. & Kraemer, D.C. (1969). Observations on the menstrual cycle, optimal mating time and pre-implantation embryos of the baboon, *Papio anubis* and *Papio cynocephalus*. — J. Reprod. Fertil. Suppl. 6: 119-128.

Henzi, S.P. & Barrett, L. (1999). The value of grooming to female primates. — Primates 40: 47-59.

Hill, D.A. (1987). Social relationships between adult male and female rhesus macaques: 1. Sexual consortships. — Primates 28: 439-456.

Hohmann, G. & Fruth, B. (2000). Use and function of genital contacts among female bonobos. — Anim. Behav. 60: 107-120.

Hohmann, G., Gerloff, U., Tautz, D. & Fruth, B. (1999). Social bonds and genetic ties: kinship, association and affiliation in a community of bonobos (*Pan paniscus*). — Behaviour 136: 1219-1235.

Hrdy, S.B. & Whitten, P.L. (1987). Patterning of sexual activity. — In: Primate societies (Smuts, B.B., Cheney, D.L., Seyfarth, R.M., Wrangham, R.W. & Struhsaker, T.T., eds). University of Chicago Press, Chicago, IL, p. 370-384.

Idani, G. (1991). Social relationships between immigrant and resident bonobo (*Pan paniscus*) females at Wamba. — Folia Primatol. 57: 83-95.

Jolly, A. (1966). Lemur behavior: a Madagascar field study. — University of Chicago Press, Chicago, IL.

Kano, T. (1982). The social group of pygmy chimpanzees (*Pan paniscus*) of Wamba. — Primates 23: 171-188.

Kano, T. (1992). The last ape: pygmy chimpanzee behavior and ecology. — Stanford University Press, Stanford, CA.

Kuroda, S. (1979). Grouping of the pygmy chimpanzees. — Primates 20: 161-183.

Kuroda, S. (1980). Social behavior of the pygmy chimpanzees. — Primates 21: 181-197.

McKenna, J.J. (1978). Biosocial functions of grooming behavior among the common Indian langur monkey (*Presbytis entellus*). — Am. J. Phys. Anthropol. 48: 503-509.

Mulavwa, M.N., Furuichi, T., Yangozene, K., Yamba-Yamba, M., Motema-Salo, B., Idani, G., Ihobe, H., Hashimoto, C., Tashiro, Y. & Mwanza, N. (2008). Seasonal changes in fruit production and party size of bonobos at Wamba. — In: The bonobos (Furuichi, T. & Thompson, J., eds). Springer, New York, NY, p. 121-134.

Mulavwa, M.N., Yangozene, K., Yamba-Yamba, M., Motema-Salo, B., Mwanza, N.N. & Furuichi, T. (2010). Nest groups of wild bonobos at Wamba: selection of vegetation and tree species and relationships between nest group size and party size. — Am. J. Primatol. 72: 575-586.

Muller, M.N., Kahlenberg, S.M., Emery Thompson, M. & Wrangham, R.W. (2007). Male coercion and the costs of promiscuous mating for female chimpanzees. — Proc. Roy. Soc. Lond. B: Biol. Sci. 274: 1009-1014.

Nunn, C.L. (1999). The evolution of exaggerated sexual swellings in primates and the graded-signal hypothesis. — Anim. Behav. 58: 229-246.

Pagel, M. (1994). The evolution of conspicuous oestrous advertisement in Old World monkeys. — Anim. Behav. 47: 1333-1341.

Paoli, T. (2009). The absence of sexual coercion in bonobos. — In: Sexual coercion in primates and humans: an evolutionary perspective on male aggression against females (Muller, M.N. & Wrangham, R.W., eds). Harvard University Press, Cambridge, MA, p. 410-423.

Paoli, T., Palagi, E., Tacconi, G. & Tarli, S.B. (2006). Perineal swelling, intermenstrual cycle, and female sexual behavior in bonobos (*Pan paniscus*). — Am. J. Primatol. 68: 333-347.

Parish, A.R. (1994). Sex and food control in the "uncommon chimpanzee": how bonobo females overcome a phylogenetic legacy of male dominance. — Ethol. Sociobiol. 15: 157-179.

Parish, A.R. (1996). Female relationships in bonobos (*Pan paniscus*): evidence for bonding, cooperation, and female dominance in a male-philopatric species. — Hum. Nat. 7: 61-96.

Parish, A.R., de Waal, F.B.M. & Haig, D. (2000). The other "closest living relative": how bonobos (*Pan paniscus*) challenge traditional assumptions about females, dominance, intra- and intersexual interactions, and hominid evolution. — Ann. NY Acad. Sci. 907: 97-113.

Pinheiro, J., Bates, D., DebRoy, S., Sarkar, D. & R Core Team (2012). nlme: linear and nonlinear mixed effects models. — R Core Team, Vienna.

Pocock, R.I. (1906). Notes upon menstruation, gestation, and parturition of some monkeys that have lived in the Society's Gardens. — Proc. Zool. Soc. Lond.: 558-570.

R Core Team (2012). R: a language and environment for statistical computing. — R Foundation for Statistical Computing, Vienna.

Reichert, K.E., Heistermann, M., Hodges, J.K., Boesch, C. & Hohmann, G. (2002). What females tell males about their reproductive status: are morphological and behavioural cues reliable signals of ovulation in bonobos (*Pan paniscus*)? — Ethology 108: 583-600.

RStudio (2012). RStudio: integrated development environment for R. — RStudio, Boston, MA.

Saayman, G.S. (1970). The menstrual cycle and sexual behaviour in a troop of free ranging chacma baboons (*Papio ursinus*). — Folia Primatol. 12: 81-110.

Sakamaki, T. (2013). Social grooming among wild bonobos (*Pan paniscus*) at Wamba in the Luo Scientific Reserve, DR Congo, with special reference to the formation of grooming gatherings. — Primates 54: 349-359.

Sakug, H., Fournier, D., Nielsen, A., Magnusson, A. & Bolker, B. (2012). Generalized linear mixed models using AD Model Builder.

Savage-Rumbaugh, E.S. & Wilkerson, B.J. (1978). Socio-sexual behavior in *Pan paniscus* and *Pan troglodytes*: a comparative study. — J. Hum. Evol. 7: 327-344.

Sbeglia, G.C., Tang-Martinez, Z. & Sussman, R.W. (2010). Effects of food, proximity, and kinship on social behavior in ringtailed lemurs. — Am. J. Primatol. 72: 981-991.

Seyfarth, R.M. (1977). A model of social grooming among adult female monkeys. — J. Theor. Biol. 65: 671-698.

Silk, J.B., Alberts, S.C. & Altmann, J. (2003). Social bonds of female baboons enhance infant survival. — Science 302: 1231-1234.

Smuts, B.B. & Smuts, R.W. (1993). Male aggression and sexual coercion of females in nonhuman primates and other mammals: evidence and theoretical implications. — Adv. Stud. Behav. 22: 1-63.

Spruijt, B.M., van Hooff, J.A.R.A.M. & Gispen, W.H. (1992). Ethology and neurobiology of grooming behavior. — Physiol. Rev. 72: 825-852.

Surbeck, M., Mundry, R. & Hohmann, G. (2011). Mothers matter! Maternal support, dominance status and mating success in male bonobos (*Pan paniscus*). — Proc. Roy. Soc. Lond. B: Biol. Sci. 278: 590-598.

Takahata, Y., Ihobe, H. & Idani, G. (1996). Comparing copulations of chimpanzees and bonobos: do females exhibit proceptivity or receptivity? — In: Great ape societies (McGrew, W.C., Marchant, L.F., Nishida, T., Goodall, J. & Itani, J., eds). Cambridge University Press, Cambridge, p. 146-155.

Thompson-Handler, N., Malenky, R.K. & Badrian, N. (1984). Sexual behavior of *Pan paniscus* under natural conditions in the Lomako Forest, Equateur, Zaire. — In: The pygmy chimpanzee (Susman, R.L., ed.). Springer, New York, NY, p. 347-368.

Tokuda, K. (1961). A study on the sexual behavior in the Japanese monkey troop. — Primates 3: 1-40.

van Schaik, C.P., Hodges, J.K. & Nunn, C.L. (2000). Paternity confusion and the ovarian cycles of female primates. — In: Infanticide by males and its implications (van Schaik, C.P. & Janson, C.H., eds). Cambridge University Press, Cambridge, p. 361-387.

Wallis, J. (1992). Chimpanzee genital swelling and its role in the pattern of sociosexual behavior. — Am. J. Primatol. 28: 101-113.

White, F.J. (1988). Party composition and dynamics in *Pan paniscus*. — Int. J. Primatol. 9: 179-193.

Wickham, H. (2002). The split-apply-combine strategy for data analysis. — J. Stat. Softw. 40: 1-29.

Wickham, H. (2009). ggplot2: elegant graphics for data analysis. — Springer, New York, NY.

Wildt, D.E., Doyle, L.L., Stone, S.C. & Harrison, R.M. (1977). Correlation of perineal swelling with serum ovarian hormone levels, vaginal cytology, and ovarian follicular development during the baboon reproductive cycle. — Primates 18: 261-270.

Wright, P.C., Izard, M.K. & Simons, E.L. (1986). Reproductive cycles in *Tarsius bancanus*. — Am. J. Primatol. 11: 207-215.

Young, C., Majolo, B., Heistermann, M., Schülke, O. & Ostner, J. (2013). Male mating behaviour in relation to female sexual swellings, socio-sexual behaviour and hormonal changes in wild Barbary macaques. — Horm. Behav. 63: 32-39.

Zeileis, A. & Hothorn, T. (2002). Diagnostic checking in regression relationships. — R News 2: 7-10.

Zinner, D.P., Nunn, C.L., van Schaik, C.P. & Kappeler, P.M. (2004). Sexual selection and exaggerated sexual swellings of female primates. — In: Sexual selection in primates (van Schaik, C.P., ed.). Cambridge University Press, New York, NY, p. 71-89.

Zuckerman, S. & Fulton, J.F. (1934). The menstrual cycle of the primates: Part VII. The sexual skin of the chimpanzee. — J. Anat. 69: 38-47.

Zuur, A.F., Ieno, E.N., Walker, N.J., Saveliev, A.A. & Smith, G.M. (2009). Mixed effects models and extensions in ecology with R. — Springer, New York, NY.

[When citing this chapter, refer to Behaviour 152 (2015) 313–334]

Sex and strife: post-conflict sexual contacts in bonobos

Zanna Clay * **and Frans B.M. de Waal**
Living Links, Yerkes National Primate Research Center and Psychology Department, Emory University, Atlanta, GA, USA
*Corresponding author's e-mail address: zannaclay@emory.edu

Accepted 15 October 2013; published online 12 February 2014

Abstract

Sexual contacts are thought to play an important role in regulating social tension in bonobos (*Pan paniscus*), and are especially common following aggressive conflicts, either between former opponents or involving bystanders. Nevertheless, research on the factors determining post-conflict sexual contacts, their effectiveness in reducing social tension and the nature of post-conflict sexual behaviour is scarce. Here, we collected data on post-conflict affiliative contacts in bonobos occurring between former opponents (reconciliation) and offered by bystanders towards victims (consolation) to investigate the role of sexual contacts in the regulation of aggressive conflicts compared to non-sexual affiliation behaviours. We tested whether post-conflict sexual contacts: (1) alleviate stress, (2) confer reproductive benefits, (3) mediate food-related conflict and (4) repair valuable social bonds. Thirty-six semi-free bonobos of all ages were observed at the Lola ya Bonobo Sanctuary, DR Congo, using standardized Post-Conflict/Matched Control methods. Consolation and reconciliation were both marked by significant increases in the occurrence of sexual behaviours. Reconciliation was almost exclusively characterized by sexual contacts, although consolation was also characterized by increases in non-sexual behaviours, such as embrace. Adults were more likely to engage in post-conflict sexual contacts than younger bonobos. Consistent with the stress-alleviation hypothesis, victims receiving sexual consolatory contact showed significantly lower rates of self-scratching, a marker of stress in primates, compared to receiving non-sexual contact. Post-conflict sexual contacts were not targeted towards valuable social partners and they did not confer obvious reproductive benefits; nor were they used to mediate food-related conflicts. Overall, results highlight the role of sex in regulating tension and social conflicts in bonobos.

Keywords

post-conflict behaviour, consolation, reconciliation, socio-sexual behaviour, tension regulation, primate.

1. Introduction

Bonobos (*Pan paniscus*) are well known for possessing a particularly rich and heightened socio-sexuality (Thompson-Handler et al., 1984; de Waal, 1987, 1995; Furuichi, 1989; Kano, 1989; White, 1996; Hashimoto, 1997; Hohmann & Fruth, 2000; Hohmann et al., 2009; Clay et al., 2011). Sex is freely incorporated into their daily life, with individuals habitually engaging in sexual interactions in all age and sex combinations. Bonobo females remain sexually active across their sexual cycles and, unlike most other primates, engage in face-to-face sexual interactions (e.g., Thompson-Handler et al., 1984; Kano, 1992; Paoli et al., 2006). Genito-genital contacts are a hallmark of their socio-sexual behaviour, during which two individuals, most commonly females, embrace ventro-ventrally, swing their hips laterally while keeping their vulva in contact (Kuroda, 1980; Hohmann & Fruth, 2000, Figure 1).

Socio-sexual contacts are thought to help regulate stress in bonobos, acting as a kind of 'social grease', to alleviate tension and to facilitate peaceful co-existence between group members, who generally lack close genetic ties (de Waal, 1987; Hohmann & Fruth, 2000; Fruth & Hohmann, 2006). Consistent with Hanby's (1977) prediction about the stress relieving function of primate socio-sexual contacts, most non-conceptive sexual behaviours in bonobos occur within socially tense periods, such as feeding, anticipation of feeding, inter-group interactions and following aggressive conflicts (Mori, 1983; de Waal, 1987; Manson et al., 1997; Hohmann & Fruth, 2000; Paoli et al., 2006; Hohmann et al., 2009).

Like any socially foraging animal, feeding is a source of contention in bonobos and sex appears to regulate feeding competition and facilitate food sharing (i.e., Parish, 1994; Hohmann et al., 2009). For instance, individuals offering sexual contacts to food possessors are more likely to gain access to the feeding source (Kuroda, 1984; Thompson-Handler et al., 1984; de Waal, 1987; Kano, 1992). With the relation between food and tension, most studies focus have focused on socio-sexual behaviour in the feeding context (e.g., Parish, 1994; Hohmann et al., 2009), consequently leaving investigation into its role in other contexts somewhat neglected. In one study however, Hohmann & Fruth (2000) showed that genital contacts in wild females increased between opponents following conflicts, a finding that warrants further investigation.

Figure 1. Socio-sexual behaviour in bonobos. (A) Female–female genito-genital contact; (B) a female bystander offering a genital touch towards a distressed victim following a conflict. Photographer: Zanna Clay at Lola ya Bonobo Sanctuary.

In response to the apparent scarcity of systematic data on bonobo socio-sexual behaviour in other contexts, this study examined the use and function of socio-sexual behaviours in another pertinent context, post-conflict interactions. As post-conflict interactions have, themselves, received considerable attention in the animal behaviour literature, it is useful to first lay out the basic framework and relevant concepts for studying post-conflict interactions and why this context is interesting for exploring the use and function of sex in bonobos.

Following aggressive conflicts, former opponents often engage in various forms of affiliative contacts, a reparative process known as reconciliation (de Waal & van Roosmalen, 1979; de Waal & Aureli, 1996; Arnold et al., 2001). In bonobos, these contacts are more often sexual in nature (de Waal, 1987, 1992; Manson et al., 1997; Hohmann & Fruth, 2000; Palagi et al., 2004), although various post-conflict behaviours are used in primates and other animals (e.g., de Waal, 1989; Fraser et al., 2008). For instance, chimpanzees reconcile primarily using embrace, kissing, 'finger in mouth' and touching (Fraser et al., 2008). In addition to reconciliation, uninvolved bystanders sometimes initiate affiliative contacts with one of the contestants, typically the former victim (de Waal & Roosmalen, 1979; de Waal & Aureli, 1996). Bystanders can accrue various direct benefits by doing so, such as protection from redirected aggression (e.g., Fraser et al., 2009). In a select number of species, however, the offering of friendly contacts appears to be more driven by a motivation to reduce the distress of a close social partner or kin-member, based on an apparent absence of purely self-serving benefits (chimpanzees, *P. troglodytes*: e.g., de Waal & van Roosmalen, 1979; Koski & Sterck, 2007; Fraser & Aureli, 2008; Romero et al., 2011; bonobos: Palagi et al., 2004; Clay & de Waal, 2013a; gorillas, *Gorilla gorilla*: Cordoni et al., 2006; crows, *C. corax*: Fraser & Bugnyar, 2010: dogs, *Canis familiaris*: Cools et al., 2008; wolves, *C. lupus*: Palagi & Cordoni, 2009: African elephants, *Loxodonta africana*: Byrne et al., 2008). This type of affiliative act, known as 'consolation', has been shown to be effective in reducing the recipient distress (e.g., Fraser et al., 2008; Clay & de Waal, 2013a). With a shift towards the other, consolation is considered an important bridge between expressions of empathy in animals and humans, as it suggests that the consoler can recognize as well as respond appropriately to alleviate anothers' distress (e.g., Preston & de Waal, 2002; Romero et al., 2010).

While sex is thought to play a key role during bonobo post-conflict interactions (i.e., de Waal, 1987; Hohmann & Fruth, 2000), there has been little systematic investigation. To address this, we used data collected on post-conflict interactions of two large social groups of semi-free bonobos to examine different hypotheses relating to the function of post-conflict sexual contacts. We explored the tension regulation hypothesis, as despite much discussion on the subject, there has been little empirical investigation into its efficiency in reducing tension as opposed to other non-sexual affiliative contacts. We compared this with three alternative hypotheses; post-conflict sexual contacts confer reproductive benefits; post-conflict sexual contacts resolve fights relating to food-competition and finally; post-conflict sexual contacts repairs valuable social bonds.

1.1. Post-conflict sexual contact alleviates stress

Despite much discussion on the topic (e.g., Kuroda, 1984; de Waal, 1987; Kano, 1989; Hohmann & Fruth, 2000; Fruth & Hohmann, 2006; Paoli et al., 2006), the tension regulation hypothesis has only been directly tested in one study (Hohmann et al., 2009), which examined the temporal relationship between salivary cortisol, a biomarker for social stress (e.g., Sapolsky, 2004), and genital contacts during a competitive feeding task. In this study, social stress was induced by providing restricted or unrestricted access to food. The results were equivocal. While the occurrence of genital contacts was temporally associated with increased cortisol and with the anticipation of food, higher rates of genital contacts did not correlate with greater reductions in cortisol. However, this study suffered from methodological limitations such as a small sample size and incomplete hormonal sampling. Moreover, it only focused on genital contacts during competitive feeding, rather than other forms of sexual behaviours and other contexts.

Applying the stress alleviation hypothesis to the post-conflict context, we predicted that rates of self-scratching of victims, a behavioural marker of social stress in primates (Maestripieri et al., 1992; Schino et al., 1996; Fraser et al., 2008), should be lower in victims that had received sexual forms of post-conflict affiliative contact rather than non-sexual ones.

1.2. Post-conflict sexual contact confers reproductive benefits

While post-conflict sexual behaviours are typically considered as social acts in bonobos, copulations may also occur following conflicts and other socially tense periods (de Waal, 1987), suggesting that individuals could also confer some reproductive benefits by offering or being offered copulations. These benefits may be further enhanced by the fact that bonobo females exhibit an especially prolonged period of oestrous compared to other primates, both in terms of extended swelling cycles and in the duration of the peak swelling phase (Thompson-Handler et al., 1984; Dahl, 1986; Blount, 1990). Under this hypothesis, copulations should therefore be expected to be a regular form of sexual contact; moreover, they should be primarily initiated by sexually mature individuals to sexually mature recipients of the opposite sex.

1.3. Sexual contact mediates food-related conflicts

Numerous studies have shown a relationship between socio-sexual behaviours and feeding competition (e.g., de Waal, 1987; Parish, 1994; White, 1996; Hohmann et al., 2009). If sex mediates food-related conflicts, sex should be more likely to occur following food-related conflicts than non-food related conflicts.

1.4. Post-conflict sexual contact repairs valuable social bonds

It has been proposed that sexual contacts consolidate and repair social bonds (de Waal, 1987, 1989; Hohmann & Fruth, 2000). This hypothesis predicts that sex would be most likely between dyads sharing a strong affiliative relationship. An alternative is that sexual contacts, more than non-sexual contacts, are facilitated by the trust that exists in dyads with strong affiliative relationships. Sexual contact in socially distant dyads may be perceived as too risky and therefore be less common.

2. Methods

2.1. Study site and subjects

Observations were conducted at the Lola ya Bonobo Sanctuary, Kinshasa, DR Congo. All information about the study site is provided in Clay & de Waal (2013a, b), which is based on the same data set used here. We conducted observations at the two largest enclosures, Group 1 and Group 2. Group 1 comprised of 25 individuals (6 adult females, 3 adult males and 16 immatures) and Group 2 comprised of 17 individuals (3 adult females, 4 adult males and 10 immatures; Clay & de Waal, 2013a). As exact birth dates for orphans were unknown, we used age estimates made by sanctuary veterinarians upon each bonobo's arrival, which were adjusted based on measurements of weight and patterns of dental emergence according to known patterns of ape development (Wobber et al., 2010; Rosati & Hare, 2012). This was validated by the known exact ages of individuals born at the sanctuary.

2.2. Data collection

From May–August 2011, observations were conducted by Z.C. and a trained assistant throughout the day ($N = 301$ and 152 observation hours recorded at Groups 1 and 2). We conducted all-occurrence observations of agonistic interactions that included at least one of the following behavioural elements: recipient fleeing and/or screaming in reaction to aggression, aggressor threat barks/grunts, directed display charge, threat arm wave, chase, hit, trample, slap, shove, poke, or bite. For each agonistic interaction, we recorded the identities of the initial recipient of the aggression, termed the 'victim', and the aggressor, as well as the identities of all visible bystanders. We also recorded the conflict context (see Clay & de Waal, 2013a, b) and the conflict intensity, which ranged from (1) threat (hand shake, bipedal swagger, threat bark, lunge); (2) directed display/charge without physical contact; (3) chase pursuit/quick poke/shove; (4) single grab/hit/slap without biting; (5) severe/multiple grab/hit or biting; to (6) injurious physical attack or biting (de Waal, 1988).

For each interaction, we conducted focal sampling of the victim using the standardized Post Conflict-Matched Control method (de Waal & Yoshihara, 1983). Post-conflict (PC) focals consisted of a 10-min focal observation of the victim immediately following the conflict. Each focal was matched with a 10-min Matched Control (MC) focal of the same victim on the following day (± 2 days) at the same or closest possible time (± 1 h). Further detail is provided in Clay & de Waal (2013a, b). We recorded all instances of affiliative contacts between the focal and the original opponent or any bystander. We coded the occurrence of nine different classes of contact behaviours: Embrace, Groom, Contact-sit, Touch, Play, Contact-peer, Pat, Hold and Sexual contact behaviours (Embrace, actor places arm around the recipient; Grooming, directed cleaning/inspection of the hair/skin; Contact-sit, sitting in physical contact; Touch, soft touch/stroke to the recipient's body, other than genitals, using a non-genital body part; Play, wrestle/run/chase/tickle with play face/laugh; Contact peer, approach close to peer at recipient, resulting in contact; Pat, tap/pat contact onto recipient's body using flat hand;

Hold, grasp/hold onto recipient (generally when walking)). Sexual contact included: Genito-genital rubbing (partners embrace ventro-ventrally and swing their hips laterally, keeping their vulvae in contact), mounting (lateral contact of the actors genitals behind the recipient and thrusting onto the recipient's behind), copulation (penile intromission and pelvic thrusting) and genital touching (touching the recipient's genitals using a non-genital body part, such as hand). We recorded the initiator of each interaction, which was the individual starting the interaction.

We collected data on rates of self-scratching per minute during all focals, a behavioural marker of social stress in primates (Maestripieri et al., 1992; Schino et al., 1996; Fraser et al., 2008). Focals were filmed using a Canon Vixia HF200 HD Camcorder. Aside from interactions involving dependent infants, interactions involving all individuals were included in the analyses.

In order to construct affinity matrices, instantaneous scan samples of all visible individuals were carried out throughout the day at 10-min intervals. The identities of all visible party members were recorded, followed by the identities of all individuals engaging in: grooming, contact sitting, sitting within arms reach, play or sexual contact (data on these state behaviours are distinct to the focal data). We collected a total of 794 and 411 scans at Groups 1 and 2, respectively. Interactions between all individuals, except dependent infants, were recorded.

2.3. *Ethical approval*

We received permission from 'Les Amis des Bonobos du Congo' (ABC) and full ethical approval from Les Amis de Bonobos du Congo (ABC) Scientific Committee and its Scientific Coordinator to conduct this study. The study complied with all legal requirements required for conducting research in the DR Congo and was approved by Emory University's IACUC committee.

2.4. *Data analysis*

In previous analyses, using the PC-MC method (de Waal & Yoshihara, 1983), on the same data set used here, we demonstrated significant occurrence of both consolation and reconciliation (Clay & de Waal, 2013a), where consolation and reconciliation were defined as increased contact affiliation spontaneously offered by the bystander toward the victim or between opponents during PC periods as compared with MCs. Having demonstrated consolation using this data set, we broadened the definition of consolation in the present analyses to be the spontaneous offering of contact affiliation by a bystander towards the victim. Likewise, reconciliation was defined as any affiliative contact occurring between former opponents during the PC period. As we were interested in affiliation offered by bystanders, any affiliations initiated by the victim itself towards a bystander were excluded.

As we were interested in the behaviours offered during consolation, we only examined PC interactions in which a bystander offered contact affiliation to the victim, which resulted in $N = 237$ events. However, as previous studies have shown that reconciliation can influence the occurrence of consolation, we removed cases where consolation co-occurred with reconciliation. This resulted in $N = 175$ interactions available for analysis ($N = 31$ victims, $N = 26$ aggressors). For each interaction, we compared the occurrence of different affiliation behaviours with those offered by bystanders to the same victim during MCs ($N = 139$ MCs in which bystanders initiated contact affiliation). The lower number of MCs reflects the lower occurrence of contact affiliations occurring during baseline periods (Clay & de Waal, 2013a).

We examined the behavioural nature of reconciliation. As there were very few cases where reconciliation occurred in the absence of consolation, we included reconciliation that occurred before consolation, resulting in $N = 65$ PC and $N = 35$ MC events.

2.4.1. *The nature of consolation and reconciliation behaviour in bonobos*

We used Generalized Linear Mixed Models (GLMM; binomial logit) to examine the extent to which PC periods influenced the offering of specific forms of affiliation behaviours by bystanders towards victims (consolation) and between opponents (reconciliation). We conducted GLMMs using the ‘lmer’ function in the R package ‘lme4’. To examine the role of sexual contacts in PC interactions, we conducted a GLMM on the occurrence of sexual contacts as the binomial dependent variable (yes/no) and context (PC or control) as the fixed factor. For consolation, these were sexual contacts initiated by the bystander to the victim and for reconciliation, the sexual contacts occurring between former opponents. All forms of sexual contacts were included in the ‘sex’ category (explored individually in subsequent analyses). We controlled for repeated sampling and inter-individual/group variation by including five random effects: the identities of Victim, Aggressor and Bystander; Group (1 or 2); and Interaction number. To compare between sexual and non-sexual contacts, we conducted a GLMM on the occurrence of Embrace, Play and Grooming during PC periods as compared to controls; behaviours all known to be especially relevant to social affiliation in bonobo sociality. We conducted GLMMs on each of these behavioural types using the same Fixed and Random Factors as above, with the occurrence of a given behaviour (yes/no) entered as the binomial dependent variable. We applied the Sidak correction in order to control for multiple comparisons ($\alpha = 0.013$). For reconciliation, we were not able to examine the frequency of these non-sexual contact behaviours (Embrace, Play, Grooming) owing to inadequate sample size of occurrence.

2.4.2. The use and determinants of post-conflict sexual contacts

In order to examine some of the determinants of PC sexual contacts, we conducted a GLMM on the occurrence of sexual contacts (yes/no) as the binomial dependent variable and the age (adult, adolescent, juvenile) and sex (male, female) of the bystander and the victim as fixed factors. We included the victim, bystander and aggressor identities as random factors, as well as the Interaction number and Group number. We conducted GLMMs for sexual contacts during consolation and during reconciliation. We computed all possible models using different combinations of the predictor variables. The best model was selected using the Akaike's information criterion (AIC), which compares multiple models and identifies the most parsimonious model that best explains the variance of the dependent variable, while penalizing for the number of variables in the model. The best model, as indicated by having the lowest AIC value, is the best model to predict values of the dependent variable in a new data set (Tabachnik & Fidell, 2001).

2.4.2.1. Does post-conflict sexual contact alleviate victim stress? Previously, using this data set, we demonstrated that consolation of any type reduced victim self-scratching as compared to receiving no consolation (Clay & de Waal, 2013a). In order to examine whether sexual contact had a more effective stress-alleviating effect compared to non-sexual contact, we used a Wilcoxon Signed Ranks test to compare mean victim self-scratch rates (per min) during PC periods in which a bystander offered sexual contact compared to non-sexual contact (i.e., touch, pat, groom, play). We removed cases in which consolation co-occurred with reconciliation in order to control for any confounding effects. We also examined whether victim self-scratch rates were lower following the occurrence of reconciliatory sexual contacts between compared to non-sexual reconciliation.

2.4.2.2. Does post-conflict sexual contact confer reproductive benefits? To examine whether PC sexual contact confers reproductive benefits, we compared the proportions of conceptive sexual contacts (copulation) with non-conceptive sexual contacts and calculated the percentage of sexual contacts that could potentially result in conception (the frequencies of copulations occurring between adult and adolescents). Adolescents were included based on the fact that bonobo adolescents have been shown to be capable of conception/insemination (e.g., San Diego Zoo, Twycross Zoo, Lola ya Bonobo).

2.4.2.3. Does sexual contact mediate food-related conflicts? To explore whether bonobos used sexual contacts to mediate food-related conflicts, we used GLMMs to examine whether the occurrence of consolatory and reconciliatory sexual contacts were more likely during food-related conflicts than non-food related conflicts. For both the consolation and reconciliation GLMMs, the occurrence of sexual contact (yes/no) was entered as the dependent variable, and Feeding Context was entered as the fixed factor (feed/non-feed). The random effects were the same as described above.

2.4.2.4. Does post-conflict sexual contact repair valuable social bonds? Previously, we showed that consolation and reconciliation were positively predicted by affiliation between victims and bystanders or, for reconciliation, between opponents (Clay & de Waal, 2013a, b). To examine whether affiliation positively predicted the occurrence of sexual contacts, we ran consolation and reconciliation GLMMs, where occurrence of sexual contacts (yes/no) was entered as the dependent variable and Affiliation was entered as the fixed factor. Individual identities, group and file number were entered as random factors.

3. Results

3.1. The nature of consolation and reconciliation behaviour in bonobos

There was considerable variation in the rates of different PC contact behaviours offered to victims by bystanders as compared to MCs (Figure 2). A χ^2-test on the pooled frequencies of each different behavioural types ($N = 9$) revealed significant differences in their relative frequencies during PC periods as compared to MCs ($\chi^2 = 101.70$, $\text{df} = 8$, $p < 0.001$). Whereas contact sitting, play, touch and grooming were typical during control periods, PC interactions showed notable increases in sexual behaviour, embrace, holding and contact peering.

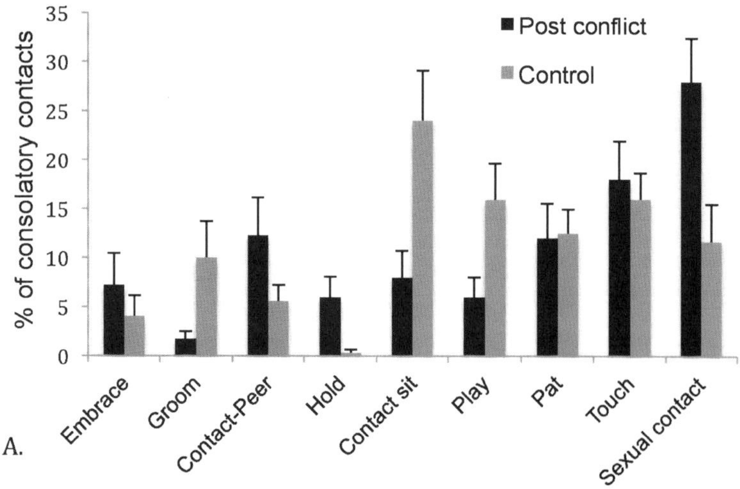

A.

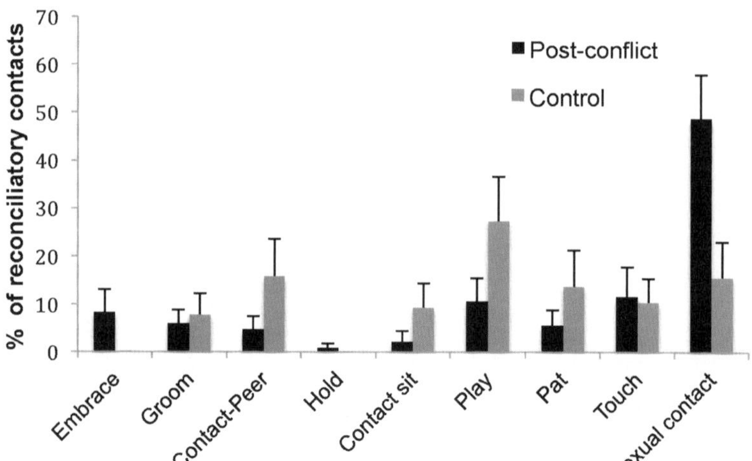

B.

Figure 2. Percentage of post-conflict contact affiliations that contained elements of one of nine behavioral categories during (A) consolation or (B) reconciliation as compared to control periods. The bar chart represents means ± SEM that were based on mean % per bystander (consolation) or per victim (reconciliation).

There was significant variation in the rates of different reconciliatory contacts occurring between former opponents as compared to MCs ($\chi^2 = 25.558$, df $= 8$, $p = 0.0013$). Again, sexual contacts were significantly more likely between former opponents during PC periods as compared to controls ($\beta = 1.876$, S.E. $= 0.594$, $Z = 3.008$, $p = 0.002$; Figure 2B). Despite interindividual variation, sexual contact was used significantly more frequently in reconciliation compared to in consolation (mean % of reconciliatory contacts that were sexual: 50 ± 43.4%; mean % consolatory contacts that were sexual = 28 ± 22%; $\chi^2 = 5.466$, df = 1, $p = 0.019$), suggesting sexual contacts are important as a form of reconciliatory behaviour in bonobos. While embracing slightly increased during PC periods, and control periods were characterized by increased play and grooming, overall sample sizes for these other behaviours were low and failed to show any significant differences (play: $\beta = -0.672$, S.E. = 0.645, $Z = -1.042$, $p = 0.297$; groom: $\beta = -0.098$, S.E. = 0.808, $Z = -0.122$, $p = 0.901$).

3.2. The use and determinants of post-conflict sexual contacts

The best fitting GLMM model for consolation only included bystander age (AIC = 332.75, $\chi^2 = 3.03$, df = 0, $p = 0.001$). This model fitted significantly better than the null model, which only included random factors ($p < 0.001$). Adults were significantly more likely to offer sexual contacts than juveniles ($\beta = 1.157$, S.E. = 0.491, $Z = 2.353$, $p \leqslant 0.019$) or adolescents ($\beta = 1.231$, S.E. = 0.628, $Z = 1.960$, $p \leqslant 0.050$). However, there was no effect of the age or sex of the bonobos these adult bystanders targeted, as indicated by the removal of factors of victim age and sex. Similarly, the model for reconciliation revealed that only victim age significantly contributed to the best fitting model (AIC = 89.99, $\chi^2 = -39.993$, df = 0, $p < 0.001$) with both juvenile and adolescent victims less likely to engage in sexual reconciliation as compared to adults (adolescents: $\beta = -3.002$, S.E. = 1.107, $Z = -2.712$, $p = 0.006$; juveniles: $\beta = -2.001$, S.E. = 1.099, $Z = -1.820$, $p = 0.068$).

3.2.1. Does post-conflict sexual contact alleviate victim stress?

Mean rates of self-scratching were significantly reduced when victims received sexual consolatory contacts (mean rate of self-scratching per victim (per min) after sexual consolatory contacts = 0.13 ± 0.21, and after non-sexual consolatory contacts = 0.36 ± 0.32; Wilcoxon signed ranks test on mean rates per individual victim: $Z = -2.614$, $N_1 = 26$, $N_2 = 23$, $p = 0.009$, Figure 3). While the offering of sexual reconciliatory contact showed a trend in the same direction (mean rate of self-scratching for sexual reconciliatory contacts = 0.21 ± 0.22, non-sexual reconciliatory contacts = 0.31 ± 0.25), sample sizes for paired analyses were very low and the result was not significant ($Z = -1.355$, $N = 7$ pairs, $p = 0.17$).

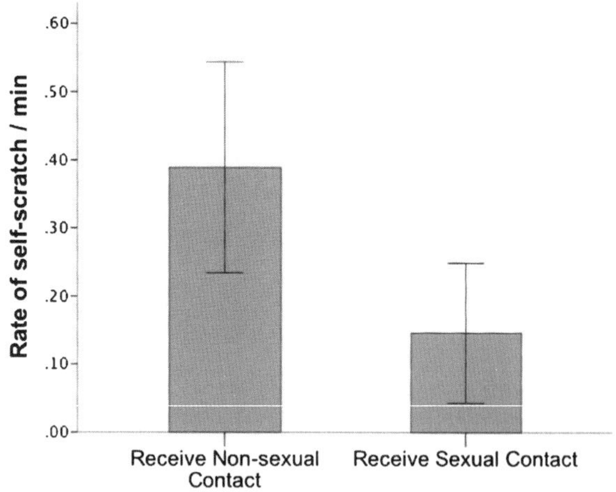

Figure 3. Mean ± SEM rate of victim self-scratching during post-conflict periods in which the victim received sexual or non-sexual consolatory contact.

3.2.2. Does post-conflict sexual contact confer reproductive benefits?

Against this hypothesis, copulations between reproductively viable partners were extremely rare, with the vast majority of PC sexual contacts non-reproductive in nature (i.e., 88% of $N = 91$ consolatory sexual contacts; 94.4% of $N = 36$ reconciliatory sexual contacts were non-reproductive). Offering genito-genital contacts outweighed all other sexual behaviours in both consolation and reconciliation (Figure 4). There were also no cases where adult male bystanders or opponents initiated copulations with female victims. In the case of consolation, copulations were only initiated by adult females and juvenile males ($N = 9$ and $N = 4$, respectively), whose targets were mostly non-adults. Only 6% of consolatory sexual contacts could potentially have led to conception ($N = 6$ copulations were offered by adult females to $N = 1$ adult male victim and to $N = 5$ adolescent male victims). The reproductively viable sexual contacts (adult females and adolescent males) are shown in Figure 5. No heterosexual copulations were observed at all for reconciliation.

3.2.3. Does post-conflict sexual contact mediate food-related conflicts?

GLMM analyses revealed no significant relationship between the occurrence of sexual behaviours and the conflict context for both consolation or reconciliation, suggesting that sexual contacts were not used particularly for mediating food-related conflicts ($\beta = -0.292$, S.E. $= 0.356$, $Z = -0.822$, $p = 0.411$).

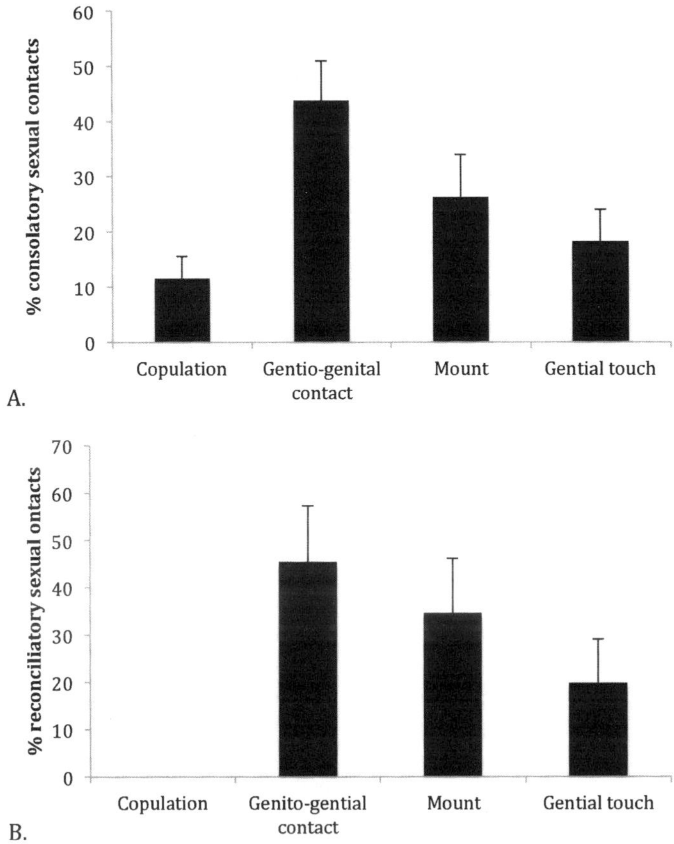

Figure 4. Percentage of (A) consolatory sexual contacts initiated by bystanders to victims and (B) reconciliatory sexual contacts occurring between former opponents. The bar chart represents mean ± SEM percentages, based on mean % of total sexual contacts offered per bystander (A) or per victim (B) ($N = 23$ for consolation, $N = 14$ for reconciliation).

3.2.4. Does post-conflict sexual contact repair valuable social bonds?

The affiliative relationship between victims and bystanders (consolation) or former opponents (reconciliation) had no significant effect on the occurrence of PC sexual contact behaviours ($\beta = -0.387$, S.E. $= 0.232$, $Z = -1.667$, $p = 0.096$). In a subsequent analysis in which all kin dyads were removed, the result still remained non-significant. Note that this analysis was specifically for sexual vs. non-sexual contacts, because overall we found affiliative relationships to be a potential factor in predicting PC contact (Clay & de Waal, 2013a).

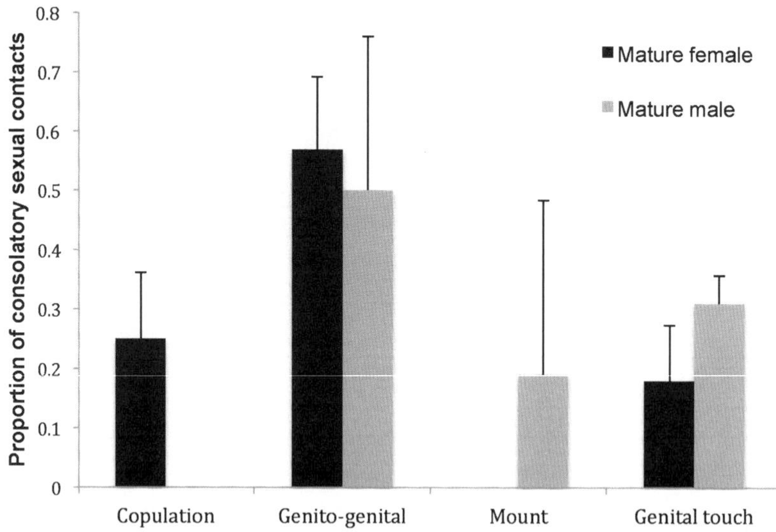

Figure 5. Percentage of sexual contacts initiated by sexually-mature males ($N = 5$ females, $N = 7$ males) and female bystander towards victims of aggression. Means ± SEM were based on mean % of sexual contacts offered per bystander per sex category (Male, Female).

4. Discussion

Following aggressive conflict, bonobos use a suite of sexual and, to a lesser extent, non-sexual behaviours to reconcile with former opponents and, as bystanders, to console distressed victims. Reconciliation and consolation were marked by pronounced increases in sexual behaviours, which included genito-genital contacts, mounting, genital touch and, to a lesser extent, copulation. Reconciliation was almost exclusively characterized by sexual contacts. While sexual contacts were also the most frequently occurring consolatory behaviour, consolation included a rise in other behaviours (embrace, touch, contact peering and holding). Grooming, play, and contact sitting occurred more frequently during baseline, suggesting these behaviours are more relevant for down-tempo social affiliation. Adults were more likely to engage in post-conflict sexual behaviours than adolescents and juveniles, indicating that the sexual nature of conflict resolution strengthens with age in bonobos and that the mechanisms underlying post-conflict behaviours are likely to vary across development.

In accordance with the tension regulation hypothesis, victims receiving sexual contact showed significantly lower rates of self-scratching compared to receiving non-sexual consolatory contact. While receiving any form of consolatory contact appears to be calming (as indicated by reduced self-scratching: Clay & de Waal, 2013a, b), the results further specify that sexual contacts are the most effective in doing so. Further research examining the physiological and behavioural effects of post-conflict sexual contacts is needed to further probe the role of sex in tension regulation in this species, as well as to verify whether other differences between sexual and non-sexual contacts, such as duration of bodily contact, may account for differences in their effects. Moreover, as the artificially close proximity of bonobos within the sanctuary setting may have resulted in increased reconciliation and consolation rates (i.e., individuals sleep together in dormitories at night, thus reducing the possibility of post-conflict avoidance, see Palagi et al., 2006), the next step is to verify whether similar effect exist in the wild.

Consistent with previous studies, genito-genital contacts were the most frequently offered form of sexual contact (de Waal, 1987; Hohmann & Fruth, 2000; Paoli et al., 2006). Mounting and genital touches were also frequently observed for both consolation and reconciliation. While, bonobos exhibit a particularly heightened level of sexuality compared to other primates, their use of genial touching and mounting during post-conflict periods appears to reflect a broader relationship for primates between sex and social tension (Hanby, 1977). In several species of macaques, for example, female–female mounting and other forms of genital touching occur frequently during post-conflict interactions (*Macaca nemestrina*: Oi, 1991; Flack & de Waal, 2007; *M. arctoides*: Call et al., 1999, 2002; *M. nigra*: Dixson, 1977). Chimpanzees frequently engage in post-conflict mounting and genital touching (i.e., testicle shakes: Arnold et al., 2001; Romero et al., 2011), as well as during other tense periods, such as during predator or inter-group encounters (Goodall et al., 1979; Herbinger et al., 2009).

We did not find support for the hypothesis that post-conflict sexual contacts are part of a reproductive strategy. Consolatory copulations were very rare, with only 6% of all sexual contacts potentially resulting in conception. Copulation was absent during reconciliation events. This is consistent with a previous study showing an increase in non-conceptive but not conceptive sexual behaviours during periods of short-term crowding and feeding (Palagi et al., 2006).

Although feeding and the anticipation of feeding are associated with socio-sexual behaviours in bonobos (Kuroda, 1984; Thompson-Handler et al., 1984; de Waal, 1987; Kano, 1992; Hohmann et al., 2009), we found no evidence that socio-sexual behaviours were used to resolve conflicts arising over food per se. Thus, while anticipation of potential competition or conflict may induce tension (Hohmann et al., 2009) and consequently, trigger sexual responses, sexual behaviours are not necessarily more effective than other contact in mediating the conflict that ensues.

While we have previously shown that affiliation predicts the overall occurrence of consolation and reconciliation in bonobos (Clay & de Waal, 2013a, b), we did not find support for the hypothesis that post-conflict sexual contacts were specifically directed towards close friends. This consequently does not support the alternative possibility that sexual interactions with socially-distant individuals are perceived as socially risky and thus avoided. The non-significant effect prevailed in a subsequent analysis, where kin dyads were removed, suggesting that this was not simply due to a correlation between affiliation and relatedness. This pattern is somewhat consistent with a study of post-conflict behaviour in stump-tailed macaques (Call et al., 2002), which showed that victims both received and directed socio-sexual behaviours to all non-kin recipients, regardless of affiliation level. The authors suggested that post-conflict socio-sexual behaviours in stump-tailed macaques, also a socially tolerant species, might reduce the risk of further or ongoing aggression, while post-conflict allo-grooming behaviours repaired damaged social relationships. However, our results did not show comparable increases in grooming during post-conflict interactions, suggesting a difference between both species.

While questions remain, our study nevertheless suggests that studying conflict dynamics in socially tolerant species, such as bonobos and stump tailed-macaques, can provide a useful means to explore the complexities and subtleties of triadic conflict interactions and to explore the role of sexual behaviour in regulating social tension.

Acknowledgements

We are very grateful to Brian Hare and Shinya Yamamoto for inviting us to contribute to the IPS Congress 2013 Symposium and this accompanying Special Issue on ‘Bonobos: The newly discovered ape’. We thank Pitshou Nsele Kayanga for assistance in data collection. We are grateful to Claudine André, Fanny Mehl, Dominique Morel, Valery Dhanani and Pierrot

Mbonzo for their collaboration and to the Ministries of Research and Environment in the DR Congo for supporting our research (research permit: MIN.RS/SG/004/2009). We thank the staff of Lola ya Bonobo for their support, particularly to Stany Mokando, Jean-Claude Nzumbi and Philippe Kunaka. We thank Brian Hare for ongoing support and Tim Eppley for valuable practical assistance. We thank two anonymous Reviewers for their comments. This study was funded by the Living Links Center, part of the Yerkes National Primate Research Center, and Emory University's College for Arts & Sciences.

References

Arnold, K. & Whiten, A. (2001). Post-conflict behaviour of wild chimpanzees (*Pan troglodytes schweinfurthii*) in the Budongo Forest, Uganda. — Behaviour 138: 649-690.

Blount, B.G. (1990). Issues in bonobo (*Pan paniscus*) sexual behavior. — Am. Anthropol. 92: 702-714.

Byrne, R., Lee, P.C., Njiraini, N., Poole, J.H. & Sayialel, K. (2008). Do elephants show empathy? — J. Conscious. Stud. 15: 10-11.

Call, J., Aureli, F. & de Waal, F.B.M. (1999). Reconciliation patterns among stumptailed macaques: a multivariate approach. — Anim. Behav. 58: 165-172.

Call, J., Aureli, F. & de Waal, F.B.M. (2002). Postconflict third-party affiliation in stumptailed macaques. — Anim. Behav. 63: 209-216.

Clay, Z. & de Waal, F.B.M. (2013a). Bonobos respond to distress in others: consolation across the age spectrum. — PLoS One 8: e55206.

Clay, Z. & de Waal, F.B.M. (2013b). Development of socio-emotional competence in bonobos. — Proc. Natl. Acad. Sci. USA 110: 18121-18126.

Clay, Z., Pika, S., Gruber, T. & Zuberbühler, K. (2011). Female bonobos use copulation calls as social signals. — Biol. Lett. 5: 513-516.

Cools, A.K.A., van Hout, A.J.M. & Nelissen, M.H.J. (2008). Canine reconciliation and third party initiated postconflict affiliation: do peacemaking social mechanisms in dogs rival those of higher primates? — Ethology 114: 53-63.

Cordoni, G., Palagi, E. & Tarli, S.B. (2006). Reconciliation and consolation in captive western gorillas. — Int. J. Primatol. 27: 1365-1382.

Dahl, J.F. (1986). Cyclic perineal swelling during the intermenstrual intervals of captive female pygmy chimpanzees (*Pan paniscus*). — J. Hum. Evol. 15: 369-385.

de Waal, F.B.M. (1987). Tension regulation and non-reproductive functions of sex in captive bonobos. — Nat. Geo. Res. 3: 318-335.

de Waal, F.B.M. (1989). Peacemaking among primates. — Harvard University Press, Cambridge, MA.

de Waal, F.B.M. (1992). Appeasement, celebration, and food sharing in the two *Pan* species. — In: Topics in primatology, Vol. 1, Human origins (Nishida, T., McGrew, W.C., Marler, P., Pickford, M. & de Waal, F.B.M., eds). University of Tokyo Press, Tokyo, p. 37-50.

de Waal, F.B.M. (1995). Sex as an alternative to aggression in the bonobo. — In: Sexual nature, sexual culture (Abramson, P.R. & Pinkerton, S.D., eds). University Press of Chicago, Chicago, IL.

de Waal, F.B.M. & Aureli, F. (1996). Consolation, reconciliation, and a possible cognitive difference between macaques and chimpanzees. — In: Reaching into thought: the minds of the great apes. Cambridge University Press, New York, NY, p. 80-110.

de Waal, F.B.M. & van Roosmalen, A. (1979). Reconciliation and consolation among chimpanzees. — Behav. Ecol. Sociobiol. 5: 55-66.

de Waal, F.B.M. & Yoshihara, D. (1983). Reconciliation and redirected affection in rhesus-monkeys. — Behaviour 85: 224-241.

Dixson, A.F. (1977). Observations on the displays, menstrual cycles and sexual behavior of the "Black ape" of Celebes (*Macaca nigra*). — J. Zool. 182: 63-84.

Flack, J.C. & de Waal, F.B.M. (2007). Context modulates signal meaning in primate communication. — Proc. Natl. Acad. Sci. USA 104: 1581-1586.

Fraser, O.N. & Aureli, F. (2008). Reconciliation, consolation and postconflict behavioral specificity in chimpanzees. — Am. J. Primatol. 70: 1114-1123.

Fraser, O.N. & Bugnyar, T. (2010). Do ravens show consolation? Responses to distressed others. — PloS One 5: 10605.

Fraser, O.N., Stahl, D. & Aureli, F. (2008). Stress reduction through consolation in chimpanzees. — Proc. Natl. Acad. Sci. USA 105: 8557-8562.

Fraser, O.N., Koski, S.E., Wittig, R.W. & Aureli, F. (2009). Why are bystanders friendly to recipients of aggression? — Commun. Integr. Biol. 2: 285-291.

Fruth, B. & Hohmann, G. (2006). Social grease for females? Same-sex genital contacts in wild bonobos. — In: Homosexual behaviour in animals: an evolutionary perspective (Sommer, V. & Vasey, P., eds). Cambridge University Press, Cambridge, p. 294-315.

Furuichi, T. (1989). Social interactions and the life history of female *Pan paniscus* in Wamba, Zaire. — Int. J. Primatol. 10: 173-197.

Goodall, J., Bandura, A., Bergmann, E., Busse, C., Matam, H., Mpongo, E., Pierce, A. & Riss, D. (1979). Inter-community interactions in the chimpanzee populations of the Gombe National Park. — In: The great apes (Hamburg, D. & McCown, E., eds). Benjamin/Cummings, Menlo Park, CA, p. 13-53.

Hanby, J.P. (1977). Social factors affecting primate reproduction. — In: Handbook of sexology (Money, J. & Mustaph, H., eds). Excerpta Medica, Amsterdam, p. 461-484.

Hashimoto, C. (1997). Context and development of sexual behaviour of wild bonobos (*Pan paniscus*) at Wamba, Zaire. — Int. J. Primatol. 18: 1-21.

Herbinger, I., Papwoth, S., Boesch, C. & Zuberbuhler, K. (2009). Vocal, gestural and locomotor responses of wild chimpanzees to familiar and unfamiliar intruders: a playback study. — Anim. Behav. 78: 1389-1396.

Hohmann, G. & Fruth, B. (2000). Use and function of genital contacts among female bonobos. — Anim. Behav. 60: 107-120.

Hohmann, G., Mundry, R. & Deschner, T. (2009). The relationship between socio-sexual behavior and salivary cortisol in bonobos: tests of the tension regulation hypothesis. — Am. J. Primatol. 71: 223-232.

Kano, T. (1989). The sexual behavior of pygmy chimpanzees. — In: Understanding chimpanzees (Heltne, G. & Marquardt, L.A., eds). Harvard University Press, Cambridge, MA, p. 176-183.

Kano, T. (1992). The last ape: pygmy chimpanzee behavior and ecology. — Stanford University Press, Stanford, CA.

Koski, S.E. & Sterck, E.H.M. (2007). Triadic postconflict affiliation in captive chimpanzees: does consolation console? — Anim. Behav. 73: 133-142.

Kuroda, S.J. (1980). Social behavior of the pygmy chimpanzees. — Primates 21: 181-197.

Kuroda, S.J. (1984). Interaction over food among pygmy chimpanzees. — In: The pygmy chimpanzee: evolutionary biology and behavior (Susman, R.L., ed.). Plenum Press, New York, NY, p. 301-324.

Maestripieri, D., Schino, G., Aureli, F. & Troisi, A. (1992). A modest proposal — displacement activities as an indicator of emotions in primates. — Anim. Behav. 44: 967-979.

Manson, J.H., Perry, S. & Parish, A.R. (1997). Nonconceptive sexual behavior in bonobos and capuchins. — Int. J. Primatol. 18: 767-786.

Mori, A. (1983). Comparison of the communicative vocalizations and behaviors of group ranging in eastern gorillas, chimpanzees and pygmy chimpanzees. — Primates 24: 486-500.

Oi, T. (1991). Non-copulatory mounting of wild pig-tailed macaques (*Macaca nemestrina nemestrina*) in West Sumatra, Indonesia. — In: Primatology today (Ehara, A., Kimura, T., Takenaka, O. & Iwamoto, M., eds). Elsevier Science, Amsterdam, p. 147-150.

Palagi, E. & Cordoni, G. (2009). Postconflict third-party affiliation in *Canis lupus*: do wolves share similarities with the great apes? — Anim. Behav. 78: 979-986.

Palagi, E., Paoli, T. & Borgognini Tarli, S. (2004). Reconciliation and consolation in captive bonobos (*Pan paniscus*). — Am. J. Primatol. 62: 15-30.

Palagi, E., Paoli, T. & Borgognini Tarli, S. (2006). Short-term benefits of play behavior and conflict prevention in *Pan paniscus*. — Int. J. Primatol. 27: 1257-1270.

Paoli, T., Palagi, E., Tacconi, G. & Tarli, S.B. (2006). Perineal swelling, intermenstrual cycle, and female sexual behavior in bonobos (*Pan paniscus*). — Am. J. Primatol. 68: 333-347.

Parish, A.R. (1994). Sex and food control in the uncommon chimpanzee — how bonobo females overcome a phylogenetic legacy of male-dominance. — Ethol. Sociobiol. 15: 157-179.

Preston, S.D. & de Waal, F.B.M. (2002). Empathy: its ultimate and proximate bases. — Behav. Brain. Sci. 25: 1-71.

Romero, T., Castellanos, M.A. & de Waal, F.B.M. (2010). Consolation as possible expression of sympathetic concern among chimpanzees. — Proc. Natl. Acad. Sci. USA 107: 12110-12115.

Romero, T., Castellanos, M.A. & de Waal, F.B.M. (2011). Post-conflict affiliation by chimpanzees with aggressors: other-oriented versus selfish political strategy. — PloS One 6: 22173.

Rosati, A.G. & Hare, B. (2012). Chimpanzees and bonobos exhibit divergent spatial memory. — Dev Sci. 15: 840-853.

Sapolsky, R.M. (2004). Social status and health in humans and other animals. — Annu. Rev. Anthropol. 33: 393-418.

Schino, G., Perretta, G., Taglioni, A., Monaco, V. & Troisi, A. (1996). Primate displacement activities as an ethopharmacological model of anxiety. — Anxiety 2: 186-191.

Tabachnick, B.G. & Fidell, L.S. (2001). Using multivariate analysis. — California State University Northridge, Harper Collins College Publishers, Northridge, CA.

Thompson-Handler, N.T., Malenky, R.K. & Badrian, N. (1984). Sexual behavior of *Pan paniscus*. — In: The pygmy chimpanzee: evolutionary biology and behavior (Susman, R.L., ed.). Plenum Press, New York, NY, p. 347-368.

White, F.J. (1996). Comparative socio-ecology of *Pan paniscus*. — In: Great ape societies (McGrew, W.C., Marchandt, L.F. & Nishida, T., eds). Cambridge University Press, Cambridge, p. 29-41.

Wobber, V., Wrangham, R. & Hare, B. (2010). Bonobos exhibit delayed development of social behavior and cognition relative to chimpanzees. — Curr. Biol. 20: 226-230.

[When citing this chapter, refer to Behaviour 152 (2015) 335–357]

Non-reciprocal but peaceful fruit sharing in wild bonobos in Wamba

Shinya Yamamoto [a,b,*]

[a] Graduate School of Intercultural Studies, Kobe University, 1-2-1 Tsurukabuto, Nada-ku, 657-8501 Kobe, Japan

[b] Wildlife Research Center, Kyoto University, Yoshida-honmachi, Sakyo-ku, 606-8501 Kyoto, Japan

[*] Author's e-mail address: shinyayamamoto1981@gmail.com

Accepted 30 December 2014; published online 29 January 2015

Abstract

Food sharing is considered to be a driving force in the evolution of cooperation in human societies. Previously postulated hypotheses for the mechanism and evolution of food sharing, e.g., reciprocity and sharing-under-pressure, were primarily proposed on the basis of meat sharing in chimpanzees. However, food sharing in bonobos has some remarkably different characteristics. Here I report details pertaining to fruit sharing in wild bonobos in Wamba based on 150 events of junglesop fruit sharing between independent individuals. The bonobos, primarily adult females, shared fruit that could be obtained individually without any cooperation or specialized skills. There was no evidence for reciprocal exchange, and their peaceful sharing seems to contradict the sharing-under-pressure explanation. Subordinate females begged for abundant fruit from dominants; this might indicate that they tested the dominants' tolerance based on social bonds rather than simply begging for the food itself, suggesting existence of courtesy food sharing in bonobos.

Keywords

bonobo, *Pan paniscus*, plant-food sharing, courtesy food sharing, begging for social bond, prosocial behaviour, reciprocity, sharing-under-pressure.

1. Introduction

Food sharing is one of the most prevalent forms of prosocial behaviour in the animal kingdom, and it has attracted much attention in investigations of the evolution of cooperation. Why does the owner relinquish food rather than keep it? This is the most basic and common question when discussing the evolution of apparently non-self-rewarding acts. Sharing with dependent related offspring is relatively common in various animal taxa (Clutton-Brock,

1991; Ydenberg, 1994; Brown et al., 2004), and its evolution can be explained by kin selection (Hamilton, 1964). In contrast, sharing among independent individuals is far less common and the benefit of sharing is often questioned or is controversial (Stevens & Gilby, 2004; Jaeggi & van Schaik, 2011). Considerable efforts have been made to find answers for these questions regarding the mechanism and evolution of food sharing by investigating non-human primates, particularly chimpanzees (*Pan troglodytes*).

To explain non-kin sharing, two main hypotheses have been proposed on the basis of previous studies involving chimpanzees: reciprocity and harassment. The reciprocity hypothesis explains that animals share in exchange for a past or future benefit (e.g., receiving the same food items or a different currency such as coalitions or mating). This hypothesis is supported by previous studies on chimpanzees, which detected statistical associations between sharing food with socially established partners (de Waal, 1989a, 1997; Mitani & Watts, 2001; Hockings et al., 2007; Gomes & Boesch, 2009). In contrast, the harassment hypothesis explains that there is an immediate benefit for the owner. Thus, when the beggar negatively affects the owner's feeding rate, the owner may give up some food but retain the majority, i.e., sharing to avoid further harassment from the beggar ('sharing-under-pressure': Wrangham, 1975; 'tolerated theft' for human sharing: Blurton Jones, 1984). This hypothesis is supported by some previous observational and experimental studies involving chimpanzees (Stevens, 2004; Gilby, 2006; Gilby et al., 2010), and it provides a more simple and parsimonious explanation for sharing behaviour (Stevens & Gilby, 2004; Stevens & Hauser, 2004). However, these two hypotheses are not exclusive and each may partially explain the sharing behaviour of chimpanzees (Jaeggi & van Schaik, 2011).

Previous studies have been mainly based on meat sharing among chimpanzees (e.g., in Gombe: Teleki, 1973; Goodall, 1986; in Mahale: Kawanaka, 1982; Nishida et al., 1992; in Kibale: Mitani & Watts, 1999, 2001; and in Tai: Boesch & Boesch, 1989; Boesch, 1994). Meat is a rare commodity and at some sites might be difficult to obtain without cooperation between individuals (Boesch, 1994). Some researchers have claimed that the contribution of an individual during cooperative hunting can predict the amount of meat they obtain during the following sharing session (Boesch, 1994). This is why sharing is often discussed in relation to hunting cooperation. In a comparative context, adult male chimpanzees at Bossou often enter 'risky' parts of their habitat to acquire large cultivated foods. In these cases, only a few 'brave' adult males could be fruit owners, and they consequently share these difficult-to-obtain foods with other community members (Hockings et al., 2007; Ohashi, 2007). The evolutionary story of human sharing cooperation has been constructed on the basis of these studies involving chimpanzees, frequently using comparisons with human hunter–gatherers.

The bonobo (*Pan paniscus*) is the sister species of the chimpanzee; however, it has been much less studied in this context, although previous studies have reported considerable differences in food sharing between these two species (Kano, 1980; Kuroda, 1980, 1984; White, 1994; Fruth & Hohmann, 2002). The unique features of sharing in bonobos are particularly characterized by their frequent fruit sharing and female participation in this social interaction. These sharing traits of bonobos are clearly different from the hunting–sharing observed in chimpanzees. A previous study involving wild bonobos in Lomako forest suggested that sharing under pressure, or mutualism, is a more plausible explanation than reciprocity (Fruth & Hohmann, 2002) because sharing increased with the number of beggars. Lomako bonobos might allow a few recipients to take pieces of food in order to avoid other begging hands; thus, both the owner and recipients seem to be mutually benefited. Meanwhile, detailed video analysis of meat and fruit sharing by a specific dyad of non-kin adult female bonobos in Wamba forest suggested that neither reciprocity nor harassment appeared to completely explain the observations (Hirata et al., 2010). More specifically, reciprocity cannot explain why the owner only tolerated the transfer of non-valuable parts, because it would be more reasonable from a psychological viewpoint to assume that the owner would share the valuable parts if he/she expected a future return benefit. Likewise, harassment cannot fully explain the differential rate of food transfer (more transfers of fruit than of meat), when the degree of pressure given by the recipient toward the two types of food remained more or less constant (Hirata et al., 2010). At present, considering that sufficient quantitative data are not available, it seems too early to make any conclusions regarding the mechanisms underlying fruit sharing in bonobos. We should also consider that the environments where animals live may influence the nature of food sharing. Unfortunately, to my knowledge, no specific study has previously compared food sharing in different wild bonobo sites. In captivity, experiments with chimpanzees and bonobos in African sanctuaries suggested that bonobos are more tolerant of each other than of chimpanzees, where they outperformed chimpanzees in a cooperative task and subsequent co-feeding (Hare et al., 2007). The bonobos even exhibited a preference for voluntary food sharing rather than eating alone with monopolisable food sources (Hare & Kwetuenda, 2010), even with unfamiliar individuals (Tan & Hare, 2013). However, an experimental study with chimpanzees and bonobos in several European zoos reported the opposite; chimpanzees shared food more tolerantly, actively and reciprocally than bonobos who demonstrated a more despotic nature (Jaeggi et al., 2010a). The environments in the zoos and sanctuaries may have affected the results, although we currently do not know what exactly might determine this difference. Therefore, bonobo–chimpanzee comparisons as well as comparisons between same-species communities are required from both captive and wild contexts to understand the complexities of ape sharing behaviours. In this sense, comparisons between bonobo field sites where food sharing can be frequently observed are important to obtain ecological insights into the nature of food sharing.

So far, however, insufficient data are available on food sharing among fully identified wild bonobos in their natural environment. Therefore, the main aim of the present study was to obtain details of fruit sharing in wild bonobos in Wamba forest, where long-term research (since 1973) has established a tracking record that began 1976 and allows for individual identification. Thus, I could investigate the detailed social relationships among sharing individuals. The excellent observational conditions and rich environment with abundant junglesop (*Anonidium mannii*) fruit also allowed me to focus on a single fruit species in the quantitative data analysis, thereby avoiding possible confusion due to different food types, which may affect the characteristics of sharing (Kano, 1980; Kuroda, 1984; White, 1994; Fruth & Hohmann, 2002; Hirata et al., 2010).

The shared item, i.e., abundant fruit and not rare meat, also raises the following question: why do independent recipients beg for fruit from other individuals when it can be obtained without any cooperation or specialized skills? The previous hypotheses, i.e., reciprocity and harassment, assume nutritional benefits to the recipients and discusses the benefits accrued to individuals that share (Stevens & Gilby, 2004). However, if the begging individual could acquire more of the same food by implementing an alternate strategy, another perspective may be required to understand this sharing behaviour (Slocombe & Newton-Fisher, 2005). The target in the present study, i.e., junglesop fruit, is often available in small quantities simultaneously, but it can still be found much more frequently and in greater quantities in Wamba than meat in the chimpanzee studies and *Treculia* fruit in the bonobo studies in Lomako. In the high season for junglesop fruits in Wamba, I counted a maximum of nine individuals who ate junglesop fruits simultaneously at a feeding site, and up to 45 individuals in total during a single observation day (Yamamoto, unpublished data), although on average 0.49 *Treculia* fruits were eaten per day in Lomako (Fruth & Hohmann, 2002). On the basis of these quantitative data, I examined some previously postulated hypotheses and propose another hypothesis from the recipient's perspective to explain this sharing behaviour and its evolution.

2. Materials and methods

2.1. Study conditions

I studied bonobos in the E1 group at Wamba (0°11′08″N, 22°37′58″E) in the northern sector of the Luo Scientific Reserve, Democratic Republic of Congo. The E1 group size varied between 27–31 individuals during the study periods, including 9–10 adult/adolescent females, 8–10 adult/adolescent males, 1–2 juvenile females, 2–3 juvenile males and 6–9 dependent infants, during the study periods. These numbers did not include some individuals from other groups who made short-term visits (<1 month at the point of study). The age–sex class was based on Hashimoto's (1997) categorization for Wamba bonobos, i.e., age < 4: infant, 4–7: juvenile, 8–14: adolescent, ⩾15: adult; and I adjusted it for the purpose of the present study. In the present study, the distinction between a dependent infant and a juvenile was based on whether an individual was carried by his/her mother or walked independently for >20 m during travel on the ground. In this sense, a juvenile can be considered as an 'independent individual' who could obtain his/her food by him/herself. The age when the transition occurred was approximately 4 years. I categorized a young female as juvenile if she remained in her natal group. Corresponding to this categorization of females, I used females' average emigration age, i.e., 8 years (Furuichi, 1989; Hashimoto, 1997; Furuichi et al., 1998, 2012; Hashimoto et al., 2008), for the categorization of males into juveniles and adults. The distinction between adolescent and adult is not that meaningful for the present study; hence, hereafter both classes have been taken together as adult. In this study, I focused on sharing among independent individuals, i.e., within and between adults and juveniles. Not all of the individuals lived in the community simultaneously (some vanished or emigrated, whereas others immigrated or matured), but 315 dyads of independent individuals were considered. For each age–sex class ratio and relevant analyses, I used the median number of individuals in each age–sex class during the study period, with a total of 22 independent individuals, i.e., 10 adult females (45.5%), nine adult males (40.9%), one juvenile female (4.5%) and two juvenile males (9.1%). All individuals in the group were well identified and habituated. Artificial provisioning was abolished in 1996. The history of the E1 group and the details of the study site have been previously described by Kano (1992), Furuichi et al. (1998, 2012), Hashimoto et al. (2008) and Idani et al. (2008).

2.2. Data collection and analysis

I collected sharing data ad libitum by direct observations and occasionally with video cameras during four field seasons (July–August 2010, June–August 2011, September–November 2012 and August–September 2013) when there was a relatively high availability of junglesop fruit. During these periods, the bonobos normally move around in a large party that comprises most group members. If they split into separate parties, I attempted to follow the larger party. In the present study, in order to avoid possible confusion caused by different food types, I focused on the sharing of junglesop fruit, although I collected sharing data for any food type as far as possible, which will be analysed more in detail in future studies. A junglesop fruit is typically 40–50 cm in length and weighs 4–6 kg (Figure 1a). This fruit is easy to process and bonobos consume soft flesh around big seeds. They normally spat out the seeds which sometimes have fresh remains around them (Figure 1b). This junglesop fruit is one of the most frequently shared food items among Wamba bonobos (Kano, 1980; Kuroda, 1984).

I defined food sharing as the unhampered transfer of food from A (the owner) to B (the recipient) (Kano, 1980; Kuroda, 1984; Feistner & McGrew, 1989). I excluded transfers in which there was no clear possession, such as collecting scraps in the vicinity of a feeding individual. A sharing event was recorded when one food item was divided between two individuals (owner–recipient dyad), which sometimes comprised two or more transfers of small portions divided from a single food item. Therefore, multiple transfer types could be observed in a single sharing event. If one food item was shared with several recipients, one event was recorded for each owner–recipient dyad. For example, when A shared a fruit with B and C, I counted two events (A–B and A–C); if A shared a fruit with B, and then B shared his/her gain with C, I counted two events (A–B and B–C).

(a)

(b)

Figure 1. (a) Junglesop (*Anonidium mannii*) fruit, and (b) its remains after a bonobo ate the fruit. Some flesh remained around the seeds (photo by Shinya Yamamoto). This figure is published in colour in the online edition of this journal, which can be accessed via http://booksandjournals.brillonline.com/content/journals/1568539x.

Based on Jaeggi et al. (2010b) and other psychological experimental studies of ape prosociality (Ueno & Matsuzawa, 2004; Warneken et al., 2007; Yamamoto et al., 2009, 2012), I examined which individual, the owner or recipient, initiated food transfer. Food transfer was categorized into three types: proactive transfer (offering), which was initiated by the owner in the absence of the recipient's begging; reactive transfer where the owner facilitated taking upon the recipient's begging; and passive transfer where the owner did not facilitate taking but they simply tolerated the recipient's taking.

I recorded 'begging' when the recipient reached for the owner's food when it was in his/her hands, feet or mouth. Close observation (identical to 'staring' in Kuroda (1984) and 'peering' in Furuchi (1989)) was often included in the category of 'begging' in previous studies (Fruth & Hohmann, 2002; Gilby, 2006). However, I excluded this from my 'begging' category because (1) close observation has almost no effect on the soliciting of sharing (Kuroda, 1984; Furuchi, 1989); (2) the animals might only be interested in the item in another individual's hand without having any intention of taking it and (3) close observation is also conspicuous in other contexts, particularly during social learning in apes (e.g., Yamamoto et al., 2013), and therefore, the function and motivation of close observation seems ambiguous. I also recorded the body part (hand or mouth) from which the recipient obtained their share, which corresponded to whether sharing occurred before or after the transferred part was chewed or licked by the owner. Even after the owner chewed or licked the fruit, some edible flesh sometimes still remained around the seed, which could be transferred to the recipient. The shared portion size was also categorized as smaller ('small portion') or larger ('large portion') than the recipient's hand. Any social interactions before, during and after sharing were analysed, where the main targets of this analysis were agonistic interactions, genito–genital (G–G) rubbing and copulation.

With respect to the social relationship between an owner and a recipient, I considered two key factors: kin relationships (mother–offspring and brothers or sisters, but not father–offspring since no genetic information is available for this at this moment) and pairwise dominance relationships. Assessments of dominance relationships were based on the outcomes of dyadic agonistic interactions and non-agonistic displacement (approach–retreat). With the help of experienced local assistants, I confirmed dominant–subordinate relationships only when their evaluations and my own observations were in complete agreement. If the evaluations did not completely agree for a given pair of individuals, I considered that the dominance relationship between them was ambiguous.

Statistical analyses were performed using R version 3.0.3 (R Core Team, 2014) except for matrix analysis which was done with MatrixTester version 2.2.4 developed by Hemelrijk (1990). I generally analysed all the sharing events between independent individuals; when I paid attention to sharing between non-kin individuals, I recorded this as 'non-kin sharing'.

3. Results

Among independent individuals in the E1 group in Wamba, I recorded a total of 178 sharing events with 17 food types including fruit, other plant food, honey and meat of small animals (Table 1). The sharing of junglesop fruit accounted for 84.3% of these events (150 sharing events); thus, I focused on interactions related to junglesop fruit.

3.1. Junglesop fruit eating

The junglesop fruit was obtained independently by individuals (i.e., not through sharing from others) 901 times in total. The field seasons of 2010, 2011 and 2013 occurred during the high season for junglesop fruit, whereas the field season of 2012 occurred after the high season (Table 2). All of the age–sex classes consumed junglesop fruit (Tables 3 and 4), although the adult females consumed them more frequently and the adult males and juveniles consumed them less than expected, based on the number of individuals

Table 1.
Shared food types among independent individuals observed in the Wamba E1 group.

Species	Local name	Food type	Size of food	No. of events
Anonidium mannii	Bolingo	Fruit	Big	150
Treculia africana	Boimbo	Fruit	Big	6
Anomalurus spp.	Itere	Meat	Big	3
Saba florida	Bossenda	Fruit	Big	2
Brachystegia laurentii	Langa	Fruit (seed)	Big (small)	2
Cola chlamydantha	Bokotikoti	Fruit	Small	2
?	Botete	Fruit	Small	2
Meliponinae spp.	Liutsu	Honey	Big	2
Isolona congolana	Bofiningo	Fruit	Small	1
Musanga cecropioides	Bombambo	Fruit	Small	1
Dialium pachyphyllum	Elimilimi	Fruit	Small	1
Dacryodes edulis	Bosou	Fruit	Small	1
Parkia bicolor	Lilembe	Fruit	Small	1
Pancovia laurentii	Botende	Fruit	Small	1
Landolphia owariensis	Batofe	Fruit	Small	1
Raphia sp.	Bolilo	Pith	Big	1
Guarea laurentii	Litoku	Pith	Big	1
Total				178

The species identification of plant food referred to Idani et al.'s (1994) plant list. Size was divided into two categories: 'big' (bigger than an individual's hand) or 'small' (smaller than an individual's hand) (see text for details).

Table 2.
Numbers of junglesop (*Anonidium mannii*) fruit eaten and shared by independent individuals during each field season.

	Observation days	No. of fruits eaten	No. of fruits shared	Shared/eaten (%)
2010	35	337	54	16.0
2011	37	347	56	16.1
2012	37	2	2	100.0
2013	20	215	13	6.1
Total	129	901	125	13.9

in each age–sex class (age–sex class, observed versus expected: adult female, 531 versus 409.5; adult male, 290 versus 368.6; juvenile female, 12 versus 41.0; juvenile male, 68 versus 81.9; $\chi^2 = 75.6$, df $= 3$, $p < 0.01$). The edible ripe fruits could normally be found on the ground. In a typical case, an individual who found a fruit on the ground first checked the inside of the fruit by breaking it, and if it was good, he/she took a part of it away (normally 1/4–3/4, but taking an entire fruit was rare) or started eating it at the site. Occasionally, another individual took some of the remaining fruit later. I observed 14 cases where two individuals (adult females in each case) arrived and touched a fruit almost simultaneously before dividing it. These cases generated a tension between the participants, resulting in G–G rubbing between them (13/14 cases), which is considered to be a tension-reducing behaviour, although physically agonistic interactions were never observed. I did not count these cases as sharing because the ownership of the fruit was ambiguous at that point and the fruit was not directly transferred between individuals.

3.2. Junglesop fruit sharing

Of the 901 junglesop fruit eaten by the bonobos, 13.9% ($N = 125$) were shared among two or more independent individuals (Table 2; Figure 2). In total, 150 sharing events were observed, and each fruit was shared with 1.2 recipients on average. There were 95 events (63.3%) between non-kin individuals (non-kin sharing) and 55 (36.7%) between independent kin individuals (sharing between mother and offspring or between brothers). Analysis of the age–sex class of the owners and recipients in all the sharing events showed that adult females account for the majority of either role (Tables 3 and 4). Adult females accounted for 92.7% of the owners, whereas adult

Table 3.
Numbers of junglesop (*Anonidium mannii*) fruit eaten by each age–sex class of independent individuals.

	Eaten individually (%)
Adult female (median $N = 10$)	531 (58.9)
Adult male (median $N = 9$)	290 (32.2)
Juvenile female (median $N = 1$)	12 (1.3)
Juvenile male (median $N = 2$)	68 (7.5)
Total	901 (100)

The number of individuals in each age–sex class in the eaten individually group is represented by the median number throughout the study periods. Adult in the present study includes adolescent and adult individuals in the categorization of Hashimoto (1997). See text for details of age–sex categorization.

males accounted for only 7.3%. Sharing never originated from juvenile individuals. Among the recipients, although juveniles were frequent recipients (juvenile female: 10.7%, juvenile male: 35.3%) given the small number of individuals (4.5% and 9.1%, respectively), the frequent adult female recipients (50.7%) and much less frequent adult male recipients (3.3%) were still notable in comparison to their population ratios (45.5 and 40.9%, respectively). In non-kin sharing events, adult females accounted for 89.5% of the owners (10.5% adult males) and 80.0% of the recipients (2.1% juvenile females; 17.9% juvenile males; there was no adult male recipient).

Table 4.
Numbers of junglesop fruit *Anonidium mannii* shared by each age–sex class of independent individuals.

Shared from	To				
	Adult female	Adult male	Juvenile female	Juvenile male	Total (%)
Adult female	72	5	16	46	139 (92.7)
Adult male	4	0	0	7	11 (7.3)
Juvenile female	0	0	0	0	0 (0.0)
Juvenile male	0	0	0	0	0 (0.0)
Total (%)	76 (50.7)	5 (3.3)	16 (10.7)	53 (35.3)	150 (100)

Adult in the present study includes adolescent and adult individuals in the categorization of Hashimoto (1997). See text for details of age–sex categorization.

Figure 2. Sharing between adult females. Sl (above) begged for a portion of junglesop (*Anonidium mannii*) fruit from Nv's mouth (video-still by Shinya Yamamoto). This figure is published in colour in the online edition of this journal, which can be accessed via http://booksandjournals.brillonline.com/content/journals/1568539x.

No significant relationship was observed between the number of times of individual eating and the number of times of receiving (Pearson's product–moment correlation: $t = 1.21$, df $= 24$, $p = 0.24$, $R = 0.24$; Figure 3). Thus, not only individuals who could not obtain junglesop fruit by themselves but also those who could obtain the fruit by themselves became recipients in sharing events.

3.3. *Relationship between owners and recipients*

No significant relationship was observed between the number of times of giving and the number of times of receiving in sharing events (Pearson's product–moment correlation: $t = -0.15$, df $= 24$, $p = 0.883$, $R = -0.03$;

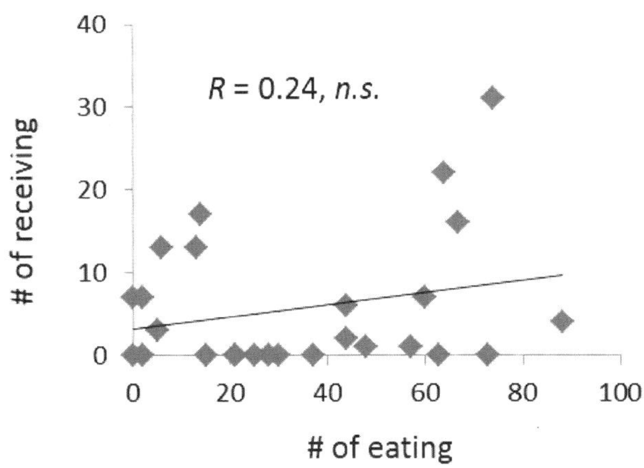

Figure 3. Numbers of times eating junglesop fruits individually and receiving them through sharing. Each dot represents an independent individual.

Figure 4). An adult female (named Hs) frequently gave and received fruits ('A' in Figure 4). However, the analysis of Hs's interacting partners showed that there was no significant overlap among individuals in her giving and receiving interactions. Hs received fruit most frequently from Yk (10 events), No (five events), Ki and Nv (four events, respectively), while she gave fruit most frequently to Ot (six events), An (three events) and Nv (two events).

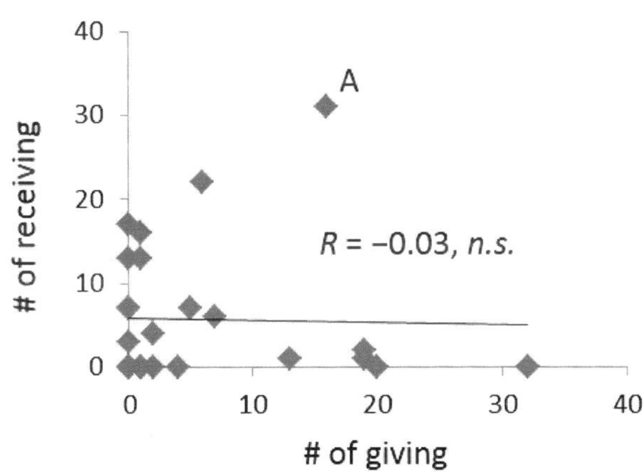

Figure 4. Numbers of times of giving and receiving. Each dot represents an independent individual. No significant correlation was observed between the two. Even for an individual (Hs, indicated by 'A' in this figure) who frequently gave and received fruits, the giving and receiving partners did not overlap (see text for details).

Matrix analysis also showed no evidence for reciprocity. No significant correlation between giving and receiving was found in analyses with all the 150 sharing events (τK_r-test: $\tau K_r = 0.068$, permutations = 2000, p_r (one-tailed probability value in the right tail of the permutation distribution) = 0.22) nor with the 95 non-kin sharing (τK_r-test: $\tau K_r = 0.074$, permutations = 2000, $p_r = 0.22$). I also calculated a giving/receiving ratio for each of the 48 dyads where sharing was observed at least once and found that a giving/receiving score of between 0.5 and 2 (i.e., 'balanced') was observed only in two pairs.

The shared fruit were predominantly transferred from dominants to subordinates. Among 150 sharing events, 129 events (86.0%) were from dominants to subordinates, 10 (6.7%) were from subordinates to dominants and 11 (7.3%) were between individuals where the dominance relationship was ambiguous.

Next, I examined whether a recipient received fruits from a specific individual. In the E1 group, 14 individuals became recipients in the sharing events: seven adult females, one adult male, two juvenile females and four juvenile males. With the exception of one juvenile male, every juvenile received fruits predominantly from his/her mother (more than half of the fruit they received). The only adult male received fruit four times only from his mother. However, the adult females who were involved in sharing as a recipient more than two times ($N = 5$) did not have such a specific donor. They each received fruits from 6.2 individuals on average (SD = 1.9).

3.4. *Food transfer types*

Analysis of the food transfer types showed that passive transfer was observed in all the 150 sharing events. Reactive transfer occurred in two sharing events (1.3%) and proactive transfer occurred in one sharing event (0.7%). Multiple transfer types could be recorded during a single sharing event because a single sharing event sometimes contained two or more transfers of small portions divided from one food item. These reactive and proactive transfers were observed only from mother to her offspring when she gave up the remaining food before travelling. Recipients obtained their share from the owners' hand (or foot) in 112 (74.7%) of the sharing events, and from the owners' mouth in 73 events (48.7%) after the owners had chewed or licked it. Transfers of small portions were observed in 140 (93.3%) of the sharing events and transfers of portions larger than the recipient's palm were involved in 16 events (10.7%).

3.5. Other social interactions

No physically agonistic interactions were observed during the sharing events. G–G rubbing between adult females was observed during 14 sharing events: two cases occurred before the first food transfer during events and 12 cases after the first transfer. Among the latter cases, five cases were observed after a large portion was taken by a recipient. Male–female copulations were observed 16 times during junglesop fruit feeding, and sharing occurred in three of these cases: one sharing event occurred from an adult female to a juvenile male before the copulation, another occurred from an adult male to an adult female during the copulation and the other occurred from an adult female to a juvenile male after the copulation. I recorded seven cases where an adult female without any food item copulated with an adult male who held a junglesop fruit in his hand, resulting in no food sharing between them except the above one case.

4. Discussion

The present study demonstrates peaceful fruit sharing among wild bonobos in Wamba, non-kin sharing of which predominantly occurred between adult females. In comparison with wild chimpanzees, fruit sharing occurred much more frequently in wild bonobos. The shared fruit was relatively abundant compared to meat, and could be obtained without any cooperation or specialized skills. Bonobos also share meat infrequently, but there is no reported evidence of group cooperation in hunting (Ihobe, 1992; Ingmanson & Ihobe, 1992; White, 1994; Fruth & Hohmann, 2002; Hohmann & Fruth, 2008; Surbeck & Hohmann, 2008; Hirata et al., 2010). Therefore, the hunting hypothesis, which assumes the successive occurrence of cooperative hunting and food sharing, is not applicable to the sharing of food that occurs among bonobos. In contrast to meat sharing in chimpanzees where a rare and precious food item is shared, more abundant fruits were shared among bonobos. Indeed, the bonobo recipients were not restricted to 'poor' individuals who could not obtain the same fruit by themselves. The bonobos, particularly adult females, could often obtain junglesop fruits by themselves, but they frequently begged for the same food items from others. More or less 'poor' adult males, compared to adult females, seldom begged for fruits from others. This clearly shows that bonobos do not share food because the recipients cannot obtain the target food by themselves. This is one of the clear differences from chimpanzees' meat sharing where individuals who could not get prey begged for the precious meat from a successful possessor.

Reciprocity is considered to explain meat sharing among chimpanzees at some sites, but it does not appear to be a good explanation for fruit sharing in bonobos. Neither individual, pairwise, nor group-level analyses of the balance of giving and receiving showed that the bonobos exchanged fruit reciprocally. Although some episodic observations in captivity and in a previous provisioning field in Wamba suggest that bonobos share food in return for sexual favours (Kuroda, 1984; de Waal, 1989b), this does not seem to be a plausible explanation for wild bonobos under natural conditions. The fruits were shared mainly between adult females or between adult females and juveniles, where sexual exchanges cannot be expected. More clearly, adult males with their hands full of fruit were observed to copulate with adult females without any fruit on seven occasions, but they did not share with their mate, except in one case. In addition, sharing-for-alliance is not a plausible explanation either. In Wamba, agonistic interactions are not frequent and supporting behaviours among females (particularly support from a subordinate to a dominant) are not conspicuous, although mothers often support their sons (Kano, 1992; Furuichi, 1997, 2011; for Lui Kotale, also see Surbeck et al., 2011; Surbeck & Hohmann, 2013; and for Lomako, also see Hohmann & Fruth, 2003). It appears that food sharing in bonobos is not directed predominantly at alliance formation to the same degree as food sharing in chimpanzees.

The present study shows that neither food-for-food, food-for-sex, nor food-for-alliance reciprocity is a plausible explanation for Wamba bonobos' food sharing, although it does not completely reject the reciprocity hypothesis. The sharers may have gained some benefit. For example, they may have occupied a central position and leadership of the group; thus, they could decide the directions of group movements. Food-for-grooming reciprocity cannot be excluded in the present study, either. However, it seems unlikely that they consciously expected benefits in the future. Hirata et al. (2010) noted that it is more reasonable to assume that the owner would share their valuable commodities if they expected a future benefit in return. The hesitancy of owners, which was represented by their passive attitude with virtually no proactive and reactive sharing, may disprove their expectation of future benefits, which is assumed in the reciprocal hypothesis.

Peaceful sharing with no physically agonistic interactions between owners and recipients also suggests that the sharing-under-pressure hypothesis does not fully explain food sharing in bonobos in Wamba. The direction of food transfer was predominantly from the dominant to the subordinate. When tensions emerged, they were resolved by G–G rubbing. This often occurred when a recipient took a large portion (still smaller than half of the item held by the owner), and the recipient adopted a posture where she laid on her back as if she was asking for a pardon. Normally, only a single independent individual begged for a single item held by an owner, and the owners could easily avoid the begging from subordinates simply by turning away. There was no obvious case in which a recipient reduced an owner's feeding rate by holding his/her arm. Based on a study in Lomako, Fruth & Hohmann (2002) suggested that the owners benefited by sharing with only a small number of recipients who assisted them in defending their possession of a *Treculia* fruit against others. However, this hypothesis does not seem to be applicable to the predominantly dyadic sharing interactions of junglesop fruits (which are smaller than *Treculia* fruits) in Wamba. Previous studies have reported that more preferred meat attracted more begging (White, 1994) or nearly constant begging (Hirata et al., 2010) from recipients, although this resulted in a lower transfer rate than fruit items. Thus, taking all of these into account, it is plausible that the owners in the present study gained little or no immediate material benefit from sharing.

The most parsimonious explanation for the majority of the fruit sharing events in bonobos in Wamba might be that they shared a surplus. The fruit was much more abundant than meat, which was suggested by the much higher frequency of eating the fruit. As well as the species differences between bonobos and chimpanzees, the present study also suggests that there are differences among the sites where wild bonobos were studied. In comparison with Lomako, Wamba seems to have greater quantities of sharable fruit. In the present study, 7.1 fruits (junglesop) were eaten on average each day during the study period in Wamba (if only the high seasons of 2010, 2011 and 2013 are considered, the number increases to 10.0) and 13.7% of these were shared among independent individuals. In contrast, 0.49 fruits (*Treculia africana*) were eaten each day and approximately half of these were shared in Lomako, where junglesop fruits were eaten and shared much less frequently than *Treculia* fruits (*Treculia* 93%, junglesop 7%) (Fruth & Hohmann, 2002). In such a rich environment, Wamba bonobos often shared only small portions, which the owners may sometimes have already chewed or licked to consume most of the edible part. In Wamba, in typical cases, bonobos who first found a junglesop fruit on the ground only took a part of it but not the whole. The bonobos did not seem to consume the greatest possible amount of the fruit. Therefore, there was little cost when large fruit owners relinquished small portions of the abundant fruit. The nature of sharing, predominantly characterized by passive transfers, also supports the ambiguous underlying psychology of the owners.

This analysis of fruit sharing in bonobos encouraged us to investigate the sharing interactions from the viewpoint of the recipients. It may be interesting to examine the reasons why recipients beg for a fruit from others when they can readily obtain it by themselves. The above hypotheses, reciprocity and sharing-under-pressure, are based on an assumption that food owners relinquish their precious food and that recipients always beg for nutritional gain; however, this does not always seem to suit the bonobos' cases. The access to junglesop fruits was not severely restricted during the study period and even subordinate males who are considered to have the least access priority to food (Furuichi, 1989, 1997; Kano, 1992; White & Wood, 2007) could eat the fruit without any expectation of sharing. The results show that individuals who consumed many fruits still begged for a share. The female-biased sharing might indicate strong female–female social bonds, as suggested in previous studies (Idani, 1991; Kano, 1992; Furuichi, 2011). It is also possible that the subordinates begged to strengthen social bonds, rather than to obtain the food itself, i.e., 'courtesy' food sharing. Of course, the food itself would still have been attractive to them and this cannot be excluded; however, the begging-for-social bond hypothesis may be a better explanation of begging by subordinates when they could find the fruit by themselves. This has also been suggested anecdotally in previous studies (Kuroda, 1984; Furuichi, 2011; for chimpanzee fruit sharing, see also Slocombe & Newton-Fisher, 2005), but insufficient qualitative and quantitative data and analyses are available at present; thus, further examinations are required to support this hypothesis, i.e., 'courtesy' food sharing characterized by begging-for-social bond.

In addition to the species differences between bonobos and chimpanzees, the differences between Wamba and Lomako sites seem to be important and merit further detailed analysis. As stated above, there appeared to be considerable differences between the two study sites in terms of the food types shared and the sharing frequency, which might have been influenced by possible differences in the fruit and tree compositions between the two forests. At present, however, no exact comparative data are available to address these differences; therefore, this should be examined in future studies. The present study also found that there were remarkable annual differences in the frequency of junglesop fruit sharing in Wamba. In the field seasons of 2010 and 2011, the bonobos shared junglesop fruits frequently (16.0 and 16.1%, respectively). In 2013, however, they shared the fruit much less frequently (6.1%). Considering that they consumed many junglesop fruits in each field season (9.6 fruit eaten per observation day in 2010, 9.4 in 2011 and 10.8 in 2013), the amount of fruit available might not be the best explanation for this difference. Other possible influential factors might include changes in food quality and/or social relationships, which were not fully examined. Further analysis of between-site differences and within-site annual changes, as well as species differences, will surely help us understand how different environments and societies facilitate the evolution of sharing cooperation in the animal kingdom, particularly in *Pan* and humans.

Acknowledgements

The field observations were certified by the government of DR Congo, and they complied with the current laws of the countries involved, as well as with Guidelines for Studying Wild Primates or Using Wild Primates in Research of Primate Research Institute, Kyoto University. Special thanks to the Research Center for Ecology and Forestry (CREF) and the Ministry of Scientific Research of DR Congo for their research permission; Takeshi Furuichi, Tetsuro Matsuzawa, Tetsuya Sakamaki, Hiroyuki Takemoto, Satoshi Hirata, Nahoko Tokuyama, Heungjin Ryu, Izabel Behncke, Kirsty Graham, Cintia Garai, Kazuya Toda, the local guides and staff at Wamba and the members of CREF and Wamba Committee for Bonobo Research (WCBR) for their fruitful discussion and helpful support during my stay in DR Congo. I am grateful to the staff of Kumamoto Sanctuary of the Wild Research Center, Kyoto University, as well as the staff at the Primate Research Institute, Kyoto University, for allowing me to conduct my field trips. I also thank Brian Hare and two anonymous reviewers for their constructive comments on an earlier version of the manuscript. The present study was supported by grants from the Japanese Society of Promotion Sciences (JSPS: Nos 22800034 and

40585767 to S. Yammaoto; No. 22255007 and Asia–Africa Science Platform Program 2009–2011 and 2012–2014 to T. Furuichi; and AS-HOPE and ITP-HOPE to T. Matsuzawa), Ministry of Education, Culture, Sports, Science, and Technology in Japan (MEXT: Nos 20002001 and 24000001 and MEXT special grant 'Human Evolution' to T. Matsuzawa), and Ministry of the Environment in Japan (The Environment Research and Technology Development Fund D-1007 to T. Furuichi).

References

Blurton Jones, N.G. (1984). A selfish origin for human food sharing: tolerated theft. — Ethol. Sociobiol. 5: 1-3.

Boesch, C. (1994). Cooperative hunting in wild chimpanzees. — Anim. Behav. 48: 653-667.

Boesch, C. & Boesch, H. (1989). Hunting behavior of wild chimpanzees in the Taï National Park. — Am. J. Phys. Anthropol. 78: 547-573.

Brown, G.R., Almond, R.E.A. & van Bergen, Y. (2004). Begging, stealing, and offering: food transfer in nonhuman primates. — Adv. Stud. Behav. 34: 265-295.

Clutton-Brock, T.H. (1991). The evolution of parental care. — Princeton University Press, Princeton, NJ.

de Waal, F.B.M. (1989a). Food sharing and reciprocal obligations among chimpanzees. — J. Hum. Evol. 18: 433-459.

de Waal, F.B.M. (1989b). Peacemaking among primates. — Harvard University Press, Cambridge, MA.

de Waal, F.B.M. (1997). The chimpanzee's service economy: food for grooming. — Evol. Hum. Behav. 18: 375-386.

Feistner, A.T.C. & McGrew, W.C. (1989). Food-sharing in primates: a critical review. — In: Perspectives in primate biology, Vol. 3 (Seth, P.K. & Seth, S., eds). Today and Tomorrow's, New Delhi, p. 21-36.

Fruth, B. & Hohmann, G. (2002). How bonobos handle hunts and harvests: why share food? — In: Behavioral diversity in chimpanzees and bonobos (Boesch, C., Hohmann, G. & Marchant, L., eds). Cambridge University Press, Cambridge, p. 231-243.

Furuichi, T. (1989). Social interactions and the life history of female *Pan paniscus* in Wamba, Zaire. — Int. J. Primatol. 10: 173-197.

Furuichi, T. (2011). Female contributions to the peaceful nature of bonobo society. — Evol. Anthropol. 20: 131-142.

Furuichi, T. (1997). Agonistic interactions and matrifocal dominance rank of wild bonobos (*Pan paniscus*) at Wamba. — Int. J. Primatol. 18: 855-875.

Furuichi, T., Idani, G., Ihobe, H., Hashimoto, H., Tashiro, Y., Sakamaki, T., Mulavwa, M.N., Yangozene, K. & Kuroda, S. (2012). Long-term studies on wild bonobos at Wamba, Luo Scientific Reserve, D.R. Congo: towards the understanding of female life history in a male-philopatric species. — In: Long-term field studies of primates (Kappeler, P.M. & Watts, D.P., eds). Springer, Berlin, p. 143-433.

Furuichi, T., Idani, G., Ihobe, H., Kuroda, S., Kitamura, K., Mori, A., Enomoto, T., Okayasu, N., Hashimoto, C. & Kano, T. (1998). Population dynamics of wild bonobos (*Pan paniscus*) at Wamba. — Int. J. Primatol. 19: 1029-1043.

Gilby, I.C. (2006). Meat sharing among the Gombe chimpanzees: harassment and reciprocal exchange. — Anim. Behav. 71: 953-963.

Gilby, I.C., Emery Thompson, M., Ruane, J.D. & Wrangham, R.W. (2010). No evidence of short-term exchange of meat for sex among chimpanzees. — J. Hum. Evol. 59: 44-53.

Gomes, C.M. & Boesch, C. (2009). Wild chimpanzees exchange meat for sex on a long-term basis. — PLoS ONE 4: e5116.

Goodall, J. (1986). The chimpanzees of Gombe: patterns of behavior. — Harvard University Press, Cambridge, MA.

Hamilton, W.D. (1964). Genetical evolution of social behaviour I. — J. Theor. Biol. 7: 1-16.

Hare, B. & Kwetuenda, S. (2010). Bonobos voluntarily share their own food with others. — Curr. Biol. 20: R230-R231.

Hare, B., Melis, A.P., Woods, V., Hastings, S. & Wrangham, R. (2007). Tolerance allows bonobos to outperform chimpanzees on a cooperative task. — Curr. Biol. 17: 619-623.

Hashimoto, C. (1997). Context and development of adult behavior of wild bonobos (*Pan paniscus*) at Wamba, Zaire. — Int. J. Primatol. 18: 1-21.

Hashimoto, C., Tashiro, Y., Hibino, E., Mulavwa, M., Yangozene, K., Furuichi, T., Idani, G. & Takenaka, O. (2008). Longitudinal structure of a unit-group of bonobos: male philopatry and possible fusion of unit-groups. — In: The bonobos: behavior, ecology, and conservation (Furuichi, T. & Thompson, J., eds). Springer, New York, NY, p. 107-119.

Hemelrijk, C.K. (1990). Models of, and tests for, reciprocity, unidirectionality and other social interaction patterns at a group level. — Anim. Behav. 39: 1013-1029 (for Matrix-Tester, see http://www.rug.nl/research/behavioural-ecology-and-self-organization/people/hemelrijk).

Hirata, S., Yamamoto, S., Takemoto, H. & Matsuzawa, T. (2010). A case report of meat and fruit sharing in a pair of wild bonobos. — Pan Africa News 17: 21-23.

Hockings, K.J., Humle, T., Anderson, J.R., Biro, D., Sousa, C., Ohashi, G. & Matsuzawa, T. (2007). Chimpanzees share forbidden fruit. — PLoS ONE 2: e886.

Hohmann, G. & Fruth, B. (2003). Intra- and inter-sexual aggression by bonobos in the context of mating. — Behaviour 140: 1389-1413.

Hohmann, G. & Fruth, B. (2008). New records on prey capture and meat eating by bonobos at Lui Kotale, Salonga National Park, Democratic Republic of Congo. — Folia Primatol. 79: 103-110.

Idani, G. (1991). Social relationships between immigrant and resident bonobo (*Pan paniscus*) females at Wamba. — Folia Primatol. 57: 83-95.

Idani, G., Kuroda, S., Kano, T. & Asato, R. (1994). Flora and vegetation of Wamba forest, Central Zaire with reference to bonobo (*Pan paniscus*) foods. — Tropics 3: 309-332.

Idani, G., Mwanza, N., Ihobe, H., Hashimoto, C., Tashiro, Y. & Furuichi, T. (2008). Changes in the status of bonobos, their habitat, and the situation of humans at Wamba in the Luo Scientific Reserve, Democratic Republic of Congo. — In: The bonobos: behavior, ecology, and conservation (Furuichi, T. & Thompson, J., eds). Springer, New York, NY, p. 291-302.

Ihobe, H. (1992). Observations on the meat-eating behavior of wild bonobos (*Pan paniscus*) at Wamba, Republic of Zaire. — Primates 33: 247-250.

Ingmanson, E. & Ihobe, H. (1992). Predation and meat eating by *Pan paniscus* at Wamba, Zaire. — Am. J. Phys. Anthropol. 35(Suppl. 14): 93.

Jaeggi, A.V., Burkart, J.M. & van Schaik, C.P. (2010b). On the psychology of cooperation in humans and other primates: combining the natural history and experimental evidence of prosociality. — Philos. Trans. Roy. Soc. B 365: 2723-2735.

Jaeggi, A.V., Stevens, J.M.G. & van Schaik, C.P. (2010a). Tolerant food sharing and reciprocity is precluded by despotism in bonobos but not chimpanzees. — Am. J. Phys. Anthropol. 143: 41-51.

Jaeggi, A.V. & van Schaik, C.P. (2011). The evolution of food sharing in primates. — Behav. Ecol. Sociobiol. 65: 2125-2140.

Kano, T. (1980). Social behavior of wild pygmy chimpanzees (*Pan paniscus*) of Wamba: a preliminary report. — J. Hum. Evol. 9: 243-260.

Kano, T. (1992). The last ape: pygmy chimpanzee behavior and ecology. — Stanford University Press, Stanford, CA.

Kawanaka, K. (1982). Further studies on predation by chimpanzees of the Mahale Mountains. — Primates 23: 364-384.

Kuroda, S. (1980). Social behavior of the pygmy chimpanzees. — Primates 21: 181-197.

Kuroda, S. (1984). Interaction over food among pygmy chimpanzees. — In: The pygmy chimpanzee: evolutionary biology and behavior (Susman, R.L., ed.). Plenum Press, New York, NY, p. 301-324.

Mitani, J.C. & Watts, D.P. (1999). Demographic influences on the hunting behavior of chimpanzees. — Am. J. Phys. Anthropol. 109: 439-454.

Mitani, J.C. & Watts, D.P. (2001). Why do chimpanzees hunt and share meat? — Anim. Behav. 61: 915-924.

Nishida, T., Hasegawa, T., Hayaki, H., Takahata, Y. & Uehara, S. (1992). Meatsharing as a coalition strategy by an alpha male chimpanzee? — In: Topics in primatology, Vol. 1: human origins (Nishida, T., McGrew, W.C., Marler, P., Pickford, M. & de Waal, F.B.M., eds). University of Tokyo Press, Tokyo, p. 159-174.

Ohashi, G. (2007). Papaya fruit sharing in wild chimpanzees at Bossou, Guinea. — Pan Africa News 14: 14-16.

R Core Team (2014). R: a language and environment for statistical computing. — R Foundation for Statistical Computing, Vienna, available online at http://www.R-project.org/.

Slocombe, K.E. & Newton-Fisher, N.E. (2005). Fruit sharing between wild adult chimpanzees (*Pan troglodytes schweinfurthii*): a socially significant event?. — Am. J. Primatol. 65: 385-391.

Stevens, J.R. (2004). The selfish nature of generosity: harassment and food sharing in primates. — Proc. Roy. Soc. Lond. B: Biol. Sci. 271: 451-456.

Stevens, J.R. & Gilby, I. (2004). A conceptual framework for nonkin food sharing: timing and currency of benefits. — Anim. Behav. 67: 603-614.

Stevens, J.R. & Hauser, M.D. (2004). Why be nice? Psychological constraints on the evolution of cooperation. — Trends. Cogn. Sci. 8: 60-65.

Surbeck, M. & Hohmann, G. (2008). Primate hunting by bonobos at LuiKotal, Salonga National Park. — Curr. Biol. 18: R906-R907.

Surbeck, M. & Hohmann, G. (2013). Intersexual dominance relationships and the influence of leverage on the outcome of conflicts in wild bonobos (*Pan paniscus*). — Behav. Ecol. Sociobiol. 67: 1767-1780.

Surbeck, M., Mundry, R. & Hohmann, G. (2011). Mothers matter! Maternal support, dominance status and mating success in male bonobos (*Pan paniscus*). — Proc. Roy. Soc. Lond. B: Biol. Sci. 278: 590-598.

Tan, J. & Hare, B. (2013). Bonobos share with strangers. — PLoS ONE 8(1): e51922.

Teleki, G. (1973). The predatory behavior of wild chimpanzees. — Bucknell University Press, Lewisburg, PA.

Ueno, A. & Matsuzawa, T. (2004). Food transfer between chimpanzee mothers and their infants. — Primates 45: 231-239.

Warneken, F., Hare, B., Melis, A.P., Hanus, D. & Tomasello, M. (2007). Spontaneous altruism by chimpanzees and young children. — PLoS Biol. 5: 1414-1420.

White, F.J. (1994). Food sharing in wild pygmy chimpanzees, *Pan paniscus*. — In: Current primatology, Vol. II: social development, learning and behavior (Roeder, J.J., Thierry, B., Anderson, J.R. & Herrenschmidt, N., eds). Universite Louis Pasteur, Strasbourg, p. 1-10.

White, F.J. & Wood, K.D. (2007). Female feeding priority in bonobos, *Pan paniscus*, and the question of female dominance. — Am. J. Primatol. 69: 837-850.

Wrangham, R.W. (1975). The behavioural ecology of chimpanzees in Gombe National Park, Tanzania. — PhD thesis, Cambridge University, Cambridge.

Yamamoto, S., Humle, T. & Tanaka, M. (2009). Chimpanzees help each other upon request. — PLoS ONE 4: e7416.

Yamamoto, S., Humle, T. & Tanaka, M. (2012). Chimpanzees' flexible targeted helping based on an understanding of conspecifics' goals. — Proc. Natl. Acad. Sci. USA 109: 3588-3592.

Yamamoto, S., Humle, T. & Tanaka, M. (2013). Basis for cumulative cultural evolution in chimpanzees: social learning of a more efficient tool-use technique. — PLoS ONE 8: e55768.

Ydenberg, R.C. (1994). The behavioral ecology of provisioning in birds. — Ecoscience 1: 1-14.

[When citing this chapter, refer to Behaviour 152 (2015) 359–374]

Can fruiting plants control animal behaviour and seed dispersal distance?

David Beaune [a,b,*], **François Bretagnolle** [b,c], **Loïc Bollache** [d,e],
Gottfried Hohmann [a] **and Barbara Fruth** [a]

[a] Max Planck Institute for Evolutionary Anthropology, Department of Primatology, Deutscher Platz 6, 04103 Leipzig, Germany

[b] Laboratoire Biogéosciences, UMR CNRS 6282, Université de Bourgogne, 6 boulevard Gabriel, 21000 Dijon, France

[c] UMR 5175 CEFE-CNRS, 1919 Route de Mende, 34293 Montpellier 5, France

[d] INRA, UMR 1347 Agroécologie, BP 86510, 21000 Dijon, France

[e] Université de Bourgogne, 6 boulevard Gabriel, 21000 Dijon, France

[*]Corresponding author's e-mail address: david.beaune@gmail.com

Accepted 31 March 2014; published online 7 May 2014

Abstract

In an Afrotropical forest, we tested the hypothesis that fleshy-fruit plants with interspecific differences in fruit quality and quantity affect ranging behaviour of their seed dispersal vector. If fruiting plants could affect their dispersal vector, the plants also affect their seed dispersal distance and eventually their plant population biology. From 2007 to 2011, we measured seed transport by georeference daily bonobo group movements via GPS. Seed dispersal distance was estimated with mechanistic model, using 1200 georeferenced dispersal events and the average seed transit time through bonobo (24.00 h). We compared dissemination for eight plant species that deal with this trade-off: attracting dispersers by means of fruit quality/quantity versus retaining them in the patch because of the same quality/quantity value that attracted them. Because fruit traits of these eight species were different, we expected a difference in seed dispersal distance. Surprisingly, seed dispersal distances induced by bonobos were not affected by fruit traits. Although fruit nutrient contents, abundance and average patch feeding duration differed between plant species, patch feeding time was not related to subsequent dispersal distances. The apes' dispersal distance survey gave an average dispersal distance estimated of 1332 $\pm$ 24 m from the parent plant (97.9% > 100 m). To conclude, feeding time invested in the patch, fruit quality and abundance had no apparent effect on bonobo seed dispersal distance. The possible effects in plant population biology are discussed.

Keywords

Africa, bonobo, Congo Basin, dispersal distance, foraging behaviour, forest structure, mutualism, optimal foraging, *Pan paniscus*, seed dispersal, zoochory.

1. Introduction

The spatial pattern of seed deposition such as dispersal distance is an aspect of dispersal ecology that may have major consequences on several aspects of plant population dynamics as well as on plant community structure and dynamics (Jordano, 1995; Levin et al., 2003; Howe & Miriti, 2004; Schupp et al., 2010). However data that quantify real dispersal patterns are scarce, particularly concerning forest species dispersed through endozoochory (Clark et al., 2005; Russo et al., 2006; McConkey & Chivers, 2007; Cousens et al., 2010). For zoochoric plants, the spatial distribution of seed deposition (i.e., seed shadows, Willson, 1993) results from the movement and behaviour of animals that feed on the fruit and transport the seeds (Westcott et al., 2005). Frugivores can shape spatial configuration of their food species, in numerous interactive ways such as by foraging decisions and by migration behaviour (Jordano et al., 2007; Spiegel & Nathan, 2007; Carlo & Morales, 2008). By that they will influence the shape of the probability distribution of the seed dispersal distance that depends on the disperser distance from the source and its seed retention time (Westcott et al., 2005; Russo et al., 2006; Cousens et al., 2010). The gut transit time of the seed is another parameter that could potentially affect the probability distribution of dispersal and very few studies show that this parameter can be affected by seed size and chemical components of the fruit that can increase or decrease seed transport time (Levey & Grajal, 1991; Gardener et al., 1993; Westcott et al., 2005; Tsuji et al., 2010a). The seed dispersal distances for animals with short gut passage time, such as birds is related to the time spent in fruiting trees (Lenz et al., 2010). The quantity and the quality of fruits produced by an individual plant as well as the level of aggregation of fruiting plants in a landscape can also affect the probability distribution of seed dispersal (Carlo & Morales, 2008). If the food patch can sustain the dispersers for a time superior to the transit time, or if the dispersers frequently return to the patch, and remain in its vicinity, the amount of seeds transported can be high while dispersal distance is low. For example, orang-utans can select large fruiting trees that they repeatedly visit staying around between feeding bouts (Leighton, 1993). In other words an optimal foraging theory would predict that food quality/quantity would negatively affect the animals' foraging effort (daily travel distance). Large and medium-sized frugivores, such as elephants or apes disperse numerous plant species (Campos-Arceiz & Blake, 2011; Forget et al., 2011). In the lowland tropical rainforest of Congo, 85% of the plant species (95% of the tree species) produce fleshy fruits and rely on animals for primary seed dispersal (Beaune et al., 2013b). The bonobos (*Pan paniscus* Schwarz) are efficient seed dispersers that transport viable seeds of several fruiting species by endozoochory (Idani, 1986; Tsuji et al., 2010b; Beaune et al., 2013c). Bonobos have a long gut passage time (24 h 00 min, Beaune et al., 2013a) and are wide-ranging animal foraging on many fruiting plants during a day although occasionally they stay around a big fruiting tree and/or frequently come back to this patch (Beaune et al., 2013a, c). Plant populations are affected by the dispersal distance, which depends on dispersal vectors, but can the plant affect the behaviour of their mutualistic partners? Here we test the hypothesis that plant species with different fruit production strategies can affect the behaviour of their disperser and, consequently, their seed dispersal distance. (a) Fruiting trees that produce large quantities and/or highly nutritive fruit can attract and maintain the disperser in place, resulting in a low seed dispersal distance. Conversely, (b) trees with limited fruit production could achieve a high dispersal distance although being less attractive.

For this, we first analyse whether or not bonobos exhibit variation in the time they spend in fruiting trees. We compare the difference in quality and quantity of the fruiting species with analysis of the fruit nutrient composition, traits and the average feeding duration of the bonobo parties (sub-groups) in the fruiting species. Secondly we develop a mechanistic estimation of seed dispersal for more than 50 zoochoric plant species incorporating seed transit time and the empirical movement behaviour of the bonobos as common dispersers for all these species. Many tropical plants have evolved fleshy fruit that are attractive to only a limited subset of frugivores (Fleming, 1979). Afrotropical forest frugivores use different canopy strata with low feeding overlap (Fleming, 1979; Clark et al., 2001; Poulsen et al., 2002). Bonobos can be considered as main seed dispersers for the fruiting species selected here, although alternative dispersers such as birds, rodents and other primates cannot be excluded (Beaune, 2012; Beaune et al., 2013a).

We compared dispersal distances for several tree species with different species traits and fruit production strategies (see Table 1). Those dispersal distances can be used to test whether fruiting plants affect their frugivore ranging behaviour and thus control their zoochoric partners for seed dispersal distance and eventually affecting their population biology.

Table 1.
Main fruiting plant species characteristics for feeding ecology (seven tree species, one liana: *Cissus dinklagei*).

Family	Species	Average feeding session		Energy (kJ/g dry) matter)	Protein (mg/g)	Sugar (mg/g)	Mean fruit weight (g)	Mean fruit diameter (mm)	Average DBH	
		min	N						cm	N
Annonaceae	*Polyalthia suaveolens*	16 ± 3	75	18.3	96.9	128.5	3.2 ± 0.2	19.4 ± 0.7	18.3 ± 0.4	408
Caesalpiniaceae	*Dialium* sp.	47 ± 3	230	16.3	145.7	101.4	0.9 ± 0.1	22.2 ± 0.5	31.3 ± 1.1	761
Moraceae	*Treculia africana*	78 ± 19	16	19.8	106.9	17.3	>8000	>800	37	1
Sapindaceae	*Pancovia laurentii*	27 ± 5	58	14.6	65.5	160.3	15.1 ± 1.2	29.0 ± 1.3	26.7 ± 1.5	31
Sapindaceae	*Placodiscus paniculatus*	46	1	16.4	125.9	101.6	2.4 ± 0.1	17.3 ± 0.5	16.7 ± 0.5	104
Sapotaceae	*Gambeya lacourtiana*	16 ± 2	81	–	–	–	207.1 ± 28.6	70.7 ± 3.7	96.3 ± 34.2	4
Tiliaceae	*Grewia* sp.	27 ± 2	89	18.9	80.2	172.4	9.2 ± 0.3	32.9 ± 0.4	22.6 ± 1.6	50
Vitaceae	*Cissus dinklagei*	22 ± 1	204	20.4	106.2	119.2	8.3 ± 0.5	26.5 ± 0.6	–	–

Average diameter at breast high (dbh) based on 12-ha plots inventory at LuiKotale, DRC; average foraging session time based on 1879 h of field observation, average fruit weight and largest diameter ($N = 10$) and mean nutritional value. Values are mean ± SE.

2. Methods

2.1. Study species and site

The bonobo is restricted to the tropical rain forest of the Democratic Republic of Congo (DRC) on the southern bank of the Congo River. The bonobo is mainly frugivores, feeding on and disseminating hundreds of plant species (Kano & Mulavwa, 1984; Tsuji et al., 2010b). Seeds of around 40% of the tree species in the forest are dispersed by bonobos (Beaune et al., 2013c). Bonobos live in polyandrogynous communities with fission of subunit groups (parties) during the day while foraging, and fusion in the nesting place at night (Fruth & Hohmann, 1993). In the Congo Basin, at the southwest fringe of the Salonga National Park, there is a habituated group of free-ranging bonobos, tracked by research teams at the LuiKotale field site (LK) (Hohmann & Fruth, 2003). The bonobo community includes 25–35 individuals identifiable by individual physical traits (genital, face, pilosity, color). The long-term project of LuiKotale with a habituated bonobo community which can be daily observed, identified, followed and georefenced by researchers, allowing us to build a mechanistic model of seed dispersal distance (Figures 1 and 2).

2.2. Ethics statement

The studied apes are free ranging bonobos observed without invasive methods, constraint, contact or any interaction with the researchers. Animal welfare had greater priority than scientific interests. The methods used to collect data in the field are in compliance with the requirements and guidelines of the Institut Congolais pour la Conservation de la Nature, and adhere to the legal requirements of the host country, the DRC.

2.3. Dispersal analysis

The probability distribution of seeds is based on empirical bonobo movements, georeferenced from 8th of July 2007 to the 21st of September 2011. With bonobo movements (i.e., group movements) recorded after observed feeding sessions in feeding trees georeferenced (Figure 1) and mean transit time of seeds known (see below), a mechanistic model of seed dispersal distance can be calculated (Westcott et al., 2005; Tsuji et al., 2010b; Côrtes & Uriarte, 2012). For example, if a bonobo group is observed feeding in an *Gambeya lacourtiana* identified and georeferenced at 8.05 a.m. and continuously followed for at least 24 h (and then including at least one or several

Figure 1. An habituated and identified bonobo of the party (here Emil) is swallowing the fruits and seeds of *Dialium* sp. at LuiKotale, DRC. The party is geolocalised at this point and continuously during 24 h (with GPS track log, including at least Emil) for an estimation of this seed dispersal event. Photo by LKBP/David Beaune. This figure is published in colour in the online edition of this journal, which can be accessed via http://booksandjournals.brillonline.com/content/journals/1568539x.

individuals followed since the record), the seed dispersal distance from the parent tree is calculated after the georeferenced position of the bonobo party at 8.05 + 24.00 h (mean transit time, see Figure 2). Whenever possible, a bonobo party was followed daily from nest to nest (approx. 05:30 to 17:30). Daily travel routes of parties were tracked with a GPS (Garmin® 60CSX) using 1 point position/5 min for georeferencing. Bonobo transit time was calculated from direct observations (24.00 h ± 9 min SE) and was not affected by the seed size or by bonobo gender (Beaune et al., 2013a). Bonobo feeding trees were georefenced when identified during group feeding sessions. The most abundant fruiting species eaten by bonobo were selected and seed dispersal calculations compared to test our hypotheses. The non-parametric dispersal distance with species effects (i.e., fruit trait) were tested with Kruskal–Wallis one-way analysis of variance.

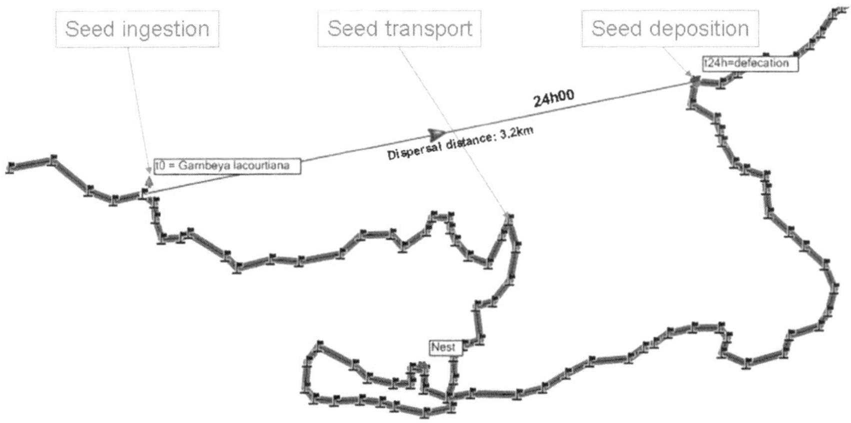

Figure 2. Illustration of the mechanistic seed dispersal model with an example of seed transport (*Gambeya lacourtiana*). Identified bonobo feeding trees were georeferenced during group observations (2007–2011, LuiKotale, DRC) and bonobo group movement daily recorded (dark track log). Theoretical seed deposition site were determined by actual bonobo group position (dark track log) after 24.00 h, corresponding to the average transit time for seed through bonobo.

2.4. Variation in feeding time session

Feeding durations were estimated by direct observation of the LK bonobo community. From September 2009 to June 2011 behavioural data of bonobos were recorded for 315 days across 22 months of observations. Bonobos are a fission-fusion society that is depending on season and time of day the community splits up into smaller foraging subgroups called parties (White, 1988). Males and female bonobos forage together, and the offspring follow their mothers, an individual effect on seed dispersal distance is by consequent not considered. As parties are largely cohesive and bonobos feed in group, synchronising their feeding sessions, we considered group feeding activity to be that of the majority of the visible animals during a continuous behavioural records (a party feeding boots was recorded when >50% of the bonobos from a party under observation entered a feeding tree and started feeding until >50% of the animal terminated feeding and the party moved on or changed its activity, i.e., group scan; Beaune et al., 2013a). As our estimates of dispersal distances are also at the party level, using this party-level feeding metric is appropriate for the purposes of this study. A total of 573 h of feeding session with fruiting species was analysed. Among these group feeding sessions, the potential correlation of 278 dispersal events linked with feeding duration from 22 different fruiting species was analysed with Spearman correlation.

2.5. *Interspecific fruit differences*

Fleshy-fruited plant species are different in fruit production and quality (Hohmann et al., 2006, 2010). The aim of this study was to test whether fruit production and quality affect the probability of seed dispersal distance. We contrasted medium-sized tree species with relatively low fruit production (i.e., *Polyalthia* (= *Greenwayodendron*) *suaveolens*, *Placodiscus paniculatus*) and large-sized fruiting trees which support and maintain dispersers for longer periods (i.e., *Dialium corbisieri* and *Gambeya lacourtiana*). To estimate the mean fruit abundance of each selected species we have calculated the diameter at breast height (dbh) (Chapman et al., 1992). The mean dbh was calculated for the main species, based on a 12-ha plot inventory (Beaune et al., 2013d). One liana, *Cissus dinklagei*, was added to the test and compared with the seven tree species, for a total of eight species analysed. Fruits from LK forest were collected for nutrient analyses (Hohmann et al., 2006; Hohmann et al., 2010). Average fruit mass and diameter were measured on at least ten mature fruits.

3. Results

Table 1 reports differences in nutritional values, fruit size, weight and dbh for adult trees, and mean feeding duration in fruiting species. Bonobo parties spent an average of 30 ± 2 min ($CI_{95\%}$ = 27–34 min; range = 1–168 min) feeding in fruiting plants ($N = 22$ species using bonobo seed dispersal service among 91 identified in another study, Beaune et al., 2013c). Although feeding times differed among species (see Table 1; three species tested: $N = 50$ *Cissus dinklagei*, 122 *Dialium* sp., 27 *Grewia* sp.; $H = 15.1015$, df = 2, $p = 0.0005257$), bonobo party feeding time did not correlate with the 24-h dispersal distance ($r = -0.07$, $p = 0.2422$; see Figure 3).

We calculated seed transport for 51 different endozoochoric plant species ($N = 1$–362 events). The average dispersal distance ± SD from the parent plant is 1332 ± 24 m, median = 1198; $CI_{95\%}$ = 1282–1380 m; range = 1–4492 m. The resulting dispersal distance kernel is a probability density function (Figure 4), characterized by a unimodal leptokurtic distribution, with a fat-tailed dispersal kernel (i.e., with exceptional seed transports at

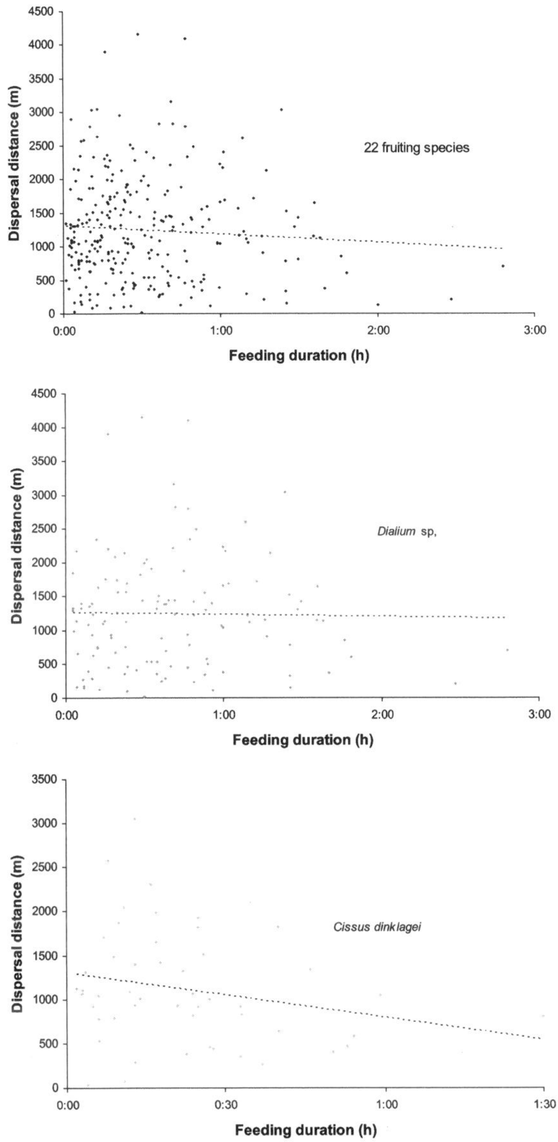

Figure 3. There is no correlation between feeding time spent by the bonobo group on the fruiting plant and the dispersal distance induced by the group after 24 h. For 22 fruiting species analysed as whole ($N = 278$, $r = -0.07$, $p = 0.2422$) or other species as *Dialium* sp. (122, $r = -0.01$, $p = 0.8572$) or *Cissus dinklagei* (50, $r = -0.22$, $p = 0.1178$) at LuiKotale, DRC. This figure is published in colour in the online edition of this journal, which can be accessed via http://booksandjournals.brillonline.com/content/journals/1568539x.

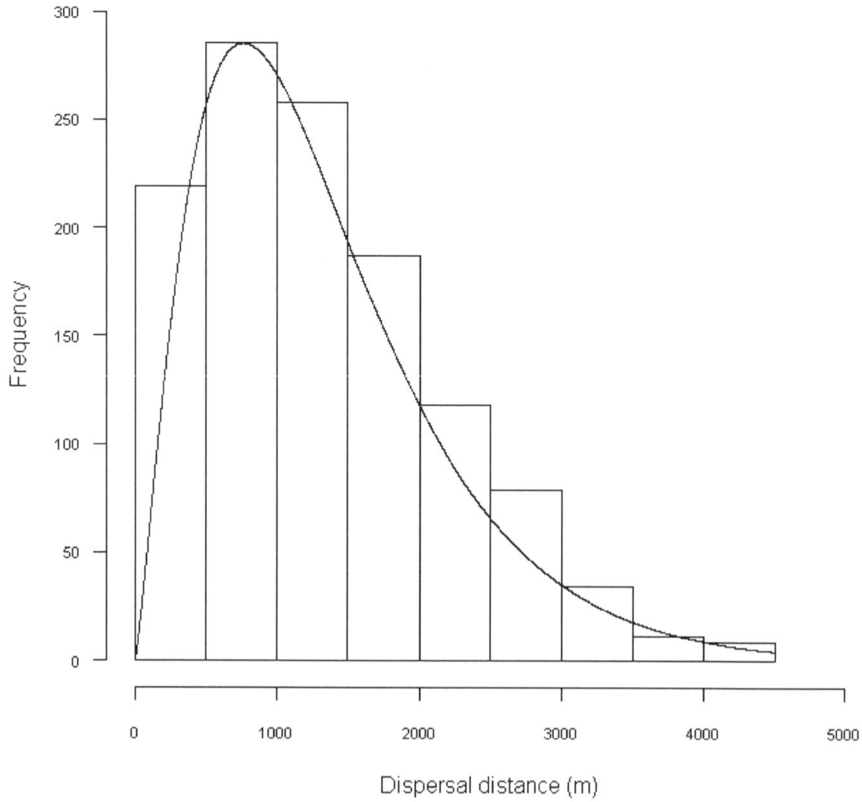

Figure 4. Seed dispersal distribution induced by bonobo based on group movement behaviour ($N = 1200$ dispersal events with all plant species at LuiKotale, DRC) and fixed bonobo transport duration (transit time = 24.00 h).

very long distance; right skewness = 0.81; Kurtosis = 3.38; see (Nathan & Muller-Landau, 2000) for the different shapes of dispersal kernels). Bonobos disperse seeds over long distances, with 97.9% of the dispersal distances longer than 100 m. There is no annual effect on bonobo dispersal distance (from 2007 to 2011, four years tested: $H = 7.6939$, df = 4, $p = 0.1035$) but the seasons seems to affect the group travel distance (September–January = 1422 m on average, March–April = 1284 m, February (dry) = 1132, May–August (dry) = 1337 m tested: $H = 17.4279$, df = 3, $p = 0.000577$). The main plant species tested (*Cissus dinklagei* ($N = 173$), *Dialium* sp. (362), *Gambeya lacourtiana* (56), *Grewia* sp. (97), *Pancovia laurentii* (73), *Placodiscus paniculatus* (48), *Polyalthia suaveolens* (61), *Treculia africana* (21)), from large *G. lacourtiana* to medium tree species (*P. paniculatus* and

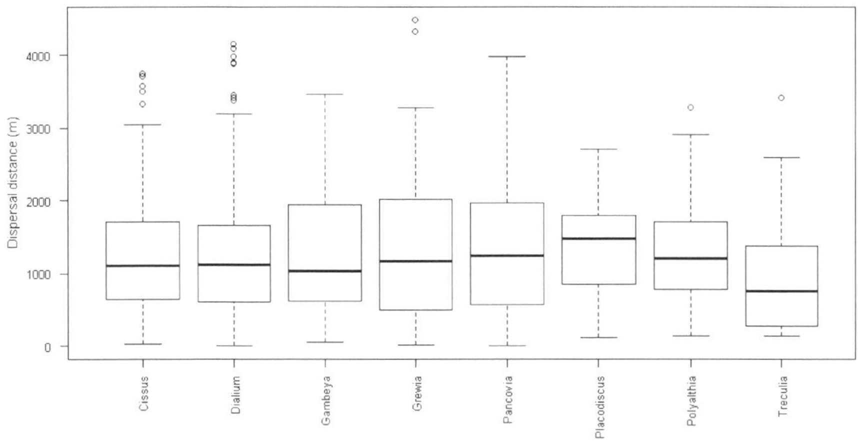

Figure 5. Seed dispersal distances induced by bonobo group for eight plant species (*Cissus dinklagei* ($N = 173$), *Dialium* sp. (362), *Gambeya lacourtiana* (56), *Grewia* sp. (97), *Pancovia laurentii* (73), *Placodiscus paniculatus* (48), *Polyalthia suaveolens* (61), *Treculia africana* (21)) at LuiKotale, DRC.

P. suaveolens) (see Table 1 for average dbh) did not significantly affect the dispersal distance induced by the bonobo ranging behaviour ($H = 8.3769$, df $= 7$, $p = 0.3005$) (Figure 5). The hypothesis on interspecific differences in seed dispersal distance was not supported.

4. Discussion

Our findings provide an analysis based on long-term data, of differences in dispersal distance among fleshy-fruited plants disseminated by the same endozoochoric dispersal vector. Surprisingly, all the fleshy-fruited species are dispersed at the same average distance, whatever the feeding time on the fruiting plant, their fruit quality and abundance: 1.3 km on average. This is a remarkable long dispersal distance with 97.9% of the seed transports superior to 100 m away from the parent plants. However, other extensive studies are required for seed dispersal comparison with other frugivores and other bonobo communities. Tsuji and his colleagues found a lowest dispersal distance induced by a Wamba community with an average distance of 783 and 777 m while foraging on two *Dialium* species (Tsuji et al., 2010b). This difference can be due to low sample size, a seasonal effect or a different community behaviour and ressource availability at Wamba). Bonobos seems to move at homogenous and regular distances from food patches. This could be due to the regular ranging behaviour and consistent travel times of bonobo groups, whatever the quality and quantity of the fruit they currently feed on. This interpretation does not fit with predictions of optimal foraging theory which suggest that animals would adapt their foraging investment (such as travel) according to food and nutrients availability. One potential explanation is the stochastic phenology of fruiting species at the site (Fruth et al., unpubl. data). This unpredictability could force bonobos to forage permanently for food and then regularly disperse the seeds at long distances. Another unverified hypothesis is differential ripening for these species. With asynchrony in ripe fruit availability, frugivores cannot forage for long sessions in the same area. However, further studies are needed on differential ripening in tropical plants. Surprisingly in another study, we found that seed size does not seem to affect transit time (Beaune et al., 2013a), unlike other animals, where a shorter gut passage is induced by smaller seed size (Westcott et al., 2005). Whatever their differences in seed/fruit size, colour, fruit quality and quantity, the fruiting-species traits did not seem to affect sufficiently the bonobo transit time and foraging behaviour for an effect on their seed dispersal distance.

Extensive seed dispersal among communities homogenises species composition, and eventually makes competitive ability dependent on global rather than local abundances, thus facilitating domination by the single most abundant species (Levin et al., 2003). This is the case here with bonobos as dispersal vectors. Seed dispersal limitation in distance (Muller-Landau et al., 2008) does not exist for different plant species sharing the same dispersers.

The assumption in the theoretical dispersal model that animals move randomly in space (Levin et al., 2003) is supported by our finding for the distance parameter, which is consistent and without any plant species effect. This is coherent with mechanistic models of zoochoric seed dispersal (Cousens et al., 2010).

However we did not explore post-dispersal fate for seed, which surely shapes species distribution (Réjou-Méchain et al., 2011). Several studies have shown that the interaction between environmental heterogeneity and the biological characteristics of species can influence distribution patterns at various spatial scales (Muller-Landau, 2004; ter Steege et al., 2006). Negative density dependence with environmental filtering contributes to community assembly (Paine et al., 2012). Nevertheless, it has also been shown that dispersal syndrome predicts spatial distribution, which is relatively dispersed for zoochoric species (Seidler & Plotkin, 2006). For species using large mammals such as the bonobo, we show that seed dispersal is long-distance but without interspecific differences; although interspecific difference in fruit characteristics is wide. Do species dispersed by the same partner share the same distribution pattern? Studies have hypothesised that spatial patterns are highly context dependent but can be predicted by dispersal syndrome (Réjou-Méchain et al., 2011) and plant traits (Muller-Landau et al., 2008). Our hypothesis goes further with spatial prediction, trait-based generalisation and modelling of seed dispersal in tropical forests, based not on fruit characteristics, but rather on the disperser variable itself (elephant, bonobo, guenon, bat, hornbill, etc.).

To conclude, fruit traits of a species (quantity and quality) did not seem to affect disperser behaviour in relation to dispersal distance. Our finding leads to new questions about possible plant adaptations to force zoochoric partners to move constantly within their range. This research challenges the validity of optimal foraging theory predictions, at least in the case of bonobo ecology, as nutrient and food availability did not affect foraging and travel investment. The seed dispersal limitation theory explains how plant biodiversity is maintained, however bonobos induce homogenous dissemination in their ecosystem and thus dominant species could overcompete other species.

Acknowledgements

We thank the Institut Congolais pour la Conservation de la Nature (ICCN) for granting permission to conduct research at Salonga National Park, and also the Lompole village for granting permission to access the forest of their ancestors. Research at LuiKotale was conducted under the auspices and with financial support of the Max Planck Society, the German Ministry of Education and Research, the Conseil Régional de Bourgogne, the SFE and SFDP (France). We thank the international and local staff of LuiKotale for their help, Carmela Chateau-Smith for proofreading the English and the reviewers for their valuable suggestions.

References

Beaune, D. (2012). Latest news from the bonobos: *Pan paniscus* myths and realities. — Rev. Primatol. 4, DOI:10.4000/primatologie.1090.

Beaune, D., Bretagnolle, F., Bollache, L., Hohmann, G., Surbeck, M., Bourson, C. & Fruth, B. (2013a). The bonobo–*Dialium* positive interactions: seed dispersal mutualism. — Am. J. Primatol. 75: 394-403.

Beaune, D., Bretagnolle, F., Bollache, L., Hohmann, G., Surbeck, M. & Fruth, B. (2013b). Seed dispersal strategies and the threat of defaunation in a Congo forest. — Biodiv. Conserv. 22: 225-238.

Beaune, D., Bretagnolle, F., Bollache, L., Bourson, C., Hohmann, G. & Fruth, B. (2013c). Ecological services performed by the bonobo (*Pan paniscus*): seed dispersal effectiveness in tropical forest. — J. Trop. Ecol. 29: 367-380.

Beaune, D., Fruth, B., Bollache, L., Hohmann, G. & Bretagnolle, F. (2013d). Doom of the elephant-dependent trees in a Congo tropical forest. — Forest Ecol. Manage. 295: 109-117.

Campos-Arceiz, A. & Blake, S. (2011). Megagardeners of the forest — the role of elephants in seed dispersal. — Acta Oecol. 37: 542-553.

Carlo, T.A. & Morales, J.M. (2008). Inequalities in fruit-removal and seed dispersal: consequences of bird behaviour, neighbourhood density and landscape aggregation. — J. Ecol. 96: 609-618.

Chapman, C.A., Chapman, L.J., Wrangham, R., Hunt, K., Gebo, D. & Gardner, L. (1992). Estimators of fruit abundance of tropical trees. — Biotropica 24: 527-531.

Clark, C., Poulsen, J. & Parker, V. (2001). The role of arboreal seed dispersal groups on the seed rain of a lowland tropical forest. — Biotropica 33: 606-620.

Clark, C., Poulsen, J., Bolker, B., Connor, E. & Parker, V. (2005). Comparative seed shadows of bird-, monkey-, and wind-dispersed trees. — Ecology 86: 2684-2694.

Côrtes, M.C. & Uriarte, M. (2012). Integrating frugivory and animal movement: a review of the evidence and implications for scaling seed dispersal. — Biol. Rev. 88: 226-254.

Cousens, R.D., Hill, J., French, K. & Bishop, I.D. (2010). Towards better prediction of seed dispersal by animals. — Funct. Ecol. 24: 1163-1170.

Fleming, T.H. (1979). Do tropical frugivores compete for food? — Am. Zool. 19: 1157-1172.

Forget, P.-M., Jordano, P., Lambert, J.E., Böhning-Gaese, K., Traveset, A. & Wright, S.J. (2011). Frugivores and seed dispersal (1985–2010); the 'seeds' dispersed, established and matured. — Acta Oecol. 37: 517-520.

Fruth, B. & Hohmann, G. (1993). Ecological and behavioral aspects of nest building in wild bonobos (*Pan paniscus*). — Ethology 94: 113-126.

Gardener, C., McIvor, J. & Jansen, A. (1993). Passage of legume and grass seeds through the digestive tract of cattle and their survival in faeces. — J. Appl. Ecol. 30: 63-74.

Hohmann, G. & Fruth, B. (2003). Lui Kotal – a new site for field research on bonobos in the Salonga National Park. — Pan Afr. News 10: 25-27.

Hohmann, G., Fowler, A., Sommer, V. & Ortmann, S. (2006). Frugivory and gregariousness of Salonga bonobos and Gashaka chimpanzees: the influence of abundance and nutritional quality of fruit. — In: Feeding ecology in apes and other primates (Hohmann, G., Robbins, M. & Boesch, C., eds). Cambridge University Press, Cambridge, p. 123-159.

Hohmann, G., Potts, K., N'Guessan, A., Fowler, A., Mundry, R., Ganzhorn, J.U. & Ortmann, S. (2010). Plant foods consumed by *Pan*: exploring the variation of nutritional ecology across Africa. — Am. J. Phys. Anthropol. 141: 476-485.

Howe, H.F. & Miriti, M.N. (2004). When seed dispersal matters. — BioScience 54: 651-660.

Idani, G. (1986). Seed dispersal by pygmy chimpanzees (*Pan paniscus*): a preliminary report. — Primates 27: 441-447.

Jordano, P. (1995). Angiosperm fleshy fruits and seed dispersers: a comparative analysis of adaptation and constraints in plant-animal interactions. — Am. Nat. 145: 163-191.

Jordano, P., García, C., Godoy, J.A. & García-Castaño, J.L. (2007). Differential contribution of frugivores to complex seed dispersal patterns. — Proc. Natl. Acad. Sci. USA 104: 3278-3282.

Kano, T. & Mulavwa, M. (1984). Feeding ecology of the pygmy chimpanzees (*Pan paniscus*) of Wamba. — In: The Pygmy chimpanzee: evolutionary biology and behavior (Susman, R.L., ed.). Plenum Press, New York, NY, p. 435.

Leighton, M. (1993). Modeling dietary selectivity by Bornean orangutans: evidence for integration of multiple criteria in fruit selection. — Int. J. Primatol. 14: 257-313.

Lenz, J., Fiedler, W., Caprano, T., Friedrichs, W., Gaese, B.H., Wikelski, M. & Böhning-Gaese, K. (2010). Seed-dispersal distributions by trumpeter hornbills in fragmented landscapes. — Proc. Roy. Soc. Lond. B: Biol. Sci. 278: 2257-2264.

Levey, D.J. & Grajal, A. (1991). Evolutionary implications of fruit-processing limitations in cedar waxwings. — Am. Nat. 138: 171-189.

Levin, S.A., Muller-Landau, H.C., Nathan, R. & Chave, J. (2003). The ecology and evolution of seed dispersal: a theoretical perspective. — Annu. Rev. Ecol. Evol. Syst. 34: 575-604.

McConkey, K.R. & Chivers, D.J. (2007). Influence of gibbon ranging patterns on seed dispersal distance and deposition site in a Bornean forest. — J. Trop. Ecol. 23: 269.

Muller-Landau, H.C. (2004). Interspecific and inter-site variation in wood specific gravity of tropical trees. — Biotropica 36: 20-32.

Muller-Landau, H.C., Wright, S.J., Calderón, O., Condit, R. & Hubbell, S.P. (2008). Interspecific variation in primary seed dispersal in a tropical forest. — J. Ecol. 96: 653-667.

Nathan, R. & Muller-Landau, H.C. (2000). Spatial patterns of seed dispersal, their determinants and consequences for recruitment. — Trends Ecol. Evol. 15: 278-285.

Paine, C.E.T., Norden, N., Chave, J., Forget, P.-M., Fortunel, C., Dexter, K.G. & Baraloto, C. (2012). Phylogenetic density dependence and environmental filtering predict seedling mortality in a tropical forest. — Ecol. Lett. 15: 34-41.

Poulsen, J.R., Clark, C.J., Connor, E.F. & Smith, T.B. (2002). Differential resource use by primates and hornbills: implications for seed dispersal. — Ecology 83: 228-240.

Réjou-Méchain, M., Flores, O., Bourland, N., Doucet, J.-L., Fétéké, R.F., Pasquier, A. & Hardy, O.J. (2011). Spatial aggregation of tropical trees at multiple spatial scales. — J. Ecol. 99: 1373-1381.

Russo, S.E., Portnoy, S. & Augspurger, C.K. (2006). Incorporating animal behavior into seed dispersal models: implications for seed shadows. — Ecology 87: 3160-3174.

Schupp, E.W., Jordano, P. & Gomez, J.M. (2010). Seed dispersal effectiveness revisited: a conceptual review. — New Phytol. 188: 333-353.

Seidler, T.G. & Plotkin, J.B. (2006). Seed dispersal and spatial pattern in tropical trees. — PLoS Biol. 4: e344.

Spiegel, O. & Nathan, R. (2007). Incorporating dispersal distance into the disperser effectiveness framework: frugivorous birds provide complementary dispersal to plants in a patchy environment. — Ecol. Lett. 10: 718-728.

ter Steege, H., Pitman, N.C.A., Phillips, O.L., Chave, J., Sabatier, D., Duque, A., Molino, J.-F., Prevost, M.-F., Spichiger, R., Castellanos, H., von Hildebrand, P. & Vasquez, R. (2006). Continental-scale patterns of canopy tree composition and function across Amazonia. — Nature 443: 444-447.

Tsuji, Y., Morimoto, M. & Matsubayashi, K. (2010a). Effects of the physical characteristics of seeds on gastrointestinal passage time in captive Japanese macaques. — J. Zool. 280: 171-176.

Tsuji, Y., Yangozene, K. & Sakamaki, T. (2010b). Estimation of seed dispersal distance by the bonobo, *Pan paniscus*, in a tropical forest in Democratic Republic of Congo. — J. Trop. Ecol. 26: 115-118.

Westcott, D.A., Bentrupperbaumer, J., Bradford, M.G. & McKeown, A. (2005). Incorporating patterns of disperser behaviour into models of seed dispersal and its effects on estimated dispersal curves. — Oecologia 146: 57-67.

White, F. (1988). Party composition and dynamics in *Pan paniscus*. — Int. J. Primatol. 9: 179-193.

Willson, M. (1993). Dispersal mode, seed shadows, and colonization patterns. — Plant Ecol. 107: 261-280.

[When citing this chapter, refer to Behaviour 152 (2015) 375–406]

Context influences spatial frames of reference in bonobos (*Pan paniscus*)

Alexandra G. Rosati *

Department of Psychology, Yale University, New Haven, CT, USA
* Author's e-mail address: alexandra.rosati@yale.edu

Accepted 15 February 2014; published online 14 March 2014

Abstract

Primates must solve complex spatial problems when foraging, such as finding patchy resources and navigating between different locations. However, the nature of the cognitive representations supporting these types of behaviors is currently unclear. In humans, there has been great debate concerning the relative importance of egocentric representations (which are viewer-dependent) versus allocentric representations (which are based on aspects of the external environment). Comparative studies of nonhuman apes can illuminate which aspects of human spatial cognition are shared with other primates, versus which aspects are unique to our lineage. The current studies therefore examined spatial cognitive development in one of our closest living relatives, bonobos (*Pan paniscus*) across contexts. The first study assessed how younger bonobos encode locations in a place-response task in which apes first learn that one of two locations is consistently baited with a reward, and then must approach the two locations from a flipped perspective. The second study examined how a larger age sample of bonobos responded to a spatial relations task in which they first experience that one location is baited, and then can generalize this learning to a new set of targets. Results indicated that while bonobos exhibited a predominantly allocentric strategy in the first study, they consistently exhibited an egocentric strategy in the second. Together, these results show that bonobos can use both strategies to encode spatial information, and illuminate the complementary contributions to cognition made by egocentric and allocentric representations.

Keywords

spatial cognition, memory, spatial relations, habit formation, allocentric, egocentric.

1. Introduction

Primates in the wild face complex spatial problems. For example, all primates must locate important items in their environment — such as patchily-distributed food resources or other conspecifics — as well as travel efficiently in their environment (Gallistel, 1990; Shettleworth, 1998; Janson & Byrne,

2007). Studies of primate foraging both in the wild (Janson, 1998, 2007; Normand & Boesch, 2009; Normand et al., 2009) and in captivity (Menzel, 1973; Menzel et al., 2002; Mendes, 2008; Rosati & Hare, 2012) suggest that primates exhibit sophisticated behaviors when solving these types of spatial problems. However, the particular cognitive capacities underlying these behaviors are currently unclear. There are two general strategies that organisms could use for goal-oriented navigation (Maguire et al., 1998; Burgess, 2008). First, an individual could move through the environment using an allocentric strategy that depends on representations of the world at large, by encoding the set of relationships between environmental cues and the relevant goal (such as a food patch). This type of enduring representation of the layout of objects in the environment has been referred to as a 'cognitive map' (Tolman, 1948). Alternatively, an individual could use an egocentric strategy that depends on representations of the organism's movements centered on their own body (e.g., turn left), sometimes called 'dead reckoning'. The relative importance of these types of strategies in human spatial cognition is a topic of debate. Moreover, some hypotheses suggest that language allows humans to use spatial information with a degree of flexibility and accuracy not seen in other species (Hermer-Vazquez et al., 2001; Wang & Spelke, 2002). Consequently, studies of nonhuman ape cognition can identify the capacities underlying their natural spatial behaviors, as well as illuminate what aspects of human spatial cognition are unique to our lineage.

How might language influence the manner in which humans encode and remember spatial information? First, some research suggests that language influences the coordinate systems that humans use when thinking about space. Whereas some languages encode spatial relations in terms of a viewer-relative system (e.g., the table is to the left or right), others use a system that encodes the absolute position of objects (e.g., the table is to the North or South). The dominant way a given language describes spatial relations is thought to influence the way mature language speakers encode spatial relations in nonverbal spatial relations tasks: children who grow up speaking language with a preponderance of relative constructions — such as English and other European languages — tend to show an egocentric biases in spatial tasks, whereas speakers of indigenous languages in Africa and other parts the world that use absolute constructions exhibit allocentric bias (Brown & Levinson, 1992, 2000; Levinson, 1996; Pederson et al., 1998; Levinson et al., 2002; Majid et al., 2004). For example, in one relevant task children watch an experimenter hide a reward under one cup in a spatial array, and then move to a second table with an identical array of cups. Their orientation relative to the array is therefore flipped 180°, and the child is asked to predict which location will have the target in this new setup. Their predictions in this context concord with their language's dominant relation construction. As apes tend to depend primarily on an allocentric strategy in the same spatial relations task (Haun et al., 2006a), one proposal stemming from this work is that allocentric encoding is phylogenetically ancestral, whereas human-specific cultural influences can override these biologically-based biases so that some humans favor egocentric strategies (see also Gentner, 2007).

An alternative proposal suggests that human spatial memory is supported primarily by egocentric representations of location, which are then combined with geometrical information about the surface of the environment (Wang & Spelke, 2002). According to this proposal, egocentric representations involve recognizing scenes based on viewer-dependent perspectives, and then updating these egocentric representations based on information about one's own movements. While pre- or nonlinguistic organisms possess geometric representations of the surface geometric of the environment, this information does not directly represent goal-relevant locations in the environment. Thus, this proposal suggests that flexible representations of space integrating both landmark and geometric information require spatial language (e.g., Hermer-Vazquez et al., 1999, 2001). Thus, younger children and animals use only geometric information, failing to integrate salient information about landmarks into their search patterns (Cheng, 1986; Hermer & Spelke, 1994). Indeed, when adults' language systems are taxed by engagement in a second simultaneous task, they perform like young children (Hermer-Vazquez et al., 1999). This hypothesis thus suggests that language allows geometric and feature information to be integrated to form these flexible spatial representations (Spelke et al., 2010).

These proposals both emphasize the importance of language in supporting human-unique forms of spatial reasoning, but differ in which type of strategy humans alone exhibit. However, current research offers a third possibility: organisms might posses multiple memory systems, with egocentric and allocentric representations coexisting in parallel (Sherry & Schacter, 1987; Burgess, 2006, 2008). That is, this hypothesis suggests that humans and animals may encode both egocentric representations of space as well as allocentric representations of the relations between objects and the environment at large, with the particular strategy used depending on the context at hand. For example, neurobiological studies indicate that parallel memory systems can influence behavior, and that these systems are supported by distinct neural circuits in the brain. One illustrative set of studies involves the responses of rats (*Rattus norvegicus*) to a memory task in which both types of strategies could be used to locate food (Packard, 2009). In the cross maze task, rodents are placed in a cross-shaped maze and initially learn that one side wing is consistently baited with food. This could be encoded using either an allocentric strategy (e.g., 'the food is in that spatial location'), or an egocentric strategy (e.g., 'turn left to find the food'). The animals' strategy is determined by reversing their start position in the maze so they must approach the side wings from a flipped orientation to assess if they search in the previously-rewarded location (allocentric strategy), or if they turn in the same direction even though this now leads to the opposite wing (egocentric strategy). In this task, early learning is supported by place-based memory such that animals initially exhibit allocentric strategies. However, with repetition animals exhibit a shift towards egocentric strategies (Packard, 2009). Moreover, temporary lesions and other neurochemical manipulations can selective inhibit or strengthen these different types of encoding strategies: the hippocampus selectively impacts allocentric encoding, whereas changes to the basal ganglia influence egocentric encoding (Packard, 1999). Together, this evidence indicates that both allocentric encoding of specific locations in space, and egocentric encoding of particular motor responses, exist in parallel (Poldrack et al., 2001; White & McDonald, 2002; Poldrack & Packard, 2003).

Developmental work further supports the claim that both types of representations can control behavior depending on context (Newcombe & Huttenlocher, 2006). For example, when infants initially experience that something interesting (such as a toy) repeatedly appears to one side, but then are flipped to view the same space from the opposite orientation, they tend to show an egocentric bias. However, this shifts to an allocentric bias over the first year of life (Acredolo, 1978, 1979). Studies using search tasks similarly show that egocentric frameworks are available at an earlier age, whereas older children can successfully use both egocentric and allocentric strategies (Newcombe et al., 1998). Importantly, aspects of the external environment can influence children's strategies: even young babies can use allocentric strategies when there are very salient landmarks in the external environment (Acredolo & Evans, 1980). Similarly, although babies tend to exhibit egocentric biases in unfamiliar experimental rooms, they are more likely to use an allocentric strategy in their own homes (Acredolo & Evans, 1980). Adults exhibit similar patterns, with a greater preference for using an allocentric frame of reference when there are salient landmark cues that situate the array in a larger spatial context (Li & Gleitman, 2002). Together, these results indicate that different sources of spatial information may be combined in a weighted fashion, with some information more important in specific contexts than others (Ratliff & Newcombe, 2008).

Overall, these hypotheses concerning human spatial cognition predict that nonlinguistic apes should favor different spatial strategies. The first proposal suggests that apes use primarily allocentric strategies, but this preference can be 'overwritten' by cultural linguistic influences in humans (Majid et al., 2004; Haun et al., 2006a; Gentner, 2007). The second view suggests that nonhumans generally use an egocentric framework to encode space, and that language enables human-unique forms of allocentric representations (Wang & Spelke, 2002). The final view suggests that nonhuman apes and humans should be both capable of using egocentric and allocentric encoding strategies, although their preference may vary across contexts (Newcombe & Huttenlocher, 2006). Consequently, evidence from apes can play an important role in evaluating the role of language in how humans represent spatial relations.

What kinds of strategies do apes use when faced with spatial problems? Several lines of evidence suggest that apes can use allocentric frameworks. For example, chimpanzees (*Pan troglodytes*) can form allocentric 'cognitive maps' of their surrounding: when chimpanzees are shown multiple hiding places for food in a large field, they later will move between those locations using an optimal route to retrieve the food with the least amount of travel distance (Menzel, 1973). Similarly, wild chimpanzee foraging patterns in tropical rainforests suggest that the apes traveled in direct routes to reach food resources (Normand & Boesch, 2009). In tasks designed to examine spatial strategies in more controlled settings, apes also exhibit allocentric strategies. For example, all four species of great apes exhibited allocentric strategies in the spatial relations task described previously (Haun et al., 2006a). Apes also exhibit allocentric strategies when they initially watch food being hidden in an array, and then can choose after their orientation is flipped (Hoffman &

Beran, 2006; Albiach-Serrano et al., 2010). Similarly, when apes' orientation towards the array was the same during both the baiting event and when making a choice (but they move in the interim), apes appear to predominately use spatial cues (e.g., proximity to the edge of the table) rather than egocentric information (left or right locations) when searching for food (Hribar & Call, 2011; Hribar et al., 2011). Finally, some tasks have examined the extent to which apes use landmark information about the features of items in the environment (such as color or shape), versus purely spatial information about specific locations in the environment. For example, in one task apes initially watched food being hidden under one container in an array where all containers had different features, but then the locations of the containers are switched before the apes could choose. Some evidence suggests that apes tend to show to choose the same location, ignoring that the container's features had changed (Haun et al., 2006b), whereas other studies show that apes exhibit a greater mixture of spatial strategies (Kanngiesser & Call, 2010). Importantly, however, apes showed a greater tendency to use feature information in tasks involving longer periods of experience (Kanngiesser & Call, 2010), which might suggest the acquisition of a more habit-based response over time. Nevertheless, current evidence overall suggests that apes have a bias towards allocentric strategy rather than an egocentric strategy.

However, current comparative data on spatial cognition in apes cannot address the relative importance of these strategies for two reasons. First, although humans are equally related to both chimpanzees (*P. troglodytes*) and bonobos (*P. paniscus*), chimpanzees are much more commonly studied than are bonobos (Hare, 2007). For example, in the previous studies that explicitly contrasted allocentric versus egocentric frameworks, few bonobos were tested (three to six across studies; Haun et al., 2006a; Albiach-Serrano et al., 2010; Hribar & Call, 2011). Importantly, differences in the feeding ecology of bonobos and chimpanzees suggest that these species face different spatial challenges while foraging. In particular, chimpanzees are thought to face more patchy, uncertain environments than bonobos (White & Wrangham, 1988; Kano, 1992; Malenky & Wrangham, 1993; White, 1998; Hare et al., 2012). This difference in feeding ecology predicts that chimpanzees would show more accurate spatial cognition and memory than bonobos due to their increased reliance of patchily-distributed foods. Cognitive evidence supports this prediction: while both chimpanzees and bonobo can successfully recall the location of hidden food in a large space, chimpanzees are more successful at recalling multiple food items than bonobos (Rosati & Hare, 2012).

Second, there is little data addressing whether apes show any ontogenetic changes in spatial cognition. This is critical for assessing hypotheses pointing to a causal role for language in shaping human spatial cognition: if language drives a particular ontogenetic change in humans, then that developmental shift should be found in our species alone (e.g., Matsuzawa et al., 2006; Matsuzawa, 2007; Rosati & Hare, 2012; Rosati et al., in press). However, it is currently unclear if nonhuman apes show any human-like shifts in spatial frameworks. For example, the study by Haun and colleagues (2006a) involved only sub-adult and adult apes, and did not examine whether apes exhibited age-dependent change in their preferred strategy. While some studies suggest improvement with age in some memory tasks (Herrmann et al., 2010; Martin-Ordas et al., 2010) these studies did not include infants in the age range where some human changes in spatial skills occur (e.g., Newcombe & Huttenlocher, 2006). Other studies of infants have been limited to small samples of three or fewer chimpanzees (Matsuzawa et al., 2006; Matsuzawa, 2007). Finally, the few previous studies of spatial cognition involving larger samples sizes of infant and juvenile apes have focused on whether apes could remember particular locations, not the specific strategy they used to do so do (e.g., Rosati & Hare, 2012; Wobber et al., 2013).

The current paper therefore examines spatial cognition in bonobos in two tasks that have both been commonly used to assess allocentric versus egocentric encoding of locations. The first study examines how bonobos encode locations in the place-response task, which has been used to disentangle allocentric, hippocampal-dependent strategies from egocentric basal-ganglia-dependent strategies. This study focuses on younger apes ($N = 10$) between the ages of 2 and 7 years. The second study uses a modified version of the spatial relations task previously developed by Haun and colleagues (2006a) to examine linguistic influences on spatial cognition. This study examines a larger sample of bonobos ($N = 19$) ranging in age from 2 to 13 years. Thus, these studies can compare the types of strategies bonobos used in these two different contexts. In addition, these studies can assess whether bonobos exhibit any developmental shift in strategies, suggesting that such shifts might reflect intrinsic maturation of spatial abilities, not the influence of language.

2. Study 1: place-response task

This study adapted the cross-maze task used in studies of rodent (Packard, 2009) and lemur memory (Rosati et al., 2014) for use with infant and juvenile bonobos. Apes were presented with two overturned bowls places on opposite sides of the room, which they could approach to search for food. As in the cross maze task, one of the locations was consistently baited with (hidden) food across sessions. The ape then completed probe trials from a flipped orientation towards the cups to assess if they had encoded the baited option from a spatial (allocentric) perspective or a habitual (egocentric) perspective.

2.1. Methods

2.1.1. Ethics statement

All behavioral studies were noninvasive. The studies had IACUC approval from Duke University (No. A078-08-03), strictly adhered to the legal requirements of the Democratic Republic of Congo, and had approval from the Ministry of Research and the Ministry of Environment in the DRC (research permit No. MIN.RS/SG/004/2009). Animal husbandry and research practices complied with the PASA Primate Veterinary Healthcare Manual and the policies of Lola ya Bonobo Sanctuary in Kinshasa, Democratic Republic of Congo. All apes were socially housed, and the vast majority semi-free-ranged in large tracts of tropical forest during the day (5–40 ha across groups). In the evening, all apes spent the night in indoor dormitories (12–160 m^2). Apes were tested individually in these familiar dormitories buildings. Following testing, most apes were released back with their larger social group outside. Apes had ad libitum access to water and were never food deprived for testing. In addition to the food the apes could eat in their forest enclosures, they were fed a variety of fruits, vegetables, and other species-appropriate food 2–4 times daily. Subjects completed no more than one test session per day, and participating in all tests were voluntary: if the ape stopped participating, the session was halted.

2.1.2. Subjects

Ten bonobos (7 females and 3 males) participated in this study, ranging in age from 3 to 7 years of age (mean age $= 4.2 \pm 0.4$ years; see Table 1 for all subject details) living at Lola ya Bonobo Sanctuary. Apes in African sanctuaries are typically born in the wild, and enter the sanctuary after being confiscated at an early age (approximately 2–3 years old) as a result of

Table 1.
Subject characteristic and performance in Study 1.

Subject	Sex	Age	Learning trials	Probe trials: allocentric
Kanaga	♀	2	0.93	1.00
Kipolo	♂	5	0.85	0.67
Kodoro	♂	3	0.85	1.00
Lukuru	♀	4	0.64	0.83
Masisi	♀	4	0.83	1.00
Muanda	♀	6	0.69	0.33
Sandoa	♀	3	0.96	0.83
Shibombo	♂	3	0.94	1.00
Tchilenge	♀	7	0.78	0.50
Waka	♀	4	0.81	0.17
Overall			0.83	0.73

Age estimate is listed in years. Learning trials indicates the proportion of correct choices across all learning trials. Probe trials indicates the proportion of test trials where subjects made an allocentric (spatial) response; they made an egocentric response on the remainder of trials.

the trade in apes for pets and bushmeat. Previous work indicates that sanctuary apes are psychologically healthy relative to other captive populations (Wobber & Hare, 2011). As exact birth dates for sanctuary apes are generally unknown, we used age estimates made by sanctuary veterinarians at arrival (typically when the infants are between 1 and 3 years old), and adjusted these age estimates using measurements of weight and patterns of dental emergence according to known patterns of ape development (following the methods described in Wobber et al., 2010, 2013; Rosati & Hare, 2012), a technique that is further validated using the known ages of individuals born in the sanctuaries. The apes had previously participated in a variety of cognitive testing involving making choices between options presented on tables (e.g., Wobber et al., 2013), but were naïve to the particular task here.

2.1.3. General setup

Apes were tested individually in a room in their familiar sleeping dormitory, with the experimenter and a caretaker present. They were presented with two identical and opaque overturned bowls (24.5 cm diameter, 15 cm tall, and red in color) that were placed 2.5 m apart; food could be hidden under the bowls. The caretaker sat at the starting position with the bonobo in front

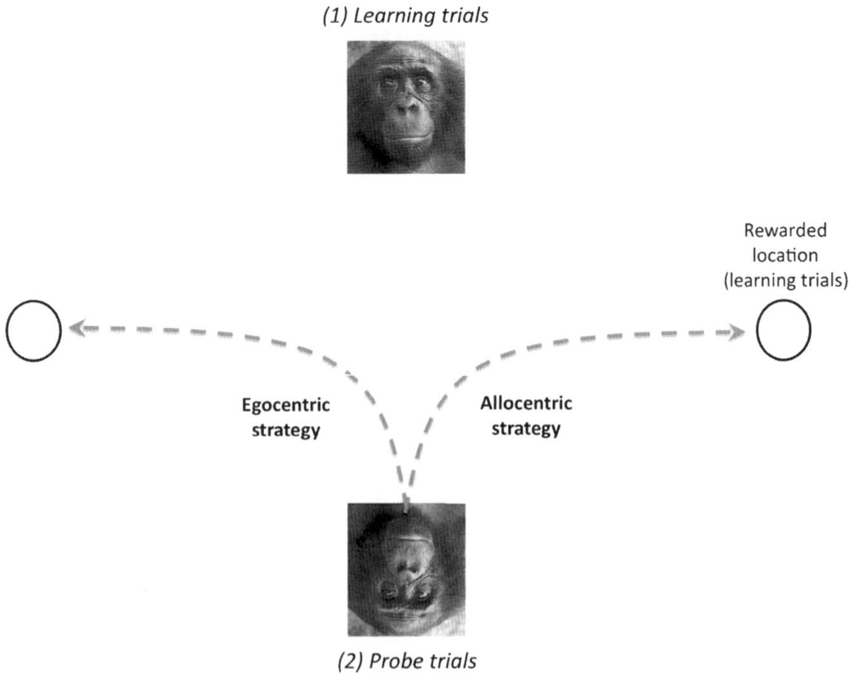

Figure 1. Setup for place-response task (Study 1). In each session, apes first completed 12 learning trials in which one of two locations (both overturned bowls) was consistently baited. In the final probe trial, the apes' starting position was rotated 180° so they faced the locations from a flipped perspective. Their responses therefore indicated if they had encoded the baited location from a viewer-dependent (egocentric) framework or a spatial (allocentric) framework. On all trials the caretaker centered the bonobo at their starting position, and the experimenter stood in the opposite position across the midline of the testing area.

of him, centering the bonobo approximately 1 m away from the midline of the two bowls (see Figure 1).

2.1.4. *Session and trial structure*

The bonobos completed 6 sessions total, consisting of three types of trials. All sessions were identical with the exception of the introductory trials in the first session (described below). A given bonobo was always tested in the same room, and the orientation of the setup in that room as well as the baited location in learning trials was kept identical across sessions, following the methods of previous work using this task in other species (Packard, 1996, 1999, 2009). In particular, the consistent setup across sessions made it possible to assess if the apes' spatial strategy changed over time, following previous work showing that rodents initially exhibit allocentric strategies but switch to egocentric strategies with increase repetition. The trial types in the sessions were as follows:

(1) Introduction trials (two trials, only in the first session): At the beginning of their first session, apes completed two introduction trials where they visibly saw the experimenter placing food under both bowls. Here, the caretaker directed the subject to the centered starting position while the experimenter baited the bowls (with bonobos typically sitting on their lap). The experimenter showed the ape two pieces of preferred food (apple), approached both bowls (in counterbalanced order), called the ape's name, and visibly placed a piece of food under each bowl while the ape watched. These trials therefore served to introduce the idea that the bowls could be baited, and allowed bonobos to successfully experience retrieving food from both. Once the experimenter finished baiting, she returned to a midline position across from the ape and looked straight ahead to avoid cuing the subject. After the experimenter said "start", the keeper released the bonobo so he or she could choose one of the bowls by approaching it. Bonobos had to approach both bowls to retrieve the food in order to proceed to the subsequent learning trials in that session; if bonobos who did not initially approach both locations on their own, the experimenter attracted their attention to the other bowl. These trials were not included in sessions two through six.

(2) Learning trials (12 in each session): These trials were similar to introductory trials, but only one bowl was baited. When the experimenter approached both bowls (in counterbalanced order within sessions), she placed her hand under the bowl while covering it with an occluder, such that the ape could not see which location had actually been baited with the food. If the subject chose incorrectly by approaching the empty bowl in learning trials, they could correct their choice (e.g., then approach the correct location) in order to equate reward experience across all subjects (following previous studies; Packard, 1996). After the bonobo retrieved the food, the caretaker attracted them back to the starting position.

(3) Probe trials (one trial at the end of each session): The procedure for probe trials was similar to that for learning trials. Prior to the trial, the caretaker walked with the ape to the opposite side of the bowls so that the ape started in the flipped position (180°) relative to the bowls and the larger spatial layout of the room. The trial then proceeded like learning trials, with two main exceptions. First, the experimenter baited both bowls (approaching

them in counterbalanced order across sessions) to avoid biasing the subjects in subsequent trials. That is, the animal always was rewarded regardless of what strategy they exhibited. Second, subjects could only approach one bowl so there was no opportunity for self-correction; the caretaker attracted the bonobo away from the bowls after they made a choice.

2.1.5. Coding and data analysis

Choices were coded live by the experimenter. A coder blind to the correct side coded 20% of sessions from videotape for reliability; agreement was excellent (Cohen's kappa = 0.99). Choice data were analyzed using two statistical approaches. The first approach used parametric statistics. For these analyses, mean proportions of choices for the correct location (in learning trials) and for allocentric responses (in probe trials) were calculated for each subject.

The second approach focused on the bonobos' responses to probe trials, the main measure of interest. This approach used generalized linear mixed models (GLMM) to analyze choices as a binary outcome variable, while accounting for correlation in responses due to repeated measures within subjects (Baayen, 2008). This analysis used the LME4 software package (Bates, 2010) in the statistics program R (R Development Core Team, 2011). The advantage of using a GLMM model, compared to ANOVA, is that it allowed trial-by-trial modeling of responses, while also controlling for various covariates and random effects (such as subject). These analyses used likelihood ratio tests to compare fit across models incorporating different factors (Bolker et al., 2008). Across models, subject was included as a random factor, and predictors included sex (as a factor), age in years (as a covariate), session number (one through six) as a covariate, and proportion of correct learning trials in that session (as a covariate). This analysis therefore could examine the degree to which age, sex, and previous learning trial experience predicted the bonobos' responses on probe trials.

2.2. Results and discussion

The first analysis focused on the bonobos' performance in learning trials. Overall, bonobos chose correctly on $82.8 \pm 3.3\%$ (mean $\pm$ SE) of all learning trials, significantly above chance overall (one sample t-test: $t_9 = 9.91$, $p < 0.001$; see Table 1 for subject means). Indeed, apes were above chance in selecting the correct location in learning trials across all six sessions (session 1: mean $\pm$ SE $= 65.8 \pm 6.3\%$, $t_9 = 2.53$, $p < 0.05$; session 2: $75.0 \pm$

5.0%, $t_9 = 5.03$, $p = 0.001$; session 3: 75.0 ± 7.8%, $t_9 = 3.22$, $p = 0.01$; session 4: 90.8 ± 3.6%, $t_9 = 11.31$, $p < 0.001$; session 5: 93.3 ± 2.7%, $t_9 = 15.92$, $p < 0.001$; session 6: 96.7 ± 1.4%, $t_9 = 34.29$, $p < 0.001$). A repeated-measures ANOVA comparing learning trial performance across sessions revealed a main effect of session ($F_{5,45} = 9.62$, $p < 0.001$) with a significant linear contrast ($F_{1,9} = 30.18$, $p < 0.001$) indicting the bonobos' recall of the baited location improved with repeated exposure to the correct location. Pair-wise comparisons confirmed this: the apes' memory in session one was less accurate than that in sessions 4–6, and session 2 performance was also lower than in session 6 (Bonferroni correction; $p < 0.05$ for all significant cases). That is, the bonobos' performance in their last sessions was significantly better than in their initial sessions.

We next examined performance on probe trials. Bonobos made an allocentric response on 73.3 ± 9.7% of trials, significantly above chance ($t_9 = 2.41$, $p < 0.05$; see Table 1 for subject means). Examining the performance of individual subjects revealed that seven of the ten subjects made predominantly allocentric choices (67–100% allocentric choices), two subjects made predominantly egocentric choices (17–33% egocentric choices), and one individual showed no overall preference (exhibiting both strategies equally across probe trials). A comparison of performance on learning compared to probe trials indicated that there was no significant difference between overall correct choices in the learning trials compared to allocentric choices in the probe trials (paired sample t-test: $t_9 = 1.09$, $p = 0.30$, n.s.; see Figure 2a). That is, overall bonobos chose the allocentric option at levels comparable to their rate of success on learning choices. This suggests that they generalized their experience from the learning trials to predict that the food was located in the same spatial location in probe trials. Finally, while there was no relationship between an individual's performance in learning trials and allocentric choices in probe trials ($r_p = 0.47$, $p = 0.17$, n.s.), there was a negative relationship between age and probe trial performance: older bonobos were less likely to make allocentric responses than younger apes ($r_p = -0.65$, $p < 0.05$). The results from the GLMM, which modeled each individual probe trial choice as a binary outcome, generally confirmed these results. The first step was to fit a basic model including subject as a random factor, sex as a fixed factor, and proportion of correct learning trials as a covariate. Learning trial performance was included in the model to control for the possibility that the bonobos' performance in the immediately preceding session

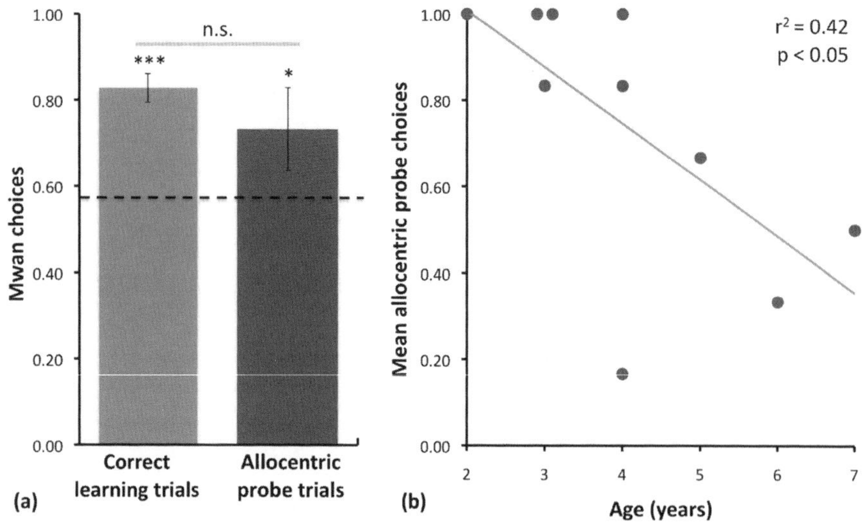

Figure 2. Results from place-response task (Study 1). (a) Bonobos' mean proportion choices for the correct location in learning trials, and for the allocentric location in probe trials. (b) Relationship between overall allocentric probe choices and the subject's age. Error bars indicate standard error. $^{*}p < 0.05$, $^{***}p < 0.001$.

predicted their probe choices (note that all bonobos had equal experience with the correct location regardless of their initial choice, as bonobos were allowed to self-correct following incorrect choices in the learning trials). In a second model, session number was included as a covariate to the more parsimonious basic model to assess whether apes showed any shifts in probe trial performance over time. Finally, in a third model age was also added as a covariate to assess any potential effects of cognitive development on the bonobos' performance. The full model revealed that age was a significant negative predictor on the apes' propensity to make an allocentric response (see Figure 2b and Table 2 for parameters from the full model). However, there was no change in probe strategies over time. Moreover, whereas comparing the basic model to the second model including session number did not improve model fit ($\chi^2 = 0.13$, df = 1, $p = 0.72$, n.s.), the third model also including age did show greater fit (model 2 versus model 3: $\chi^2 = 3.89$, df = 1, $p < 0.05$). This highlights the importance of age as a predictor of apes' allocentric responses.

Overall, these results indicate that the bonobos primarily used an allocentric spatial strategy in this context. The bonobos quickly learned the location

Table 2.
Factors influencing the apes' likelihood for making an allocentric response on probe trials in Study 1.

Factor	Estimate	SE	Z	p
Sex	1.363	1.302	1.047	0.30
Age	−0.824	0.406	−2.031	<0.05*
Session	−0.036	0.284	−0.127	0.90
Correct learning trials	−0.100	2.750	−0.036	0.97

The full generalized linear mixed model assessed the influence of the following factors: sex (female as baseline), age (in years), session number (1–6) and learning trial performance (proportion correct in the preceding learning trials that session).
* Significant value.

that was baited in learning trials, showing a preference for that location from the first session. When the bonobos' orientation was flipped so they approached the bowls from the opposite direction, they generalized their experience in learning trials to consistently select the allocentric location. That is, they approached the same location in space that had been baited previously, rather than using a viewer-dependent frame of reference. This generalization did not change across sessions, indicating the repeated experience with the correct location on learning trials did not shift the bonobos from using an allocentric strategy towards an egocentric strategy, as has been previously demonstrated in other animals (Packard, 2009). However, probe trial performance was negatively related to age — younger bonobos were more likely to make allocentric responses on probe trials than were older bonobos — and a model including age as a factor was a better fit in the GLMM analysis. It is important to point out, however, that the age range and sample size in this study were small. Thus, the second study examined spatial cognition in a large sample of bonobos ranging from infants to adults. In addition, this study used a different task that involved choices between three options presented on a smaller scale to examine if the bonobos' allocentric strategies were stable across contexts.

3. Study 2: spatial relations task

The second study examined how bonobos process spatial relations by presenting them with two tables with identical arrays of three overturned opaque cups on top, following the basic procedure from (Haun et al., 2006a). Apes

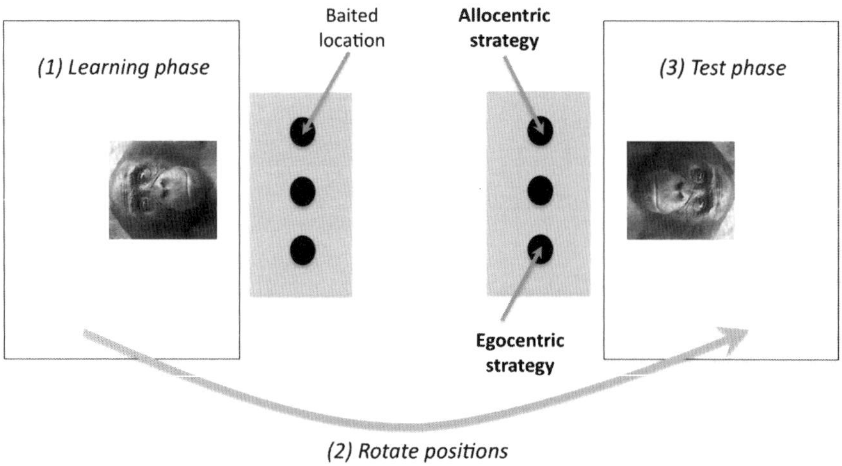

Figure 3. Setup for relations task (Study 2). Apes first completed a learning phase in which one location (always either the left or right side) was consistently baited. After apes met a learning criterion indicating they consistently chose the correct side, they moved 180° into a new room for the test phase. In the last 10 trials, apes faced an identical table from a flipped perspective to assess if they had encoded the baited location from a viewer-dependent (egocentric) framework or a spatial (allocentric) framework. The experimenter sat centered behind the table at both locations.

first made choices at one of the tables and learned that one of the outer cups (left or right) always contained food. After demonstrating that they would consistently pick that cup, apes moved to the second table where apes' orientation was flipped 180° (see Figure 3). In this second test phase, bonobos could use either use an allocentric strategy (choosing the cup with the same orientation in the broader spatial layout), use an egocentric strategy (choosing the cup with the same relation to their own body), or fail to generalize their previous experience by choosing the center location.

While this study followed the same basic procedure as that by Haun and colleagues (2006a), there were some adjustments to the methods to test a larger age range of subjects (in particular, younger bonobos). First, each individual only completed one session where one location was baited, to see how this experience generalized in the test phase, whereas in the previous study apes were trained on all possible locations across three different sessions. Second, bonobos initially completed several types of warm-up trials to ensure they were proficient with choosing all the various cup locations, accounting for the fact that the infants were less familiar with these kinds of table-based tasks than are older apes. Finally, we did not include the initial baseline trials used by Haun and colleagues (2006), used there to detect if the apes had some initial bias for one of the locations in the absence of any relevant experience in the study. As apes in the current study had to demonstrate proficiency with choosing all three locations in the warm-up phase, the current analyses therefore focused only on mean choices in the test phase, and did not involve a difference score measuring change between the baseline and test phase.

3.1. Methods

3.1.1. Subjects

Nineteen bonobos (9 females and 10 males) from the same population of bonobos living at Lola ya Bonobo Sanctuary participated in this study. Individuals ranged from 3 to 13 years of age (mean age = 6.6 ± 0.7 years; see Table 3 for all subject characteristics). Seven individuals had participated in Study 1 approximately 1 year previously.

3.1.2. General setup

Apes were tested in two rooms that faced each other across a hallway. The basic setup consisted of two tables (each 80 cm wide, 40 cm deep, 40 cm high) with sliding tops that allowed the objects on the tables to be set up and then moved into the reach of the apes. Each table was placed in the hallway and facing one of the rooms. An identical array of three overturned cups (diameter 5 cm, height 10 cm) was placed on each table. During testing, apes sat in one of the rooms facing the array of three cups, and the experimenter sat across the table from the apes, separated by bars. Apes were tested individually; adults were alone in the test rooms, whereas infants were tested with the human caretaker present to ensure that they were comfortable. The caretaker sat at the back of the room behind the bonobo, and did not interact with the ape or provide any cues. Apes always began the study in one room (for the learning phase) facing one of the tables, and then moved to the second room (for the test phase) via an overhead tunnel connecting them (juvenile and adult bonobos), or by being walked across the hallway by their caretaker (infant bonobos). The width of the hallway between the learning phase room and the testing phase room (as well as the overhead tunnel that adults utilized) was 2 m. Thus, the two tables were also 2 m apart.

In contrast to the first study, this setup made it possible to disentangle whether some apes do not generalize their initial learning to the test phase, versus whether some bonobos exhibit a mixed strategy involving both

Table 3.
Subject characteristic and performance in Study 2.

Subject	Sex	Age	Learning trials	Test trials	
				Allocentric	Egocentric
Dilolo	♂	10	30	0.0	0.8
Fizi	♂	11	21	0.0	1.0
Kalina	♀	13	11	0.0	0.0
Kanaga*	♀	3	17	0.2	0.0
Kasongo	♂	9	29	0.0	0.9
Katako	♀	7	13	0.0	0.9
Kikwit	♂	12	20	0.0	0.9
Kindu	♀	3	23	0.0	0.4
Kodoro*	♂	4	20	0.0	0.9
Lomako	♂	4	21	0.0	1.0
Lukuru*	♀	5	12	0.0	0.9
Mabali	♂	8	19	0.0	0.1
Maliaka	♀	4	22	0.0	1.0
Masisi*	♀	5	18	0.0	0.9
Matadi	♂	11	31	0.0	1.0
Oshwe	♂	3	18	0.0	1.0
Sandoa*	♀	4	12	0.0	1.0
Shibombo*	♂	4	27	0.0	0.4
Waka*	♀	5	30	0.0	0.0
Overall			20.7	0.01	0.69

Age estimate is listed in years. Learning trials indicates the total number of learning trials that the subject completed to meet criterion. Test trial columns indicate the proportion of allocentric and egocentric choices in the test phase; the remainder of choices were for the center option. *Subjects that previously participated in Study 1.

egocentric and some allocentric choices, due to the presence of the center location. Moreover, these methods could be used with older apes as the experimenter was in a different room than the subject.

3.1.3. Session and trial structure

In the initial learning phase, apes learned that one cup (always one of the outer cups, left or right side assignment counterbalanced across subjects) provided food. Here, apes completed four types of trials, including various warm-up trials to familiarize the apes to the setup, as well as the learning trials where apes learned that one particular cup was consistently baited. Across all trials only one cup was baited with food on a given trial. After meeting a criterion indicating they consistently chose the baited location, apes were switched 180° to the other table, faced with an identical array of cups for the test phase. In those trials all three cups were baited to examine how the apes' generalized their previous learning to the new array. All trials were completed as part of the same session. The order of trials was as follows:

(1) Visible food trials (one per cup location in randomized order, three trials total): In the first part of the learning phase, bonobos completed a basic introduction to making choices on the table and familiarize them to the procedure (as some of the infant apes were naïve to this test setup). In particular, food was placed on top of one of the cups and thus was visible at the time the ape made their choice. These trials introduced the bonobos to the procedure of choosing cups to get the food, and were repeated if the bonobo initially chose incorrectly to ensure that all bonobos were proficient with this basic skill.

(2) Hidden food trials (two per cup location in randomized order, six trials total): Next, apes saw the experimenter place the food under the correct cup, but it was hidden under the cup at the time of choice. These trials introduced the idea of choosing concealed food, and were also repeated if the bonobo initially chose incorrectly. Thus, these warm-up trials ensured that bonobos had to demonstrate an ability to correctly choose hidden food items before proceeding to the learning phase.

(3) Help trials (four trials, always involving the same cup as in the learning trials): After the warm-up phase was completed, apes learned that only one of the cups (left or right) consistently contained food. In the four initial help trials, the experimenter placed the food visibly under the baited side cup. These trials were similar to the previous hidden food trials, except that only the one cup was baited; in addition, if the ape chose incorrectly, they were shown that their choice was empty as well as the location of the correct choice, without being allowed to repeat the trial to obtain the food.

(4) Learning trials (as main trials necessary to meet criterion, up to 32 trials total): In the last type of trial, the experimenter hid the food under the baited cup behind an occluder (the same side as had been baited in the help trials). The experimenter touched all three cups in a counterbalanced order behind the occluder so as to not give any auditory or movement cues about the location of the food. As in help trials, if the ape chose incorrectly, they were shown that their choice was empty as well as the location of the food, but were not allowed to correct their choice. Apes had to choose the correct location in 10 of the previous 12 trials to meet the criterion and move onto

the test phase. They had to meet this criterion within 32 trials or the session was halted without proceeding to the test phase. Three bonobos did not meet this criterion within 32 trials, so they were given a break of several days and then repeated the session (all three then met the criterion). The limit was set after relatively few trials to ensure that the infant bonobos would remain food-motivated to continue for the duration of the session, as they would become disinterested if the session continued for an extended period.

(5) Test trials (10 in total, all three locations baited): After meeting the learning criterion, apes were switched 180° to the other table, faced with an identical array of cups during the test phase. As in learning trials, the experimenter baited the cups behind an occluder and touched all three cups in counterbalanced order to avoid giving any clues to the food's location. In contrast to the learning trials, however, here the experimenter baited all three locations so as to avoid biasing the ape's choices on subsequent trials. Once the ape made a choice, they received the food from their chosen cup and were not shown the contents of the other cups. This allowed us to assess if the apes were using an allocentric strategy (chose the cup in the same spatial direction as in the learning trials), and egocentric strategy (chose the same cup from their own perspective, left or right), or did not generalize their previous learning to this new context. The side of the hallway assigned as the learning versus test room was counterbalanced across subjects to ensure that any external spatial cues were counterbalanced.

3.1.4. Coding and data analysis

Choices were coded live by the experimenter. A coder blind to the correct side coded 20% of sessions from videotape for reliability; agreement was 100%. Choice data was analyzed using the same approaches as in the first study. The first approach used parametric statistics to assess mean proportions of choices for the correct location (in learning trials) and for allocentric versus egocentric responses (in test trials) for each subject. The second approach used generalized linear mixed models (GLMM) to analyze test choices on a trial-by-trial basis.

3.2. Results and discussion

The first analysis focused on the bonobos' performance in learning trials. Bonobos took an average of 20.7 ± 1.5 learning trials to reach the criterion of choosing the last ten of twelve trials correctly. Overall, they chose correctly on $66.1 \pm 3.2\%$ of all learning trials, significantly above chance overall (with chance at 33.3%; one sample t-test: $t_{18} = 10.38$, $p < 0.001$; see Table 3 for subject means). A univariate GLM with sex as a between-subjects factor and age as a covariate revealed that female bonobos reached criterion in fewer trials than did males (females requires an average of 17.6 ± 2.1 trials compared to 23.6 ± 1.5 trials for males; $F_{1,16} = 4.66$, $p < 0.05$). However, there was no effect of age on the number of learning trials ($F_{1,16} = 0.03$, $p = 0.88$, n.s.), indicating that younger and older bonobos learned the baited location at similar rates.

The next analysis examined the bonobos' performance on test trials, where they faced a set of cups from a flipped perspective (see Figure 4a and Table 3 for subject means). Across all subjects, bonobos made an allocentric responses on only 1.1 ± 1.1% of trials, significantly below chance (at 33.3%; $t_{18} = -30.66$, $p < 0.001$). In contrast, bonobos showed a clear pattern of making egocentric generalizations, choosing that option on 69.0 ± 9.0% of trials, significantly above chance ($t_{18} = 3.93$, $p = 0.001$; see Table 3 for subject means). A comparison of egocentric versus allocentric responses indicated that apes were more likely to use an egocentric strategy ($t_{18} = -7.12$, $p < 0.001$). Apes chose the middle option 30.0 ± 8.7%, at chance levels ($t_{19} = -0.39$, $p = 0.71$, n.s.); these responses represent fail-

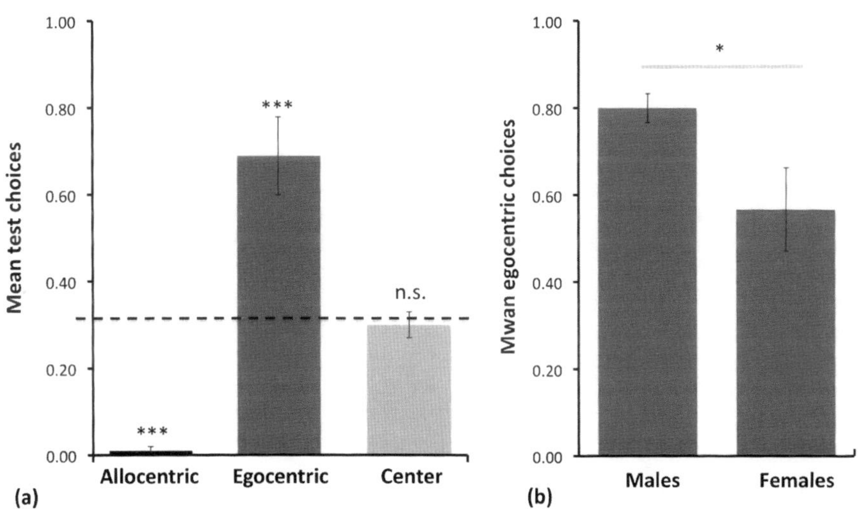

Figure 4. Results from the relations task (Study 2). (a) Bonobos' mean proportion choices in the test phase for the allocentric option, the egocentric option, and the middle option. (b) Breakdown of egocentric choices in test phase by the subjects' sex. Error bars indicate standard error. $^{*}p < 0.05$, $^{***}p < 0.001$.

ures to generalize their previous experience to the test phase. Examining the performance of individual subjects revealed that 13 of the 19 subjects made predominantly egocentric choices (80–100% egocentric choices), but no individuals preferred the allocentric option. Indeed, the remaining six individual bonobos predominantly chose the center location, suggesting a lack of generalization (they made 60–100% of choices for the center cup during the test phase). Overall, this suggests that bonobos tended to generalize their experience from the learning trials to predict that the food was located in the locations with the same relationship to their own body (e.g., they encoded the baited location based on whether it was to their own left or right, not based on external spatial cues). Finally, the number of learning trials need to reach criterion was not correlated with either the proportion of egocentric test responses that bonobo exhibited in the test phase ($r_p = 0.01$, $p = 0.99$, n.s.), nor allocentric test responses ($r_p = -0.14$, $p = 0.56$, n.s.).

The GLMM analysis modeled each test trial choice in terms of whether or not the bonobo made an egocentric response (as that was the dominant type of response). The first step fit a basic model including subject as a random factor; sex as a fixed factor; and number of learning trials necessary to reach criteria as a covariate to assess whether the bonobos' learning patterns predicted their test choices. In a second model, test trial number was included as a covariate to the more parsimonious basic model to ascertain whether apes showed any shifts in their test preferences over time. Finally, in a third model age was also added as a covariate to assess any potential effects of cognitive development on the bonobos' performance. The full model revealed that male bonobos showed more egocentric choices than females (at $80.0 \pm 9.7\%$ for males, but only $56.7 \pm 15.4\%$ for females; see Figure 4b). This was the case even when controlling for their differential learning patterns in the learning phase, as learning trials to criterion was included as a covariate. In addition, apes' egocentric choices increased over the course of the ten test trials (see Table 4 for parameters from the full model). Moreover, while the second model including trial number did improve model fit in comparison to the basic model (model 1 versus model 2: $\chi^2 = 4.59$, df = 1, $p < 0.05$), the third model including age did not improve fit in terms of predicting whether apes' made an allocentric response (model 2 versus model 3: $\chi^2 = 0.94$, df = 1, $p = 0.33$, n.s.). Thus, there was no effect of age on the bonobos' patterns of choices in the test phase.

Table 4.
Factors influencing the apes' likelihood for making an egocentric response on test trials in Study 2.

Factor	Estimate	SE	Z	p
Sex	4.422	2.219	1.993	<0.05*
Age	−0.286	0.294	−0.972	0.33
Test trial number	0.202	0.096	2.099	<0.05*
Learning trials to criterion	−0.163	0.163	−1.001	0.32

The full generalized linear mixed model assessed the influence of the following factors: sex (female as baseline), age (in years), trial number (1–10) and number of learning trials necessary to reach criterion in preceding learning phase.
* Significant value.

Overall, these results indicate that the bonobos primarily used an egocentric strategy in this context. That is, after the bonobos learned which cup was baited in the learning trials, the test trials revealed that they had encoded this location in terms of a viewer-dependent framework. Indeed, bonobos rarely made an allocentric response: in the vast majority of trials they did not make an egocentric response, they chose the middle cup (suggesting lack of generalization from the previous trials). Finally, The GLMM analysis indicated that bonobos tended to increase their choices for the egocentric location over the course of the ten test trials, but this reflected a decrease in center cup choices, not a change in their likelihood of making an allocentric response. This analysis further indicated that male bonobos were more likely to make this egocentric generalization; females in contrast showed greater rates of choosing the middle cup, suggesting those individuals did not generalize their previous experience to the test phase. One possibility is that males' slower learning rates in the learning phase meant that they in fact received more feedback about the correct location, and therefore more likely to generalize this experience in the test phase (although, the number of learning trials apes took to reach criterion was not a significant predictor of their probe trial performance). Regardless, no bonobo of either sex exhibited a bias towards an allocentric strategy in the test trials; in fact there was only one individual who ever chose it.

4. General discussion

The current studies had two major results. First, across two commonly used but distinct tasks to assess spatial frameworks, bonobos were able to use both egocentric and allocentric strategies. In the first study, apes experienced that one of two locations in a larger spatial area was consistently baited with food, and then their strategy was assess by flipping their orientation towards the testing arena. Here, the bonobos showed allocentric inferences, tending to go to the same location (rather than making the same egocentric body movement) when searching for the food. In the second study, apes learned that food was delivered under one of three cups. When their orientation was flipped and they faced an identical set of three cups, however, apes tended to show an egocentric strategy — choosing the cup with the same position relative to their own body (e.g., to the left or right) but not in the same relationship to the external environment. Second, the results from these studies provide little evidence for systematic developmental change in the frames of reference that bonobos use. There was some suggestion of increases in use of egocentric frame in Study 1, but that study had a small sample size. In contrast, the results from Study 2 with a larger sample size did not reveal any age-related change in strategies. Overall, these results indicate that bonobos can used both allocentric and egocentric frameworks, but there was not consistent evidence for age-related shifts in their use of these strategies.

One major question raised by these results is why the bonobos exhibited different strategies in the place-response task versus the spatial relations task. Importantly, studies of multiple memory systems suggest that many diverse contextual factors can bias organisms toward the use a given strategy in dual-solution tasks that can be solved in either fashion (Packard & Goodman, 2013). For example, one factor that can shift control of behavior between hippocampal-based allocentric strategies and basal-ganglia-based egocentric strategies is the amount of training involved in the task. Rats that experience extended periods of training in the plus-maze task show a switch in their response to probe trials over time: although initial responses tend to be allocentric, over time rats show more egocentric choices (Packard, 1996, 1999). However, training effects are unlikely to have driven the difference between the two studies presented here for two reasons. First, bonobos exhibited allocentric strategies in the place-response task, which actually involved more training over a longer time period (e.g., several sessions across multiple days) compared to the spatial relations task (which took place in one day). Second, there was no consistent evidence for shifts from allocentric to egocentric strategies in either task: in Study 1 there was no change in probe responses across days, and an increase in egocentric responses in Study 2 was due to a decrease in center-cup responses, not a shift from allocentric responses.

Another major factor that appears to influence animals' performance in the plus maze concerns the characteristics of the external environment (see Packard & Goodman, 2013 for a review). In particular, an external environment with many varied visual cues and landmarks tends to bias animals towards using an allocentric strategy, likely because the presence of these landmarks fosters the formation of a cognitive map of the surrounds. In contrast, a more homogenous external environment with limited visual cues appears to bias animals towards an egocentric strategy based on their own movements. Similarly, the presence or absence of salient visual landmarks can bias the strategies that infants and young children exhibit in a variety of spatial tasks (Acredolo, 1978, 1979; Acredolo & Evans, 1980; Newcombe & Huttenlocher, 2006). However, differences in the external environment are also not a likely explanation for the current results, as the bonobos were either tested in the same building for both tasks, or in similarly constructed buildings (depending on the group the bonobo lived in at the sanctuary at the time of testing). These buildings all had similar visual features that could potentially serve as landmarks (e.g., windows serving as light sources, the presence of bars versus tiles on the walls, a hallway for humans to move through the building), and were fairly uniform rooms. However, variation in the external environment might be one factor driving the difference between the results in the spatial relations task (Study 2), where the bonobos exhibited egocentric strategies, and the results of a previous study in which apes exhibited primarily allocentric strategies (Haun et al., 2006a). For example, there may be differences in the visual cues across the testing facilities in these two studies, with more possible visual landmarks in the previous study. Additionally, one important difference between the two studies concerns the position of the experimenter relative to the cup arrays. In particular, in Study 2 the experimenter could not serve as a landmark as she was always centered relative to the cup array, sitting behind the table to administer the task (see Figure 3). However, in the previous study the experimenter sat to one side of the testing tables, and thus potentially could have served as a consistent landmark (e.g., the experimenter was physically closer to one side of the testing tables; see Figure 3 in Haun et al., 2006a). Thus, the presence of the human might have biased the apes towards using an allocentric strategy in that experiment.

One noticeable difference between the place-response task used in Study 1 and the spatial relations task used in Study 2 is the size of testing 'arena'. Namely, bonobos had to make choices concerning locations that were 2.5 m apart in the first study, but chose between an array of cups on a table less than a meter total in width in the second study. Importantly, research using reorientation paradigms indicates that both children and adults utilize landmark-dependent allocentric strategies more often when faced with locating a target in a large space than in a smaller but otherwise-identical space (Learmonth et al., 2008; Ratliff & Newcombe, 2008). One reason is that landmarks are more distal and salient in larger spaces, and therefore more useful cues for solving spatial problems. For example, in the current studies the positions of any available landmarks in the testing rooms might have been more relevant when the bonobos were making decisions between locations separated at larger distances in Study 1, than when they chose between the array of cups on the table in Study 2. This difference in size between the two studies also meant that the type of response the apes made to indicate their choice differed across the two studies. In particular, apes had to approach the locations by moving with their entire body over a distance of several meters in the place-response task, whereas in the spatial relations task the apes sat in front of a table and were within arms-reach of all the locations when making their choices. Some of the research on spatial strategies in children suggests that room size impacts strategies not only because of the relative salience of landmarks, but also due to how children can move around in the space. When children are tested in a room of a given size, children who are able to move throughout that room's space are more likely to use landmark information than are children who have identical visual access to the landmarks, but whose movements are confined to a smaller portion of the room (for example, if a smaller portion of the room is delineated by a line on the floor and children are asked to remain inside those lines; Learmonth et al., 2008). Notably, influence of room size or movement are not important factors in theories that focus on the role of language in spatial cognition (Hermer-Vazquez et al., 1999; Wang & Spelke, 2002; Majid et al., 2004; Haun et al., 2006a; Lee & Spelke, 2010). In contrast, these types of contextual factors are thought to be important in adaptive-combination theories of spatial cognition (Newcombe & Ratliff, 2007) in which the likelihood of using strategies is dependent on how different sources of information is weighted across different situations, for example due to the salience or usefulness of different sources of spatial information in particular environment.

Another issue raised by these results concerns why the bonobos did not exhibit age-related changes in spatial strategies. In particular, while there was some suggestion of increases in use of egocentric frame in Study 1, the smaller sample size in that study makes it difficult to assess how meaningful this age-related difference is. In contrast, the results from Study 2 with a larger sample size did not reveal any age-related change in strategies. Importantly, the sample size in that study was sufficient to allow detection of a sex difference in learning and generalization (e.g., females were quicker to learn which cup was baited in the learning phase, but then less likely to generalize this learning to produce either strategy in the test phase), but there was no indication that older and younger bonobos differed in their initial learning or in responses in the test phase. Thus, while developmental shifts in cognition are important in generating mature spatial competency in humans (Newcombe & Huttenlocher, 2006), the current studies provide little evidence that bonobos' spatial frameworks exhibit any age-related change.

Notably, this aligns with results of previous research which suggest that chimpanzees show greater changes in cognitive skills across development than do bonobos (Wobber et al., 2010, 2013; Rosati & Hare, 2012; Rosati et al., in press). For example, one study examined apes' spatial memory skills in a naturalistic foraging task in which animals had to recall the locations of many hidden rewards in a large-scale space (Rosati & Hare, 2012). Even though chimpanzees and bonobos are very closely related, their developmental trajectories and mature levels of performance were quite different in this task: the bonobos exhibited no developmental shifts in accuracy of recall between infancy and sub-adulthood, whereas chimpanzees exhibit substantial improvement in the same age range and therefore outperformed bonobos as older ages. Thus, one possibility is that chimpanzees may in fact show shifts in spatial frameworks in the tasks used here with bonobos. For example, chimpanzees' improved performance in the foraging task may have been due to an overall increased reliance on allocentric frameworks that are more useful in a natural context with many salient, distal landmarks. Furthermore, some evidence suggests that primate species can differ in the cognitive strategies they use in these types of tasks: more frugivorous lemur species are more likely to use allocentric encoding compared to more folivorous lemur species (Rosati et al., 2014). Consequently, future studies should also examine whether chimpanzees show similar patterns compared to bonobos, or exhibit age-related changes in use of spatial frameworks.

What do these results mean for claims about human-unique spatial cognition? First, bonobos are able to use both egocentric and allocentric frameworks, which does not support claims that either allocentric (Haun et al., 2006a) or egocentric (Wang & Spelke, 2002) frameworks alone are phylogenetically ancient. It also suggests that language is not necessary to flexibly utilize both types of frameworks. Rather, this supports views suggesting that parallel memory systems support nonhuman and human behavior (Burgess, 2006; Newcombe & Huttenlocher, 2006). These results also indicate that some types of developmental change in spatial cognition seen in humans may not be present in bonobos. This suggests that differences in human and nonhuman patterns of development could result in human-unique cognitive capacities in adulthood. However, differences in the developmental trajectories of chimpanzees and bonobos suggests that these species might in fact provide contrasting views of ancestral developmental traits in humans, for the developmental trajectories and mature levels of performance in chimpanzees and bonobos differ for other spatial skills (Rosati & Hare, 2012). Thus, models of the evolution of human cognitive development, including theories about the role of language in shaping spatial cognition, should integrate data on the development of both chimpanzees and bonobos in order to draw strong inferences about what traits are unique to our lineage.

Acknowledgements

Thanks to Brian Hare for advice on study design, Laura Lewis for assistance with coding and Felix Warneken for comments on a previous version of the manuscript. I also thank Claudine Andre, Dominique Morel, Valery Dhanani, Fanny Mehl and Pierrot Mbonzo and the animal caretakers for their support of this research at Lola ya Bonobo, in collaboration with the Ministry of Research and the Ministry of Environment in the Democratic Republic of Congo. This work was supported in by a L.S.B. Leakey Grant.

References

Acredolo, L.P. (1978). Development of spatial orientation in infancy. — Dev. Psychol. 14: 224-234.

Acredolo, L.P. (1979). Laboratory versus home: the effect of environment on the 9-month-old infant's choice of spatial reference system. — Dev. Psychol. 15: 666-667.

Acredolo, L.P. & Evans, D. (1980). Developmental changes in the effects of landmarks on infant spatial behavior. — Dev. Psychol. 16: 312-318.

Albiach-Serrano, A., Call, J. & Barth, J. (2010). Great apes track hidden objects after changes in the objects' position and in subject's orientation. — Am. J. Primatol. 72: 349-359.

Baayen, R.H. (2008). Analyzing linguistic data. A practical introduction to statistics. — Cambridge University Press, Cambridge, MA.

Bates, D. (2010). The LME4 package: linear mixed-effects models using S4 classes. — Available online at http://www.R-project.org.

Bolker, B.M., Brooks, M.E., Clark, C.J., Geange, S.W., Poulsen, J.R., Stevens, M.H.H. & White, J.S.S. (2008). Generalized linear mixed models: a practical guide for ecology and evolution. — Trends Ecol. Evol. 24: 127-135.

Brown, P. & Levinson, S. (1992). 'Left' and 'right' in Tenejapa: investigating a linguistic conceptual gap. — Z. Phonetik Sprachwissensch. Kommunikationsforsch. 45: 590-611.

Brown, P. & Levinson, S.C. (2000). Frames of spatial reference and their acquisition in Tenejapan Tzeltal. — In: Culture, thought, and development (Nucci, L., Saxe, G. & Turiel, E., eds). Lawrence Erlbaum, Mahwah, NJ, p. 167-197.

Burgess, N. (2006). Spatial memory: how egocentric and allocentric combine. — Trends Cogn. Sci. 10: 551-557.

Burgess, N. (2008). Spatial cognition and the brain. — Ann. NY Acad. Sci. 1124: 77-97.

Cheng, K. (1986). A purely geometric module in the rat's spatial representation. — Cognition 23: 149-178.

Gallistel, C.R. (1990). The organization of learning. — Bradford Books/MIT Press, Cambridge, MA.

Gentner, D. (2007). Spatial cognition in apes and humans. — Trends Cogn. Sci. 11: 192-194.

Hare, B. (2007). From nonhuman to human mind: what changed and why. — Curr. Dir. Psychol. Sci. 16: 60-64.

Hare, B., Wobber, V. & Wrangam, R. (2012). The self-domestication hypothesis: evolution of bonobo psychology is due to selection against aggression. — Anim. Behav. 83: 573-585.

Haun, D.B.M., Rapold, C.J., Call, J., Janzen, G. & Levinson, S.C. (2006a). Cognitive cladistics and cultural override in Hominid spatial cognition. — Proc. Natl. Acad. Sci. USA 103: 17568-17573.

Haun, D.B.M., Call, J., Janzen, G. & Levinson, S.C. (2006b). Evolutionary psychology of spatial representations in the Hominidae. — Curr. Biol. 16: 1736-1740.

Hermer, L. & Spelke, E. (1994). A geometric process for spatial reorientation in young children. — Nature 370: 57-59.

Hermer-Vazquez, L., Spelke, E. & Katnelson, A.S. (1999). Sources of flexibility in human cognition: dual-task studies of space and language. — Cogn. Psychol. 39: 3-36.

Hermer-Vazquez, L., Moffet, A. & Munkholm, P. (2001). Language, space, and the development of cognitive flexibility in humans: the case of two spatial memory tasks. — Cognition 79: 263-299.

Herrmann, E., Hare, B., Call, J. & Tomasello, M. (2010). Differences in the cognitive skills of bonobos and chimpanzees. — PLoS One 5: e12438.

Hoffman, M.L. & Beran, M.J. (2006). Chimpanzees (*Pan troglodytes*) remember the location of a hidden food item after altering their orientation to a spatial array. — J. Comp. Psychol. 120: 389-393.

Hribar, A. & Call, J. (2011). Great apes use landmark cues over spatial relations to find hidden food. — Anim. Cogn. 14: 623-635.

Hribar, A., Haun, D. & Call, J. (2011). Great apes' strategies to map spatial relations. — Anim. Cogn. 14: 511-523.

Janson, C.H. (1998). Experimental evidence for spatial memory in wild brown capuchin monkeys (*Cebus apella*). — Anim. Behav. 55: 1229-1243.

Janson, C.H. (2007). Experimental evidence for route integration and strategic planning in wild capuchin monkeys. — Anim. Cogn. 10: 341-356.

Janson, C.H. & Byrne, R. (2007). What wild primates know about resources: opening up the black box. — Anim. Cogn. 10: 357-367.

Kanngiesser, P. & Call, J. (2010). Bonobos, chimpanzees, gorillas, and orangutans use feature and spatial cues in two spatial memory tasks. — Anim. Cogn. 13: 419-430.

Kano, T. (1992). The last ape: pygmy chimpanzee behavior and ecology. — Stanford University Press, Stanford, CA.

Learmonth, A.E., Newcombe, N.S., Sheridan, N. & Jones, M. (2008). Why size counts: children's spatial reorientation in large and small enclosures. — Dev. Sci. 11: 414-426.

Lee, S.A. & Spelke, E. (2010). Two systems of spatial representation underlying navigation. — Exp. Brain Res. 206: 179-188.

Levinson, S.C. (1996). Frames of reference and Molyneux's question: cross-linguistic evidence. — In: Language and space. Language, speech, and communication (Bloom, P., Peterson, M.A., Nadel, L. & Garrett, M.F., eds). MIT Press, Cambridge, MA, p. 385-436.

Levinson, S.C., Kita, S., Haun, D.B.M. & Rasch, B.H. (2002). Returning the tables: language affects spatial reasoning. — Cognition 84: 144-188.

Li, P. & Gleitman, L. (2002). Turning the tables: language and spatial reasoning. — Cognition 83: 265-294.

Maguire, E.A., Burgess, N., Donnett, J.G., Frackowiak, R.S.J., Frith, C.D. & O'Keefe, J. (1998). Knowning where and getting there: a human navigation network. — Science 280: 921-923.

Majid, A., Bowerman, M., Kita, S., Haun, D. & Levinson, S. (2004). Can language restructure cognition? The case for space. — Trends Cogn. Sci. 8: 108-114.

Malenky, R.K. & Wrangham, R.W. (1993). A quantitative comparison of terrestrial herbaceous food consumption by *Pan paniscus* in the Lomako Forest, Zaire, and *Pan troglodytes* in the Kibale Forest, Uganda. — Am. J. Primatol. 32: 1-12.

Martin-Ordas, G., Haun, D., Colmenares, F. & Call, J. (2010). Keeping track of time: evidence for episodic-like memory in great apes. — Anim. Cogn. 13: 331-340.

Matsuzawa, T. (2007). Comparative cognitive development. — Dev. Sci. 10: 97-103.

Matsuzawa, T., Tomonaga, M. & Tanaka, M. (eds) (2006). Cognitive development in chimpanzees. — Springer, Tokyo.

Mendes, N. (2008). Spatial memory in chimpanzees: single-trial learning and 24 hour and 3 month long-term memory. — Folia Primatol. 79: 283-304.

Menzel, C.R., Savage-Rumbaugh, E.S. & Menzel, E.W. (2002). Bonobo (*Pan paniscus*) spatial memory and communication in a 20-hectare forest. — Int. J. Primatol. 23: 601-619.

Menzel, E.W. (1973). Chimpanzee spatial memory organization. — Science 182: 943-945.

Newcombe, N.S. & Huttenlocher, J. (2006). Development of spatial cognition. — In: Handbook of child psychology, Vol. II: cognition, perception, and language (Damon, W., Lerner, R.M., Kuhn, D. & Siegler, R.S., eds). Wiley, New York, NY, p. 734-776.

Newcombe, N.S. & Ratliff, K.R. (2007). Explaining the development of spatial reorientation: modularity-plus-language versus the emergence of adaptive combination. — In: Emerging landscapes of mind: mapping the nature of change in spatial cognitive development (Plumert, J. & Spencer, J., eds). Oxford University Press, New York, NY, p. 53-76.

Newcombe, N.S., Huttenlocher, J., Drummey, A. & Wiley, J.G. (1998). The development of spatial location coding: place learning and dead reckoning in the second and third years. — Cogn. Dev. 13: 185-200.

Normand, E., Ban, S.D. & Boesch, C. (2009). Forest chimpanzees (*Pan troglodytes verus*) remember the location of numerous fruit trees. — Anim. Cogn. 12: 797-807.

Normand, E. & Boesch, C. (2009). Sophisticated Euclidean maps in forest chimpanzees. — Anim. Behav. 77: 1195-1201.

Packard, M.G. (1996). Inactivation of hippocampus or caudate nucleus with lidocaine differentially affects expression of place and response learning. — Neurobiol. Learn. Mem. 65: 65-72.

Packard, M.G. (1999). Glutamate infused posttraining into the hippocampus or caudate-putamen differentially strengthens place and response learning. — Proc. Natl. Acad. Sci. USA 96: 12881-12886.

Packard, M.G. (2009). Exhumed from thought: basal ganglia and response learning in the plus-maze. — Behav. Brain Res. 199: 24-31.

Packard, M.G. & Goodman, J. (2013). Factors that influence the relative use of multiple memory systems. — Hippocampus 23: 1044-1052.

Pederson, E., Danziger, E., Wilkins, D.G., Levinson, S.C., Kita, S. & Senft, G. (1998). Semantic typology and spatial conceptualization. — Language 74: 557-589.

Poldrack, R.A., Clark, M.A., Pare-Blagoev, E.J., Shohamy, D., Creso Moyan, J., Myers, C., et al. (2001). Interactive memory systems in the human brain. — Nature 414: 546.

Poldrack, R.A. & Packard, M.G. (2003). Competition among multiple memory systems: converging evidence from animal and human brain studies. — Neuropsychologia 41: 245-251.

R Development Core Team (2011). A language and environment for statistical computing. — http://www.R-project.org.

Ratliff, K.R. & Newcombe, N.S. (2008). Reorienting when clues conflct: evidence for an adaptive-combination view. — Psychol. Sci. 19: 1301-1307.

Rosati, A.G. & Hare, B. (2012). Chimpanzees and bonobos exhibit divergent spatial memory development. — Dev. Sci. 15: 840-853.

Rosati, A.G., Rodriguez, K. & Hare, B. (2014). The ecology of spatial memory in four lemur species. — Anim. Cogn., in press, DOI:10.1007/s10071-014-0727-2.

Rosati, A.G., Wobber, V., Hughes, K. & Santos, L.R. (in press). How is human cognitive development unique? — Comp. Dev. Psychol.

Sherry, D.F. & Schacter, D.L. (1987). The evolution of multiple memory systems. — Psychol. Rev. 94: 439-454.

Shettleworth, S.J. (1998). Cognition, evolution, and behavior. — Oxford University Press, New York, NY.

Spelke, E., Lee, S.A. & Izard, V. (2010). Beyond core knowledge: natural geometry. — Cogn. Sci. 1: 1-22.

Tolman, E.C. (1948). Cognitive maps in rats and men. — Psychol. Rev. 55: 189-208.

Wang, R.F. & Spelke, E.S. (2002). Human spatial representation: insights from animals. — Trends Cogn. Sci. 6: 376-382.

White, F.J. (1998). Seasonality and socioecology: the importance of variation in fruit abundance to bonobo sociality. — Int. J. Primatol. 19: 1013-1027.

White, F.J. & Wrangham, R.W. (1988). Feeding competition and patch size in the chimpanzee species *Pan paniscus* and *Pan troglodytes*. — Behaviour 105: 148-163.

White, N.M. & McDonald, R.J. (2002). Multiple parallel memory systems in the brain of the rat. — Neurobiol. Learn. Mem. 77: 125-184.

Wobber, V. & Hare, B. (2011). Psychological health of orphan bonobos and chimpanzees in African sanctuaries. — PLoS One 6: e17147.

Wobber, V., Wrangham, R. & Hare, B. (2010). Bonobos exhibit delayed development of social behavior and cognition relative to chimpanzees. — Curr. Biol. 20: 226-230.

Wobber, V., Herrmann, E., Hare, B., Wrangham, R. & Tomasello, M. (2013). Differences in the early cognitive development of children and great apes. — Dev. Psychobiol., in press, DOI:10.1002/dev.21125.

[When citing this chapter, refer to Behaviour 152 (2015) 407–423]

The influence of testosterone on cognitive performance in bonobos and chimpanzees

Victoria Wobber [a,b] **and Esther Herrmann** [c,*]

[a] Department of Psychology, Harvard University, William James Hall, 33 Kirkland Street, Cambridge, MA 02138, USA

[b] Department of Human Evolutionary Biology, Harvard University, Cambridge, MA 02138, USA

[c] Department of Developmental and Comparative Psychology, Max Planck Institute for Evolutionary Anthropology, Leipzig, Germany

[*]Corresponding author's e-mail address: eherrman@eva.mpg.de

Accepted 23 March 2014; published online 7 May 2014

Abstract

Levels of the steroid hormone testosterone have been found to impact diverse features of cognition from spatial memory to decision-making regarding risk, both in humans and other animals. However less is known about whether closely-related species differ in their testosterone-cognition relationships in line with pressures shaping each species' cognitive evolution. We therefore examined relationships between testosterone and cognition in two-closely related species that differ markedly in their social behaviour, cognition, and patterns of testosterone production: bonobos (*Pan paniscus*) and chimpanzees (*Pan troglodytes*). We presented individuals of both species with a battery of 16 cognitive tasks and determined whether performance on these tasks correlated with average testosterone level. We found that among male chimpanzees, high levels of testosterone correlated with higher performance in numerous tasks, including tasks assessing spatial cognition and physical cognitive abilities more broadly. Meanwhile, in male bonobos we found no correlations between testosterone and performance on the cognitive tasks, and found no correlations in females of either species. Building on prior comparative research, these results suggest that bonobos and chimpanzees differ critically in the proximate mechanisms influencing their cognitive capacities, and that in particular the role of testosterone in shaping behaviour and cognition differs dramatically between the two species.

Keywords

cognition, testosterone, chimpanzees, bonobos.

1. Introduction

Understanding the proximate mechanisms that influence social behaviour and cognition can give us a deeper understanding of these traits. Steroid hormones in particular have been found to have diverse effects on behaviour and cognition, aside from their roles in facilitating reproductive and metabolic functions (Nelson, 2000; Luine, 2008; Eisenegger et al., 2011). Levels of testosterone, for example, have been found to correlate with decision-making in situations involving risk and cooperation (Burnham, 2007; Apicella et al., 2008; Stanton et al., 2011), abilities to discriminate emotion in faces (van Honk et al., 1999; Wirth & Schultheiss, 2007; Derntl et al., 2009), and capacities pertaining to mental rotation (Hooven et al., 2004; Alexander & Son, 2007). Meanwhile, administration of testosterone in men has been demonstrated to directly affect decision-making in cooperative interactions (Zak et al., 2009), avoid/approach responses to emotional face stimuli (Volman et al., 2011), and success in spatial memory and navigation tasks (Janowsky et al., 1994; Cherrier et al., 2001). Though research targeting the mechanisms underlying these hormone-cognition relationships is ongoing, potential means by which testosterone might influence behaviour include its effects on the amgydala and orbitofrontal cortex, specifically in reducing the links between these two areas (Mehta & Beer, 2010; van Wingen et al., 2010; Eisenegger et al., 2011).

Many of the relationships found between testosterone and cognition in humans appear to have deep roots within our evolutionary history. Relationships between testosterone and spatial cognition have been well-documented within rodents, with testosterone enhancing performance in various types of maze tasks and even aiding in spatial memory over a temporal delay (reviewed in Leonard & Winsauer, 2011), and more recent work suggests that testosterone mediates approach/avoidance responses to socio-emotional stimuli in male rhesus macaques, particularly in the area of increased vigilance and alertness in interaction with novel objects and response to social playbacks (Lacreuse et al., 2010, 2012). In addition, the role of testosterone in aggressive behaviour has been extensively catalogued across numerous animal taxa (Wingfield et al., 1990; Archer, 2006). Testosterone has been found to increase in the presence of reproductively active females, and facilitates behaviours related to male dominance hierarchies such as displays and agonistic interactions (Muller & Wrangham, 2004; Bartos et al., 2010). Indeed, several studies have revealed that individual and species differences in aggressive behaviour may be critically mediated by the influence of testosterone on patterns of decision-making during competitive interactions (Fuxjager & Marler, 2010; Wobber et al., 2010).

Despite the broad research on the relationship between testosterone and cognition across taxa, few studies have directly examined whether testosterone-cognition relationships vary among closely related species in line with differing ecological pressures that those species face. Here, we propose that bonobos (*Pan paniscus*) and chimpanzees (*Pan troglodytes*) provide an ideal comparison to suit this purpose. These two species, while closely related (Won & Hey, 2005), differ markedly in aspects of their social behaviour, cognition, and the functionality of testosterone across development and in facilitating mating effort (Muller & Wrangham, 2004; Hermann et al., 2010; Wobber et al., 2010, 2013; Hare et al., 2012; Surbeck et al., 2012). Bonobos show more fluid dominance hierarchies, greater aversion to risk in decision-making, and differ in their development of spatial memory capacities (Heilbronner et al., 2008; Rosati & Hare, submitted), indicating species differences in numerous areas where testosterone has a known influence, as discussed above. These differences between bonobos and chimpanzees have been proposed to derive from fundamental shifts in feeding and social ecology between these two species, in particular with the more predictable feeding environment in the Democratic Republic of Congo facilitating heightened associations between females and corresponding reductions in male aggressive tendencies in bonobos relative to chimpanzees (Wrangham & Pilbeam, 2001; Hare et al., 2012). We therefore hypothesized that bonobos and chimpanzees might differ significantly in their relationship between testosterone and cognition.

In this paper we discuss the results of a study examining the relationship between levels of testosterone and performance on a diverse array of cognitive tasks (the Primate Cognition Test Battery (PCTB); Herrmann et al., 2007) in bonobos and chimpanzees. We examine these relationships separately within chimpanzees and bonobos, determining the degree to which testosterone correlates with cognitive performance within each sex in each species.

2. Methods

2.1. Subjects

We collected saliva samples from 27 chimpanzees (13 males and 14 females; 3 to 18 years of age) and 30 bonobos (19 males and 11 females; aged 5 to 22 years) who participated in the battery of cognitive tests described below and elsewhere (Herrmann et al., 2010). The chimpanzees lived at the Tchimpounga Chimpanzee Sanctuary, Republic of Congo and the bonobos at Lola ya Bonobo sanctuary, Democratic Republic of Congo. All apes came to the sanctuaries as orphans as a result of the illegal bushmeat trade and were raised together with peers, living in social groups that have access to large forest enclosures (see Wobber & Hare (2011) for site description), except for one ape in the sample who was born on-site and mother-reared. The behavioural data discussed here have been published previously in Herrmann et al. (2010).

2.2. Procedure

2.2.1. Cognitive measures: Primate Cognition Test Battery (Herrmann et al., 2007)

Subjects were tested on the PCTB that comprised 16 physical and social cognitive different tasks (see Table 1, see original study (Herrmann et al., 2007, 2010)). The physical cognitive tasks consisted of problems concerning space, quantities and causality. The space scale comprised tasks in which the ape had to show an understanding of spatial relations by locating a reward (spatial memory), tracking a reward after invisible displacement (object permanence), tracking a reward after a rotation manipulation (rotation) or tracking a reward after visible changes in location (transposition). The quantities scale was divided into problems in which the ape had to discriminate between two quantities (relative numbers) or had to discriminate between quantities added to other quantities (addition numbers). The causality scale included tasks in which the ape had to show causal understanding either of noise produced by a hidden reward (noise), or a change in appearance produced by a hidden reward (shape). In addition, apes were presented with a task where they needed to use a stick in order to retrieve a reward which was out of reach (tool use), and a task where they needed to discriminate between a functional and a non-functional tool (tool properties). The social cognitive tasks consisted of problems concerning social learning, communication and 'theory of mind'. The social learning scale comprised a task in

Table 1.
Primate Cognition Test Battery (PCTB) including domains, scales and tasks.

Domain	Scale	Task	Description
Physical	Space	Spatial memory	Locating a reward
		Object permanence	Tracking of a reward after invisible displacement
		Rotation	Tracking of a reward after a rotation manipulation
		Transposition	Tracking of a reward after location changes
	Quantities	Relative numbers	Quantity discrimination
		Addition numbers	Quantity discrimination with added quantities
	Causality	Noise	Causal understanding of produced noise by hidden rewards
		Shape	Causal understanding of appearance change by hidden rewards
		Tool use	Using a stick in order to retrieve a reward which is out of reach
		Tool properties	Understanding of functional and non-functional tool properties
Social	Social learning	Social learning	Solving a simple but not obvious problem by observing a demonstrated solution
	Communication	Comprehension	Understanding communicative cues indicating a reward's hidden location
		Pointing cups	Production of communicative gestures in order to retrieve a hidden reward
		Attentional state	Choice of communicative gestures considering the attentional state of the recipient
	Theory of mind	Gaze following	Following an actor's gaze direction to a target
		Intentions	Understanding what an actor intended to do (unsuccessfully)

Table originally published in Herrmann et al. (2007).

which subjects were required to imitate another's solution to a problem (in three separate tasks of this type). The communication scale comprised one task in which the ape had to understand communicative cues indicating a reward's hidden location (comprehension) and two tasks in which subjects had to produce communicative gestures in order to retrieve a hidden reward (pointing cups and attentional state). The theory of mind scale was divided into a task in which the ape had to follow an experimenter's gaze to a target (gaze following) and a second task in which the subject had to infer the location of a hidden reward based on the intention-guided actions of an experimenter (intentions).

Given the known relationship between testosterone and spatial memory, we will outline the rotation task in more detail since this in fact represented a critical test for whether testosterone correlated with cognitive ability in either species. In the rotation task, apes were presented with an array of three cups sitting on a plastic platform, with a food reward hidden under one of these three cups in full view of the subject. After the food reward was hidden, the platform was rotated either 180 or 360 degrees (depending on the trial) in view of the subject. The platform was then pushed towards the subject so that he or she could identify the location of the hidden reward. If the subject did so successfully, he was rewarded; if not, he was shown the location of the food reward but not given the item. Subjects received 9 trials of this task, 3 trials with a reward hidden under the middle cup and 180 degree rotations, 3 trials with a reward hidden under either the left or the right cup and 360 degree rotations and 3 trials with a reward hidden under the either the left or right cup and 180 degree rotations, with the trials presented in a consistent order across all subjects.

2.2.2. Hormonal measures

To determine baseline testosterone levels of individuals participating in PCTB, we collected saliva samples from these individuals using methods outlined previously (Wobber et al., 2010, 2013) (range 1–22 samples per individual, mean 8.6 samples). Samples for an individual baseline were all collected within a 2-month period. This two-month period fell within 1 year of when subjects participated in the PCTB. Samples were collected between 8:00 AM and 5:00 PM, with the time of day (morning versus afternoon) counterbalanced within-individual to prevent circadian rhythms from biasing the average testosterone values. Our sampling regime also avoided the early morning peak in testosterone levels upon waking seen in chimpanzees and other animals (Muller & Lipson, 2003).

Saliva samples were collected according to procedures described elsewhere using cotton rounds and Sweet Tarts as stimulants (Wobber et al., 2010, 2013). Fifty microliters of 0.1% sodium azide solution was added to samples immediately after collection to prevent contamination and to allow samples to be kept at room temperature until they were returned to the laboratory (Lipson & Ellison, 1989). The saliva samples were analysed in the Reproductive Ecology Laboratory at Harvard University. Salivary testosterone measurements were made using an ^{125}I-based radioimmunoassay kit (No. 4100, Diagnostic Systems Laboratories, Webster, TX, USA) with the following modifications: standards were prepared in assay buffer and run at six concentrations from 2 to 375 pg/ml. Samples were added in 100 μl amounts together with 300 μl of assay buffer. First antibody (20 μl) and labelled steroid (50 μl) were added to each tube to yield a total reaction volume of 470 μl per tube. After overnight incubation at 4°C, 500 μl of second antibody was added to each reaction tube. Reaction tubes were subsequently centrifuged for 45 min; after aspiration of the supernatant, tubes were counted in a gamma counter for 2 min. In pilot assays, the ape testosterone values using the standard aliquot for human assays (200 μl) were too high to be readable in the assay range. Thus, we used only 100 μl of the chimpanzee and bonobo saliva for the T assays, with the same standard curve as employed in the human testosterone radioimmunoassay protocol. Validation of this sample collection and analysis procedure is discussed in Wobber et al. (2013).

2.3. Analyses

We first computed an average testosterone value for each individual. Then, we log-transformed these average testosterone values to normalize our data and enable the use of parametric statistics. Below, we first provide descriptive statistics on the testosterone values, to confirm that there was comparable variance in between sexes in each species within this subsample taken from a larger project examining testosterone production throughout development in bonobos and chimpanzees (Wobber et al., 2013). Similarly, we discuss overall variance in the cognitive measures. Second, we discuss the results of Pearson correlations measuring the relationship between an individual's average testosterone level and his or her performance on (1) each broad division of the test battery (physical versus social) and (2) each of the six cognitive scales described above: space, quantities, tools and causality, social learning, communication, and 'theory of mind'. For the scales where we found a significant relationship between testosterone and performance, we analysed task-by-task correlations with testosterone to determine which tasks might best predict the overall relationship. Finally, we performed partial correlations to ensure that age did not mediate any relationships between testosterone and cognitive performance, given that our previous findings have established that testosterone and cognitive task performance increase more markedly with age in chimpanzees than in bonobos (Wobber et al., 2013, unpublished data). Given the known sex differences in testosterone levels in mammals, we performed these analyses separately for males and females within each species.

3. Results

Descriptive statistics showed that variability in testosterone levels was comparable between males of the two species, as was variation in performance across the two main cognitive divisions (Table 2). Variance in testosterone levels was higher among female chimpanzees due to one female with extremely high testosterone levels. Finally, variance in performance on the cognitive tasks was comparable between the two species, with the only difference that female bonobos showed slightly more variance in cognitive performance than female chimpanzees (meaning that if anything, we would be more likely to find a relationship between testosterone and cognitive performance in female bonobos given their greater variance on these measures) (Table 2). Average testosterone levels in each species and sex were as follows: male chimpanzees 0.27 ng/ml, male bonobos 0.22 ng/ml, female chimpanzees 0.18 ng/ml, female bonobos 0.17 ng/ml. Note that we performed statistical analyses with log-transformed testosterone values, with the rationale discussed in our 'Analyses' section above.

Examining the results of Pearson correlations between log average testosterone and performance on the cognitive measures, within male chimpanzees we found several significant relationships between log average testosterone values and performance on the cognitive tasks. Males with higher testosterone on average performed more skilfully in the physical cognition division of tasks ($r = 0.795$, $p = 0.001$, $N = 13$), though they did not perform more skilfully within the social cognition division (Figure 1, Table 3). Looking at the cognitive scale level, male chimpanzees with higher testosterone levels performed more skilfully in space ($r = 0.620$, $p = 0.024$, $N = 13$) as well as causality ($r = 0.588$, $p = 0.035$, $N = 13$). Within these two scales, task-by-task analysis only revealed one task-level correlation between performance and testosterone level, with higher levels of testosterone correlating

Table 2.
Descriptive statistics for testosterone levels and performance on the two major cognitive divisions (physical versus social cognition) and 6 cognitive scales for each species and sex.

	Males				Females			
	Chimpanzees		Bonobos		Chimpanzees		Bonobos	
	SE	Range	SE	Range	SE	Range	SE	Range
Log testosterone	0.07	0.76	0.07	0.75	0.09	1.11	0.05	0.48
Physical cognition	0.02	0.22	0.01	0.24	0.01	0.16	0.02	0.24
Social cognition	0.02	0.26	0.02	0.32	0.02	0.24	0.03	0.35
Space	0.03	0.39	0.03	0.53	0.02	0.25	0.03	0.36
Quantities	0.03	0.36	0.03	0.42	0.03	0.30	0.03	0.34
Causality	0.03	0.37	0.02	0.32	0.04	0.42	0.04	0.36
Social learning	0.04	0.33	0.03	0.33	0.03	0.25	0.05	0.33
Communication	0.04	0.41	0.03	0.52	0.03	0.42	0.06	0.73
Theory of mind	0.03	0.39	0.03	0.56	0.03	0.47	0.03	0.39

with more skilful performance on the rotation task ($r = 0.601$, $p = 0.030$, $N = 13$). Meanwhile, there was also one scale where higher levels of testosterone were correlated with poorer performance in male chimpanzees: the theory of mind scale ($r = -0.598$, $p = 0.031$, $N = 13$), and within that scale on the gaze-following task ($r = -0.596$, $p = 0.030$, $N = 13$) (Table 3).

We next examined the potential co-variation of age with these relationships between testosterone and performance in male chimpanzees. We performed partial correlations between log average testosterone and performance on the scales/tasks where there was a significant relationship in the Pearson correlations and controlled for the effect of age as a covariate. In these partial correlations, only the positive relationships between testosterone and physical cognition (partial correlation with age, $r = 0.696$, $p = 0.012$, df $= 10$) and between testosterone and the rotation task (partial correlation with age, $r = 0.616$, $p = 0.033$, df $= 10$) remained significant. These correlations indicate that testosterone had an independent effect on physical cognition and rotation in particular, even when taking into account any effect that learning or heightened experience in interacting with the world might have conferred in terms of advantages on these tests for older individuals.

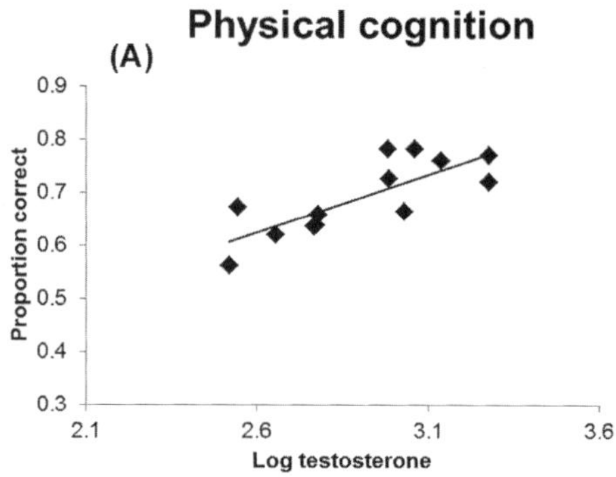

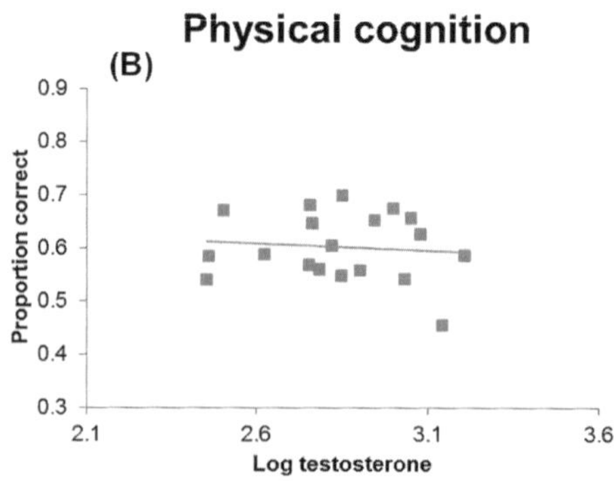

Figure 1. Regression of average log testosterone levels and average performance across 10 physical cognition tasks in (A) chimpanzee males and (B) bonobo males. Each point represents one individual (either a chimpanzee, as depicted by the black diamond, or a bonobo, as depicted by the grey squares), and that individual's average testosterone level (log-transformed) as well as their average proportion correct across the physical cognition tasks. Regression lines denoting for the relationship between testosterone and proportion correct within the physical cognition division are shown for each species.

Turning to male bonobos, we found no significant relationships between log average testosterone and performance on any divisions or scales of the cognitive test battery, despite our larger sample of male bonobos ($N = 19$) relative to male chimpanzees ($N = 13$). Similarly, no significant relationships emerged even when controlling for the effect of age with partial correlations. These findings suggest a stronger role of testosterone in association with performance on cognitive tasks in chimpanzees relative to bonobos, despite there being similar levels of variance in task performance and testosterone level in males within our sample (Table 2).

Among females, there was only one significant relationship between testosterone and performance in either species. Female chimpanzees showed

Table 3.
Relationship between testosterone and cognitive performance in male chimpanzees ($N = 13$), male bonobos ($N = 19$), female chimpanzees ($N = 14$) and female bonobos ($N = 11$).

	Males				Females			
	Chimpanzees		Bonobos		Chimpanzees		Bonobos	
	r	p	r	p	r	p	r	p
Physical cognition	*0.795*	*0.001**	−0.095	0.698	−0.356	0.211	0.098	0.774
Social cognition	−0.140	0.648	0.092	0.708	0.363	0.202	−0.503	0.115
Space	*0.620*	*0.024*	−0.162	0.509	−0.448	0.108	−0.357	0.282
Quantities	0.199	0.516	−0.045	0.853	0.016	0.957	0.192	0.571
Causality	*0.588*	*0.035*	0.112	0.649	−0.168	0.567	0.282	0.400
Social learning	−0.055	0.865	0.196	0.422	0.065	0.848	−0.311	0.352
Communication	−0.200	0.512	0.318	0.184	*0.577*	*0.031**	−0.263	0.434
Theory of mind	*−0.598*	*0.031*	−0.343	0.151	−0.166	0.570	−0.496	0.121

Pearson correlation values (r) are indicated for the relationships between log average testosterone and the overall cognitive task divisions (physical versus social cognition), as well as each of the 6 cognitive scales, along with their corresponding p-values. Italicised values indicate a positive correlation.

* Results of a partial correlation controlling for the effect of age were significant.

a positive correlation between log average testosterone and performance in the communication scale ($r = 0.577$, $p = 0.031$, $N = 14$) (Table 3), which was likely influenced by a strong relationship between testosterone and performance on the attentional state task ($r = 0.665$, $p = 0.009$, $N = 14$). These relationships remained significant even after controlling for age (partial correlations, communication scale: $r = 0.569$, $p = 0.042$, df $= 11$, attentional state task: $r = 0.755$, $p = 0.003$, df $= 11$). Meanwhile, there were no significant relationships between testosterone and performance on any division or scale of the test battery in female bonobos, even after partialling out the effects of age — though this was the group in our analysis with the smallest sample size.

4. Discussion

Our results reveal significant differences between bonobos and chimpanzees in the relationship between testosterone and performance on cognitive tasks, particularly among males. Among chimpanzee males, higher levels of testosterone correlated with higher levels of performance in the physical cognition division on the whole, as well as with the space and causality scales. In addition, among male chimpanzees, higher levels of testosterone correlated with lower levels of theory of mind performance. Meanwhile among bonobo males there were no relationships between testosterone and performance on the cognitive tasks, despite having a larger sample size of male bonobos relative to male chimpanzees and the two groups showing comparable levels of variance in these measures. Finally, among female chimpanzees there was a positive relationship between testosterone level and performance on the communication scale, largely driven by higher levels of testosterone correlating with more skilful performance in the attentional state task. Among female bonobos there were no significant relationships between testosterone and cognitive performance.

These findings indicate that in a number of areas where testosterone has been implicated in human and rodent cognition including spatial cognition (Williams & Meck, 1991; Silverman et al., 1999; Hooven et al., 2004; Leonard & Winsauer, 2011), chimpanzees also show a significant role of testosterone level in accounting for individual variance. This relationship was particularly apparent in the rotation task, where individuals needed to track a reward after the reward locations were rotated geometrically. Females of both species showed fewer significant relationships between testosterone and performance — the association between testosterone and behaviour/cognition has been less studied in females than in males, with greater inconclusiveness of findings (Bateup et al., 2002), though see Beehner et al. (2005), prompting further inquiry in this area.

Importantly, male chimpanzees and bonobos differed in their relationship between testosterone and performance on the cognitive tasks administered here. These results align with the finding that testosterone plays a significant role in numerous aspects of chimpanzee social life — with males differing in baseline testosterone level in correlation with dominance rank (Muehlenbein et al., 2004; Muller & Wrangham, 2004), showing changes in testosterone levels when females are most fecund (Muller & Wrangham, 2004), and showing more responsiveness in levels of testosterone surrounding competitive interactions (Wobber et al., 2010; Sobolewski et al., 2012). In contrast, in bonobos these behaviours have been found to correlate less well with levels of testosterone, with male bonobos showing a weaker association between testosterone and rank (Marshall & Hohmann, 2005; Surbeck et al., 2012), little change in testosterone level depending on female fecundity (Surbeck et al., 2012), minimal change in testosterone level surrounding varying types of competitive interaction (Wobber et al., 2010), and even lesser change in testosterone level throughout development relative to chimpanzees (Wobber et al., 2013). On the whole then, these findings suggest that testosterone may play a greater role in chimpanzee social behaviour and cognition than found among bonobos, suggesting striking differences in the proximate mechanisms underlying behaviour and cognition in these two species. In line with the broader cognitive differences between chimpanzees and bonobos then (Herrmann et al., 2010), the associations found here between testosterone and cognition in chimpanzees, but not bonobos, suggests one potential mechanism by which the behavioural and cognitive variation between these two ape taxa might be maintained.

Nonetheless, our results have demonstrated a correlative, rather than a causal relationship; hence, it is possible that testosterone's effect on male chimpanzees' performance on specific cognitive tasks is mediated via changes in visual attention or motivation to participate in these tasks. However, if that was the case, it would not explain why testosterone correlated only with specific tasks and scales rather than performance overall. Furthermore, we were also able to rule out age as a potential covariate driving the relationships between testosterone and at least some of the cognitive scales in male chimpanzees. It is also possible that greater ability in certain areas, such as tracking potential feeding trees in space, in fact enables males to attain higher rank and therefore to sustain a higher level of testosterone. It is therefore possible that heightened cognitive abilities drive attainment of rank and increases in testosterone levels, rather than testosterone levels facilitating differing cognitive acuity, as we have postulated here. Further inquiry experimentally inducing testosterone increases and measuring their potential effects on cognitive performance in male apes can tease apart elements of causality in the testosterone-cognition relationship.

This study provides greater insight into the role of testosterone of these two closely-related species, but additional research is necessary (1) to clarify the mechanisms by which testosterone might influence cognition, including manipulations of testosterone level to determine whether this might directly impact performance (Zak et al., 2009; Volman et al., 2011) and (2) to identify why relationships between testosterone and cognition might be less significant among bonobos relative to chimpanzees, in line with the differences in social behaviour, cognition and endocrinology between the two species (Herrmann et al., 2010; Hare et al., 2012; Wobber et al., 2013). Moreover, work with additional steroid hormones such as oestradiol and androstenedione can clarify the specific mechanisms by which hormone levels mediate individual differences in cognition across species, including humans (Azurmendi et al., 2005; Luine, 2008).

Acknowledgements

We are grateful to both sanctuaries for hosting our research. We are thankful to L. Pharoah, R. Atencia, K. Brown and the Jane Goodall Institute USA and staff of Tchimpounga Sanctuary and Claudine Andre, Dominique Morel, Crispin Kamate Mahamba, and Pierrot Mbonzo and the staff Lola ya Bonobo Sanctuary for their enthusiasm and support. In particular, we appreciate the hard work of the animal caregivers: J. Maboto, B. Moumbaka, A. Sitou, M. Makaya, B. Bissafi, C. Ngoma, W. Bouity, J.A. Tchikaya, L. Bibimbou, A. Makosso, C. Boukindi, G. Nzaba, B. Ngoma, J.C. Nzumbi, S. Mokando, C. Paluku, A. Kisungu, P. Kunaka, N. Luvualu and K. Manzambi. We also thank the Ministry of Environment in the Democratic Republic of Congo and the Congolese Ministere de la Recherche Scientifique et de l'Innovation Technique in the Republic of Congo. Finally, we thank Peter Ellison and Susan Lipson for their help in performing the endocrine analyses. This research was approved by the Institutional Care and Use Committees at Harvard University, and complied with the laws of the countries in which it was performed. This work was supported in part by an L.S.B. Leakey Foundation Grant, National Science Foundation BCS-0851291 (DDIG), and a Wenner-Gren Foundation Grant to V.W. as well as NSF grants BCS-08-27552 and BCS-25172 to B.H.

References

Alexander, G.M. & Son, T. (2007). Androgens and eye movements in women and men during a test of mental rotation ability. — Horm. Behav. 52: 197-204.

Apicella, C.L., Dreber, A., Campbell, B., Gray, P.B., Hoffman, M. & Little, A.C. (2008). Testosterone and financial risk preferences. — Evol. Hum. Behav. 29: 384-390.

Archer, J. (2006). Testosterone and human aggression: an evaluation of the challenge hypothesis. — Neurosci. Biobehav. Rev. 30: 319-345.

Azurmendi, A., Braza, F., Sorozabal, A., Garcia, A., Braza, P., Carreras, M., Munoz, J., Cardas, J. & Sanchez-Martin, J. (2005). Cognitive abilities, androgen levels, and body mass index in 5-year-old children. — Horm. Behav. 49: 187-195.

Bartos, L., Schams, D., Bubenik, G.A., Kotrba, R. & Tomanek, M. (2010). Relationship between rank and plasma testosterone and cortisol in red deer males (*Cervus elaphus*). — Physiol. Behav. 101: 628-634.

Bateup, H., Booth, A., Shirtcliff, E. & Granger, D. (2002). Testosterone, cortisol, and women's competition. — Evol. Hum. Behav. 23: 181-192.

Beehner, J., Phillips-Conroy, J. & Whitten, P. (2005). Female testosterone, dominance rank, and aggression in an Ethiopian population of hybrid baboons. — Am. J. Primatol. 67: 101-119.

Burnham, T.C. (2007). High-testosterone men reject low ultimatum game offers. — Proc. Roy. Soc. Lond. B: Biol. Sci. 274: 2327-2330.

Cherrier, M.M., Asthana, S., Plymate, S., Baker, L., Matsumoto, A.M., Peskind, E., Raskind, M.A., Brodkin, K., Bremner, W., Petrova, A., LaTendresse, S. & Craft, S. (2001). Testosterone supplementation improves spatial and verbal memory in healthy older men. — Neurology 57: 80-88.

Derntl, B., Windischberger, C., Robinson, S., Kryspin-Exner, I., Gur, R.C., Moser, E. & Habel, U. (2009). Amygdala activity to fear and anger in healthy young males is associated with testosterone. — Psychoneuroendocrinology 34: 687-693.

Eisenegger, C., Haushofer, J. & Fehr, E. (2011). The role of testosterone in social interaction. — Trends Cogn. Sci. 15: 263-271.

Fuxjager, M. & Marler, C. (2010). How and why the winner effect forms: influences of contest environment and species differences. — Behav. Ecol. 21: 37-45.

Hare, B., Wobber, V. & Wrangham, R. (2012). The self-domestication hypothesis: bonobo psychology evolved due to selection against aggression. — Anim. Behav. 83: 573-585.

Heilbronner, S., Rosati, A., Stevens, J., Hare, B. & Hauser, M. (2008). A fruit in the hand or two in the bush? Divergent risk preferences in chimpanzees and bonobos. — Biol. Lett. 4: 246-249.

Herrmann, E., Call, J., Hernandez-Lloreda, M., Hare, B. & Tomasello, M. (2007). Humans have evolved specialized skills of social cognition: the cultural intelligence hypothesis. — Science 317: 1360-1366.

Herrmann, E., Hare, B., Call, J. & Tomasello, M. (2010). Differences in the cognitive skills of bonobos and chimpanzees. — PLOS One 5: e12438.

Hooven, C., Chabris, C., Ellison, P. & Kosslyn, S. (2004). The relationship of male testosterone to components of mental rotation. — Neuropsychologia 42: 782-790.

Janowsky, J.S., Oviatt, S.K. & Orwoll, E.S. (1994). Testosterone influences spatial cognition in older men. — Behav. Neurosci. 108: 325-332.

Lacreuse, A., King, H.M., Kurdziel, L.B., Partan, S.R., Caldwell, K.M., Chiavetta, M.R., Millette, M.M., Meyer, J.S. & Grow, D.R. (2010). Testosterone may increase selective attention to threat in young male macaques. — Horm. Behav. 58: 854-863.

Lacreuse, A., Gore, H.E., Chang, J. & Kaplan, E.R. (2012). Short-term testosterone manipulations modulate visual recognition memory and some aspects of emotional reactivity in male rhesus monkeys. — Physiol. Behav. 106: 229-237.

Leonard, S.T. & Winsauer, P.J. (2011). The effects of gonadal hormones on learning and memory in male mammals: a review. — Curr. Zool. 57: 543-558.

Luine, V.N. (2008). Sex steroids and cognitive function. — J. Neuroendocrinol. 20: 866-872.

Marshall, A. & Hohmann, G. (2005). Urinary testosterone levels of wild male bonobos (*Pan paniscus*) in the Lomako forest, Democratic Republic of Congo. — Am. J. Primatol. 65: 87-92.

Mehta, P.H. & Beer, J. (2010). Neural mechanisms of the testosterone-aggression relation: the role of orbitofrontal cortex. — J. Cogn. Neurosci. 22: 2357-2368.

Muehlenbein, M., Watts, D. & Whitten, P. (2004). Dominance rank and fecal testosterone levels in adult male chimpanzees (*Pan troglodytes schweinfurthii*) at Ngogo, Kibale National Park, Uganda. — Am. J. Primatol. 64: 71-82.

Muller, M. & Lipson, S. (2003). Diurnal patterns of urinary steroid excretion in wild chimpanzees. — Am. J. Primatol. 60: 161-166.

Muller, M. & Wrangham, R. (2004). Dominance, aggression and testosterone in wild chimpanzees: a test of the 'challenge hypothesis'. — Anim. Behav. 67: 113-123.

Nelson, R. (2000). An introduction to behavioral endocrinology. — Sinauer Associates, Sunderland, MA.

Silverman, I., Kastuk, D., Choi, J. & Phillips, K. (1999). Testosterone levels and spatial ability in men. — Psychoneuroendocrinology 24: 813-822.

Sobolewski, M., Brown, J. & Mitani, J.C. (2012). Territoriality, tolerance and testosterone in wild chimpanzees. — Anim. Behav. 84: 1469-1474.

Stanton, S.J., Liening, S.H. & Schultheiss, O.C. (2011). Testosterone is positively associated with risk taking in the Iowa Gambling Task. — Horm. Behav. 59: 252-256.

Surbeck, M., Deschner, T., Schubert, G., Weltring, A. & Hohmann, G. (2012). Mate competition, testosterone and intersexual relationships in bonobos, *Pan paniscus*. — Anim. Behav. 83: 659-669.

van Honk, J., Tuiten, A., Verbaten, R., van den Hout, M., Koppeschaar, H., Thijssen, J. & de Haan, E. (1999). Correlations among salivary testosterone, mood, and selective attention to threat in humans. — Horm. Behav. 36: 17-24.

van Wingen, G., Mattern, C., Verkes, R.J., Buitelaar, J. & Fernandez, G. (2010). Testosterone reduces amygdala-orbitofrontal cortex coupling. — Psychoneuroendocrinology 35: 105-113.

Volman, I., Toni, I., Verhagen, L. & Roelofs, K. (2011). Endogenous testosterone modulates prefrontal-amygdala connectivity during social emotional behavior. — Cerebr. Cort. 21: 2282-2290.

Williams, C.L. & Meck, W.H. (1991). The organizational effects of gonadal-steroids on sexually dimorphic spatial ability. — Psychoneuroendocrinology 16: 155-176.

Wingfield, J., Hegner, R., Dufty, R. & Ball, G. (1990). The "challenge hypothesis": theoretical implications for patterns of testosterone secretion, mating systems, and breeding strategies. — Am. Nat. 136: 829-846.

Wirth, M. & Schultheiss, O. (2007). Basal testosterone moderates responses to anger faces in humans. — Physiol. Behav. 90: 496-505.

Wobber, V. & Hare, B. (2011). Psychological health of orphan bonobos and chimpanzees in African sanctuaries. — PLOS One 6: e17147.

Wobber, V., Hare, B., Maboto, J., Lipson, S., Wrangham, R. & Ellison, P. (2010). Differential changes in steroid hormones before competition in bonobos and chimpanzees. — Proc. Natl. Acad. Sci. USA 107: 12457-12462.

Wobber, V., Lipson, S., Hare, B., Wrangham, R. & Ellison, P. (2013). Different ontogenetic patterns of testoterone production reflect divergent male reproductive strategies in chimpanzees and bonobos. — Physiol. Behav. 116-117: 44-53.

Won, Y. & Hey, J. (2005). Divergence population genetics of chimpanzees. — Mol. Biol. Evol. 22: 297-307.

Wrangham, R. & Pilbeam, D. (2001). African apes as time machines. — In: All apes great and small (Galdikas, B., Briggs, N., Sheeran, L., Shapiro, G. & Goodall, J., eds). Kluwer Academic/Plenum Publishers, New York, NY, p. 5-18.

Zak, P.J., Kurzban, R., Ahmadi, S., Swerdloff, R.S., Park, J., Efremidze, L., Redwine, K., Morgan, K. & Matzner, W. (2009). Testosterone administration decreases generosity in the ultimatum game. — PLOS One 4: 7.

[When citing this chapter, refer to Behaviour 152 (2015) 425–460]

Why do wild bonobos not use tools like chimpanzees do?

T. Furuichi [a,*,**], **C. Sanz** [b,c,**], **K. Koops** [d,**], **T. Sakamaki** [a], **H. Ryu** [a], **N. Tokuyama** [a] **and D. Morgan** [c,e]

[a] Primate Research Institute, Kyoto University, Inuyama, Aichi 484-5806, Japan

[b] Department of Anthropology, Washington University, 1 Brookings Drive, Saint Louis, MO 63130, USA

[c] Congo Program, Wildlife Conservation Society, B.P. 14537, Brazzaville, Republic of Congo

[d] Department of Archaeology and Anthropology, University of Cambridge, Fitzwilliam Street, Cambridge CB2 1QH, UK

[e] Lester E. Fisher Center for the Study and Conservation of Apes, Lincoln Park Zoo, North Clark Street, Chicago, IL 60614, USA

[*]Corresponding author's e-mail address: furuichi@pri.kyoto-u.ac.jp

[**]Co-first authors

Accepted 17 June 2014; published online 1 September 2014

Abstract

One of the most conspicuous behavioural differences among great apes is the paucity of tool use among wild bonobos (*Pan paniscus*) in comparison to chimpanzees (*Pan troglodytes*) who are one of the most prolific and skilled tool users in the animal kingdom. This is in spite of the fact that bonobo tool use repertories are as large and diverse as chimpanzees' in captive settings. In this study, we compared tool using behaviours and potential drivers of these behaviours in the Wamba bonobo population located in central Democratic Republic of Congo with the Goualougo chimpanzee population of northern Republic of Congo. The tool use repertoire of wild bonobos was comprised of only 13 behaviours, compared to 42 for chimpanzees. However, the number of tool behaviours observed in each study site was similar between bonobos and chimpanzees, and many types of tool use for social, self-grooming/stimulation, and comfort/protection functions were commonly used by both species. A marked difference is that 25 of 42 tool behaviours exhibited by chimpanzees are performed for feeding, in contrast to a single report of bonobos using a leaf sponge to drink water. We examined whether the differences in tool use repertoires can be explained by the necessity, opportunity, relative profitability, or invention hypotheses. We found that habitat composition and fluctuation of fruit production at these two sites were similar, particularly when compared with variation observed between sites within each species. Thus it was unlikely that the necessity hypothesis explains the lack of tool use for feeding in bonobos. Though further study at Wamba is needed, we did not identify any obvious differences in prey availability that would indicate differences in tool using opportunities between the sites. This study could not test the relative profitability hypothesis, and further research is needed on whether tool use is the most efficient means of calorie or protein intake for wild apes. Bonobos at Wamba formed much larger and stable parties than chimpanzees at Goualougo, which was contrary to the prediction by the invention hypothesis. Another explanation is that differences in tool use behaviour between bonobos and chimpanzees might not be explained by the current ecological or social conditions, but rather by circumstances during the Pleistocene Epoch. The observed species differences might also reflect divergent behavioural predispositions, rather than actual differences in cognitive abilities.

Keywords
bonobo, chimpanzee, *Pan paniscus*, *Pan troglodytes*, tool use, environmental condition.

1. Introduction

Comparisons of bonobo (*Pan paniscus*) and chimpanzee (*Pan troglodytes*) lifestyles have revealed several intriguing differences between these closely related apes (Kano, 1992; Fruth et al., 1999). These allopatric species reside in the equatorial forests of Africa, separated by the Congo River. They both live in multi-male multi-female groups with male philopatry and a fission-fusion social structure (Kano, 1982a). One of the most conspicuous behavioural differences between these apes is the rarity of tool use among wild bonobos in comparison to chimpanzees who are one of the most prolific and skilled tool users in the animal kingdom (McGrew, 1992; Sanz & Morgan, 2007; Shumaker et al., 2011). An increased knowledge of the ecology and sociality of wild bonobos and the central subspecies of chimpanzee (*P. t. troglodytes*) now makes it possible to examine the specific environmental and social factors which may elicit tool use among chimpanzees and bonobos. Such a comparison is particularly informative because chimpanzees inhabiting the Congo Basin exist within habitat types that are more similar to bonobos than other chimpanzee subspecies residing at the edges of the species range. In this study, we compare tool use behaviours and potential drivers of these behaviours in the Wamba bonobo population located in central Democratic Republic of Congo with the Goualougo chimpanzee population of northern Republic of Congo.

Despite years of field research at multiple study sites, there are few reports of tool use by bonobos in natural settings (Kano, 1982b; Ingmanson, 1996; Hohmann & Fruth, 2003a). As we will show in this paper, the species repertoire is comprised of 13 tool use behaviours, compared to more than forty for chimpanzees (Sanz & Morgan, 2007). In striking contrast to chimpanzees, there are no reports of bonobos using tools in a feeding context except using a leaf sponge to drink water. This is despite the fact that bonobos are capable of using tools to obtain food, as shown by reports from captive populations (Jordan, 1982; Gold, 2002; Gruber et al., 2010; Shumaker et al., 2011; Boose et al., 2013).

Ecological and social factors are typically examined in isolation, but it is likely that a combination of these factors shape the emergence and maintenance of tool use in natural settings. Recent studies addressing the drivers of tool use cite some iteration of the Necessity and Opportunity Hypotheses, which state that tool use is a behavioural response to the absolute abundance of ecological resources or opportunities in the environment (Spagnoletti et al., 2012; Koops et al., 2013; Sanz et al., 2013b). More specifically, the Necessity Hypothesis posits that tool use is a response to resource scarcity which drives its practitioners to exploit novel food items during times of low resource availability (Fox et al., 1999). The Opportunity Hypothesis proposes that repeated exposure to appropriate conditions, such as encounters with target items and availability of tool materials, prompts the emergence and/or maintenance of tool use behaviours (Fox et al., 1999). Shifting the emphasis away from absolute abundance, the Relative Profitability Hypothesis suggests that tool-assisted feeding strategies targeted at embedded food items would be expected when it is more profitable than conventional methods of gathering more accessible foods (Rutz et al., 2010; Rutz & St Clair, 2012). Representing the social forces in the evolution of tool use, the Invention Hypothesis states that "behaviours such as tool use are rarely invented, and that the spread and maintenance of such behaviours requires sufficient opportunities for observational learning via social proximity to conspecifics" (Fox et al., 2004: p. 163). Recent research in both wild and captive settings has led to increasing recognition of the role of social transmission in maintaining technological traditions (Lonsdorf, 2006; Whiten et al., 2007).

The aim of this study is to review differences in bonobo and chimpanzee tool use repertoires and assess to what extent these differences can be attributed to necessity, opportunity, relative profitability, or invention (Table 1). We address these hypotheses using behavioural observations and ecological data from bonobos at Wamba in Democratic Republic of Congo and chimpanzees of the Goualougo Triangle in Republic of Congo. (1) The absence

Table 1.
Hypotheses to explain the difference in tool use for foraging.

Hypothesis	Prediction	Data to be examined	Results
(1) Necessity hypothesis	Preferred foods are more abundant at Wamba	Total basal area of trees per unit area	Not supported (basal area was similar between two sites)
	Bonobos at Wamba experience a lesser degree of seasonal resource scarcity	Pattern and degree of fluctuation of fruit availability Feeding behaviour during the season of fruit scarcity	Not fully supported (lesser degree of fluctuation of fruit availability, but the difference was small; feeding behaviour during fruit scarcity was similar).
(2) Opportunity hypothesis	Tool use opportunities are absent or far less abundant at Wamba	Availability and abundance of foods potentially eaten with tools	Not fully supported (similar foods potentially eaten with tools existed in both sites, but species level identification or quantitative data for comparison were not available).
(3) Relative profitability hypothesis	Energetic gains of tool-assisted strategies exceed that of conventional foraging in Goualougo	Per-unit-time energetic gain of tool-assisted and conventional feeding behaviours	Not examined in this study.
(4) Invention hypothesis	Parties of chimpanzees in Goualougo are larger than those of bonobos at Wamba	Mean party size	Not supported (party size in Goualougo was smaller than that at Wamba).

of tool use for feeding by bonobos would be explained by the Necessity Hypothesis if their preferred foods are more abundant and if they experience a lesser degree of seasonal resource scarcity than chimpanzees. (2) To explain the absence of feeding tool use by bonobos, one might also postulate that tool use opportunities are absent or far less abundant in bonobo habitats than chimpanzee habitats, which would support the Opportunity Hypothesis. (3) The Relative Profitability Hypothesis predicts that the energetic gains of tool-assisted strategies exceed that of conventional feeding. (4) With regard to the Invention Hypothesis, one could predict that chimpanzee parties are larger and thus provide more opportunities for social transmission of technological information among independent individuals. In this manuscript, we also summarize and propose additional possible explanations and evolutionary scenarios for the differences in tool use between *Pan* species.

2. Methods

2.1. Wamba, Democratic Republic of the Congo

The Wamba study area is located in the northern section of the Luo Scientific Reserve (22°34′E, 0°01′N), Democratic Republic of the Congo. This area is included in the Congo-equatorial climatic zone. The main habitat type is primary forest, including some areas of old secondary forest. This habitat type occurs on terra firma and is characterized by species of the Caesalpiniaceae family, with some narrow bands of monodominant *Gilbertiodendron dewevrei* forest occurring along the boundaries of swamp forests. The second main habitat type is swamp forest that exists along the Luo River and its tributaries. Most of the swamp forest is inundated all year, but a part of it near the terra firma is seasonally inundated. This habitat type is characterized by a complex mosaic of species, mainly representing the Caesalpiniaceae and Euphorbiaceae families. The northern section of the Luo Scientific Reserve contains homesteads, and so this area also includes agricultural complexes and young secondary forest (Idani et al., 1994; Hashimoto et al., 1998).

To monitor fruit abundance, we used five line transects and reconnaissance paths, the total length of which was 22.550 km (Mulavwa et al., 2008). Each trail was surveyed twice a month. We recorded the number of clusters of fallen fruit that were found within 1 m on each side of the trail, the number of fruits in each cluster, species of fruit, and whether they were ripe or unripe. The validity of this approach was established by Furuichi et al. (2001) who showed that the abundance of chimpanzee fruit foods at Kalinzu in Uganda was sufficiently evaluated by the number of clusters of ripe fallen fruits. Furthermore, Mulavwa et al. (2008) compared the numbers of ripe-fruit clusters of all species and the food species of bonobos at Wamba, and found that these numbers fluctuated proportionately. Therefore, in this study, we evaluated abundance of fruit foods based on the number of clusters of ripe fallen fruits of all species per km of transect.

A main study group of bonobos, group E, was habituated to the presence of researchers and has been observed since 1976. The group has subsequently split into two groups, E1 and E2. Another group, P, was also habituated and has been observed since the 1980's. Observations were made using artificial provisioning during some part of each year until 1996, but currently groups E1 and P are being observed from nest to nest under the natural conditions (Kano, 1992; Furuichi et al., 2012). In November 2012, group E1 consisted of 31 individuals, including 7 adult males and 9 adult females, and an eastern subgroup of P that is also continuously monitored by researchers consisted of 26 individuals, including 5 adult males and 7 adult females.

Tool use behaviours have been documented through direct observation throughout the study period (Kano, 1982b; Ingmanson, 1996). In this study, we report previously unpublished observations of tool use behaviours by bonobos at Wamba that were made before the end of 2012.

We employed the definition of the 1-h party size proposed by Hashimoto et al. (2001) for evaluating party sizes of chimpanzees. While following a party, we recorded the names of all bonobos in sight at the beginning of each hour and continued recording those bonobos that appeared in the party throughout the rest of the hour. Thus, the 1-h party represents the minimum number of bonobos that were present in the party during each 1-h observation. We also recorded the observation time in minutes in each 1-h segment. We obtained the daily 1-h party size by averaging all those observations recorded in a day, excluding those with less than 30 min of observation.

2.2. *Goualougo Triangle, Republic of Congo*

The Goualougo Triangle study area is located along the southern boundary of the Nouabalé-Ndoki National Park (16°51′–16°56′E, 2°05′–3°03′E), Republic of Congo. The climate in northern Republic of Congo can be described as transitional between the Congo-equatorial and sub-equatorial climatic zones. The lowland tropical forests of northern Congo are part of the regional centre of endemism Guinea–Congolian that ranges from Nigeria to the Congo Basin (White, 1986). The different habitat types in the Goualougo Triangle have been assessed by ground surveys and satellite imagery classification (Devos et al., 2008). The main habitat types are mixed-species forest (semi-evergreen forest with high heterogeneity of species composition and occurs on terra firma), monodominant *G. dewevrei* forest (single-species formation of *G. dewevrei* which occurs along watercourses as well as on interfluvial plateaus), and gallery/swamp forest (consists of diverse flora associated with watercourses, that may be permanently or seasonally inundated).

Relative abundance of preferred chimpanzee foods was systematically assessed through establishment of two trail networks to monitor the phenological states of tree species and strangler figs known to be consumed by apes (Chapman et al., 1994). A total of 607 trees representing 47 species were monitored each month in the Goualougo Triangle. The ground under each monitored tree was systematically surveyed to determine the abundance of mature fruit fall. Relative abundance of fruit was estimated on a scale of 0 to 4, with higher scores indicating more fruit. In this study, we evaluated abundance of fruit foods based on the proportion of monitored trees with mature fruit fall.

Direct observations of the chimpanzees in the Goualougo Triangle have been ongoing since February, 1999. Individual chimpanzees were identified from their distinct physical characteristics and these data compiled in a population history database. The main study group is the Moto community which consisted of 71 individuals, including 17 adult males and 24 adult females (Morgan, 2007).

Tool use behaviours have been documented through direct observation during reconnaissance surveys of chimpanzees since the initiation of research at this site. In 2003, we also began remote video monitoring of tool use sites. For all observations of tool use, observers record the identification of the chimpanzee, type of object used, target of object, actions, context and/or goal of the tool use behaviour, and the outcome.

20-min group scans of party composition and behaviour were recorded during chimpanzee encounters. Chimpanzee parties were considered to be all individuals travelling, feeding, resting or socializing within 50 m of one another (definition adopted from (Wrangham et al., 1992; Wilson et al., 2001)). For this study, we included information from the first scan conducted each hour so as to provide a direct comparison with the 1-h sampling of bonobo party size at Wamba. Daily values of party size were based on average sizes of all recorded 20-min group scans per day, excluding those with only orphans present.

3. Results

3.1. Habitat types and climate

Bonobos and chimpanzees inhabit a wide range of habitats across equatorial Africa (Table 2). The habitat types of bonobos range from savannah with patches of forest with 1666–1778 mm annual rainfall at Lukuru to rain forest with 2733 mm rainfall at Wamba. On the other hand, the habitat types of chimpanzees range from grassland and woodland with 954 mm rainfall at Assirik to rain forest with 3244 mm rainfall at Seringbara, including semi-evergreen forest with 1690 mm rainfall in the Goualougo Triangle. Thus, the habitat conditions of bonobos and chimpanzees largely overlap with an exception of the driest extreme in chimpanzees.

Habitat classification of satellite imagery showed that the Wamba forest is comprised mainly of primary and old secondary forest dominated by representatives of the Caesalpiniaceae family (65.5%), with swamp forest (18.9%) and young secondary forest (15.6%) representing a smaller proportion of the bonobo range (Hashimoto et al., 1998; Terada et al., unpublished data). Mixed species semi-evergreen forest (71.5%) was found to be the dominant habitat in the Goualougo Triangle, with monodominant *Gilbertiodendron dewevrei* (Caesalpiniaceae) evergreen (21.7%) and swamp forest (6.1%) also represented (Devos et al., 2008).

3.2. Ape density, home range and habitat use

Ape densities were nearly identical between the two study sites. At Wamba, bonobo densities were reported to be 1.4–2.5 individuals/km^2 (excluding infants) based on home range estimates (Hashimoto et al., 1998). Chimpanzee density in the Goualougo Triangle was estimated to be 1.5 chimpanzees/km^2 (excluding infants) from line transect surveys and 2.2 chimpanzees based on home range estimates (Morgan et al., 2006). The home ranges of bonobo communities at Wamba were estimated to be 12.3–17.8 km^2 for the E1 study group and 22.5–31.5 km^2 for the E2 study group (Hashimoto et al., 1998). The Moto chimpanzee community range was estimated to be 17.3 to 19.2 km^2 (Morgan et al., 2006). The home ranges of both species were heterogeneous in habitat composition, but with some evidence of preference for dry forest habitats by both bonobos and chimpanzees. Both direct observations of habituated groups and nest surveys indicated that use of swamps by bonobos may exceed visitation to inundated habitats by chimpanzees (Hashimoto et al., 1998; Morgan et al., 2006; Mulavwa et al., 2010).

Table 2.
Environmental conditions of study sites.

Species	Site	Country	Habitat type	Altitude (m)	Annual mean rainfall (mm)	Days of rain	No. of dry months (<100 mm/month)	Temperature (°C)		Source
								Mean max.	Mean min.	
Pan paniscus	Lac Tumba-Lac Mai Ndombe hinterland	DRC	Mixed mature swampy, mixed mature seasonally inundated, mixed mature terra firma, old-secondary, and young secondary forest, circular patch of a forest–savanna mosaic	300–700	Around 1770 in the northern part/1500–1600 in the southern part		Minor dry seasons in February and July			Inogwabini et al. (2008); Serckx et al. (2014)
	Lui Kotale	DRC	Primary evergreen lowland tropical rain forest		above 2000		Short dry season in Feb, longer one between May–Aug	28	21	Beaune et al. (2012, 2013); Hohmann & Fruth (2003b)
	Lomako	DRC	Primary climax, evergreen, polyspectific rain-forest, with smaller areas of secondary growth, slope forest and swamp forest	390	1960		No conspicuous dry season	27.3	21.2	White (1992); Myers Thompson (2002)

Table 2.
(Continued.)

Species	Site	Country	Habitat type	Altitude (m)	Annual mean rainfall (mm)	Days of rain	No. of dry months (<100 mm/month)	Temperature (°C)		Source
								Mean max.	Mean min.	
	Wamba	DRC	Predominant primary and old secondary, with young secondary, agricultural fields, and swamp	around 390	2733 ± 283 (in 2004–2006, 2008, 2010–2012)		1 to 2 months mostly around December and January	Monthly mean 31 ± 1.1	Monthly mean 21.1 ± 0.4	Furuichi et al., unpublished data for 2004–2012
	Lukuru	DRC	Dry forest and grassland	564	long-term data (1941–1963) at proximate sites: 1666 (Ilebo), 1778 (Lodja)			35.5	19.5	Myers Thompson (2002)
Pan troglodytes verus	Assirik	Senegal	Grassland and woodland with small amounts of riverine forest and bamboo thicket	100–311	954 (824–1224) (in 1976–1979)	84	7	35	23	McGrew et al. (1981); Hunt & McGrew (2002)

Table 2.
(Continued.)

Species	Site	Country	Habitat type	Altitude (m)	Annual mean rainfall (mm)	Days of rain	No. of dry months (<100 mm/month)	Temperature (°C)		Source
								Mean max.	Mean min.	
	Seringbara	Guinea	Primary tropical forest, montane forest, riverine forest, high-altitude grasslands	600–1752	3244 (in 2008)		4	24.5	18.7	Koops (2011); Koops et al. (2012, 2013)
	Bossou	Guinea	Mosaic of primary, secondary, riverine, and scrub forest, as well as farmland	550 (village)	2230 (in 1995)		4	26.1 (1.5 m high in the forest)	21.0 (1.5 m high in the forest)	Yamakoshi, 1998; Takemoto, 2004
	Taï	Côte d'Ivoire	Evergreen lowland rain forest (with only the swamp forest along the rivulet less densely forested)		1829		3 in Figure 1.3 (Boesch & Boesch, 2000)	30	24	Boesch & Boesch (2000); Hunt & McGrew (2002)
Pan troglodytes troglodytes	Lopé	Gabon	Tropical rain forest		1531					Hunt & McGrew (2002)

Table 2.
(Continued.)

Species	Site	Country	Habitat type	Altitude (m)	Annual mean rainfall (mm)	Days of rain	No. of dry months (<100 mm/month)	Temperature (°C)		Source
								Mean max.	Mean min.	
	Noua-bale-Ndoki	Congo	Swamp forest, *Gilbertiodendron* forest, and mixed species forest		1430		December–February			Nishihara (1995)
	Goua-lougo Triangle	Congo	Predominant mixed species semi-evergreen forest, with monodiminant *Gilbertiondendron dewevrei* evergreen and swamp forest		1690 ± 27 (in 2007–2011)		2–4 months around December and January	23.7 ± 0.8	21.5 ± 0.6	Morgan et al. (2006); Sanz & Morgan (2013b)
Pan troglodytes schweinfurthii	Kahuzi-Biega	DRC	Bamboo forest (37%), primary montane forest (28%), secondary montane forest (20%), swamp (7%)	2050–2350	1800					Yamagiwa et al. (1996); Hunt & McGrew (2002)

Table 2.
(Continued.)

Species	Site	Country	Habitat type	Altitude (m)	Annual mean rainfall (mm)	Days of rain	No. of dry months (<100 mm/month)	Temperature (°C)		Source
								Mean max.	Mean min.	
	Kalinzu	Uganda	Medium altitude moist evergreen forest	1000–1500	1584 (years: 1997–1998)		6 months (Jan–Mar, Jun–Aug)	25–28	14–17	Howerd (1991); Hashimoto et al. (1999)
	Kibale	Uganda	Moist semi-deciduous forest with grassland, woodland and swamp forest	1110–1590/ 1470–1750 (Kanyawara)	1475 (in 1970–1991), 1749 (in 1990–2001) (Kanyawara); 1393 ± 200 (in 1998–2009) (Ngogo)			23	16	Hunt & McGrew (2002); Watts (2012)
	Budongo	Uganda	Moist, semi-deciduous tropical rain forest and grassland	1100 (average)	1600 (range 1240–2187, in 1993–2000)		between mid-December and mid-February (50 mm or less per month); 4 months with <100 mm	32	19	Wallis (2002); Reynolds (2005)

Table 2.
(Continued.)

Species	Site	Country	Habitat type	Altitude (m)	Annual mean rainfall (mm)	Days of rain	No. of dry months (<100 mm/month)	Temperature (°C)		Source
								Mean max.	Mean min.	
	Semliki	Uganda	Dry *Combretum ghasalense* savanna and *Borassus aethiopum* palm savannah with gallery forest		1206	151	5	34	20	Hunt & McGrew (2002); Samson & Hunt (2012)
	Gombe	Tanzania	Alternating thick riverine forests, deciduous woodland, and hilltop grassland		1775	152	6 (dry season from May to October)	28	19	Hunt & McGrew (2002); Wallis (2002)
	Mahale	Tanzania	Semi-deciduous savannah woodland with montane forest and continuous gallery forest	780–1300	1836	143	5	29	18	Nishida (1990); Takasaki et al. (1990)

Table 2.
(Continued.)

Species	Site	Country	Habitat type	Altitude (m)	Annual mean rainfall (mm)	Days of rain	No. of dry months (<100 mm/ month)	Temperature (°C)		Source
								Mean max.	Mean min.	
	Kasakati	Tanzania	Woodland (59%), savannah (16%), grassland (16%), forest (10%)	750	962 (in 1941–1970 (Kigoma)		6 months of <60 mm (1941–1970)	28 (1941–1970 at Kigoma)	19 (1941–1970 at Kigoma)	McGrew et al. (1981); Moore (1992)
	Ugalla	Tanzania	Woodland (91%), swamp (5%), bamboo (3%)	900–1800	1012.3 ± 139.1 (in 1973–1988) (Uvinza); 955 (Issa)		5.2 ± 0.8 months with <60 mm (1973–1988); May–Oct with <100 mm	Average daily max: 34 (Aug)–28 (Nov)	Average daily max: 17.2 (Jan)–14.4 (Aug)	McGrew et al. (1981); Moore (1992); Hernandez-Aguilar (2009)

3.3. *Tool use repertoire*

The bonobo tool use repertoire was comprised of 13 different types of tools (Table 3), with 10 types of tool use exhibited by the Wamba population.

Table 3.
Compilation of tool using behaviours observed in wild bonobos.

Context	Behaviour	Code	Description	Sex of performer	Observed sites	
					Wamba	Lomako
Social	Play start	2*	Taking a branch, small leafy twig or fruit in hand or mouth and chasing of others in a play context. Possession of object sometimes changes (sharing the same object)	M, F	+	+
	Drag branch	3*	Dragging a branch toward one or more target individual(s) in threat. Targets sometimes ignore this agonistic approach. Dragging a branch before departure or during ranging without targeting conspecifics. Usually performed by males to propose direction of ranging.	M, F	+	+
	Drop twigs	–	Clipping twigs or small branches and dropping them from tree to get attention from opposite sex (may be to solicit copulation)	M, F**	+	–
	Aimed-throw	49	Throwing sticks and branches at human observers and at other bonobos. Bonobos at Wamba just drop twigs or branches to human observers.	M, F	+	+
	Leaf-clip mouth	53	Clipping leaves from herbs or trees by mouth to get attention from play partner.	M, F	–	+

Table 3.
(Continued.)

Context	Behaviour	Code	Description	Sex of performer	Observed sites	
					Wamba	Lomako
	Leaf-clip fingers	54	Clipping leaves from herbs or trees by finger to solicit mates.	F	−	+
Comfort, protection	Leaf-umbrella or rain-hat	–	Detaching small branches or twigs and put them on their head under heavy rain	M, F	+	−
	Leaf cover	–	Detaching small branch or twig and using it as a cover over their nest. Common during the rainy season, possibly for thermoregulation. At Wamba one adult male bent a small branch to cover his body on his day nest	M, F	+	+
	Fly-whisk	47	Swatting or whisking away sweat bees by waving and shaking small, leafy twigs	F	+	+
Self-groom, stimulate	Leaf-napkin	50	Using leaves to wipe faeces or urine from one's own body	M, F	+	−
	Toothpick	–	After feeding, using a small twig to remove debris from between teeth	M	+	−
	Stick scratch	–	Scratching one's own back with small twig	M, F	+	−
Feeding	Leaf-sponge	4*	Dipping water from a tree hole using moss. Seemed to be equivalent of leaf sponge behavior by chimpanzees	F	−	+

Contexts refer to those in Sanz & Morgan (2007). Names and code numbers of behaviours refer to those in Whiten et al. (1999), except for what are not observed in chimpanzees (marked –). Source of information for Wamba: Kano (1982b, 1997); Ingmanson (1996); this study; for Lomako: Hohmann & Fruth (2003a).

* Universal chimpanzee tool use behaviours (Whiten et al., 1999).

** Performance of this behaviour by females was first reported by this study.

Table 4.
Types of tool behaviours observed in chimpanzees and bonobos.

	Number of all types	Number of types at each site	Context				
			Social	Comfort, protection	Self-groom, stimulate	Feeding	Other
Chimpanzee	42	6 (Assirik)–22 (Goualougo)	5	4	5	25	3
Bonobo	13	8 (Lomako)–10 (Wamba)	6	3	3	1	0

Data for chimpanzees from Sanz & Morgan (2007).

Chimpanzees in the Goualougo Triangle exhibited 22 different types of tool use, which is approximately half of the species repertoire (Sanz & Morgan, 2007). The number of tool behaviours found at chimpanzee study sites varied from 22 in Goualougo to 6 in Assirik. Hence, the numbers of tool behaviours at bonobos study sites (10 in Wamba and 8 in Lomako) are within the range of variation in chimpanzees (Table 4) (Sanz & Morgan, 2007). Chimpanzee tool use was most common in feeding contexts, but was also exhibited in self grooming/stimulation, comfort/protection, and social situations. Bonobo tool use mainly had social (self-grooming/stimulation and comfort/protection) functions, except for one behaviour for feeding (leaf sponge to drink water). Among 13 tool behaviours, 8 were common in chimpanzees and 5 were uniquely found in bonobos. Out of 4 chimpanzee universal tool behaviours, 3 were observed in bonobos (play start, drag branch, leaf-sponge) but 1 was not observed (investigatory probe).

A marked difference in tool behaviours between chimpanzees and bonobos is the very limited tool use for feeding in bonobos (Table 4). In chimpanzees, 25 of 42 tool behaviours were performed for feeding (Sanz & Morgan, 2007). By contrast, tool use for feeding was observed only once in Lomako, in the use of a leaf-sponge (Hohmann & Fruth, 2003a). Therefore, what we really need to examine is why wild bonobos do not use tools for feeding.

3.4. Abundance of preferred resources

Similar to other bonobo and chimpanzee populations, the diet of apes at Wamba and Goualougo was primarily comprised of ripe fruits (Kano & Mulavwa, 1984; Morgan & Sanz, 2006). As an indication of the overall abundance of food resources, we compared the total basal area of all trees with DBH greater than 5 cm in Wamba and of all trees with DBH greater than 10 cm in Goualougo. The basal area for Wamba is 33.27 m^2 per hectare for primary forest (including old secondary forest), 28.50 for swamp forest, 21.90 for young secondary forest (calculated from DBH2 in Idani et al., 1994) and that for Goualougo is 34.04 m^2 per hectare (Morgan & Sanz, 2006). Due to the difference in sampling (inclusion of a larger sample of trees than Goualougo), the per-hectare total basal area of trees at Wamba must be somewhat overestimated as compared to Goualougo. Therefore, we can conservatively conclude that the total basal area in Wamba is not greater than at Goualougo.

Though we do not have comparable data for absolute abundance of ripe fruit, we compared the pattern of seasonal fluctuation in abundance using parameters that represent proportion of trees with ripe fruit (Figure 1a). The monthly proportion of trees with ripe fruit on the ground showed similarly stochastic fluctuations between the sites, though the degree of fluctuation was greater in Goualougo (Wamba: 5.5 ± 2.0 clusters per km of transect, CV = 0.35, $N = 61$; Goualougo: 5.6 ± 3.2% of monitored trees, CV = 0.56, $N = 67$). When fruits were scarce, bonobos at Wamba increased their intake of seeds, leaves, and terrestrial herbs (Kano & Mulavwa, 1984), and chimpanzees in the Goualougo Triangle similarly compensated by increasing their intake of leaves (Morgan & Sanz, 2006; Sanz & Morgan, 2013b).

Overall, although the degree of fluctuation in availability of ripe fruit was greater in Goualougo, similar per-hectare tree basal areas and similar tendencies of food shift during fruit scarcity, together with similar density of bonobos and chimpanzees mentioned in the previous section, suggests that the differences in the food conditions between Wamba and Goualougo were much smaller than differences among sites of each species (Table 2).

3.5. Ecological opportunities for feeding with tools

In the Goualougo Triangle, tools are used to harvest several species of termites (*Macrotermes muelleri* and *M. lilljeborgi*), army ants (*Dorylus mayri*, *D. rubellus*, *D. sjoestedti* and *D. wilverthi*), and the honey of stingless bees (*Trigona* spp.). Although we have not yet assessed the availability and distribution of these different insect species at Wamba, we confirmed that these insects (or their products) are present. Seemingly, there are ample ecological opportunities (i.e., termites, army ants, bees and suitable raw materials)

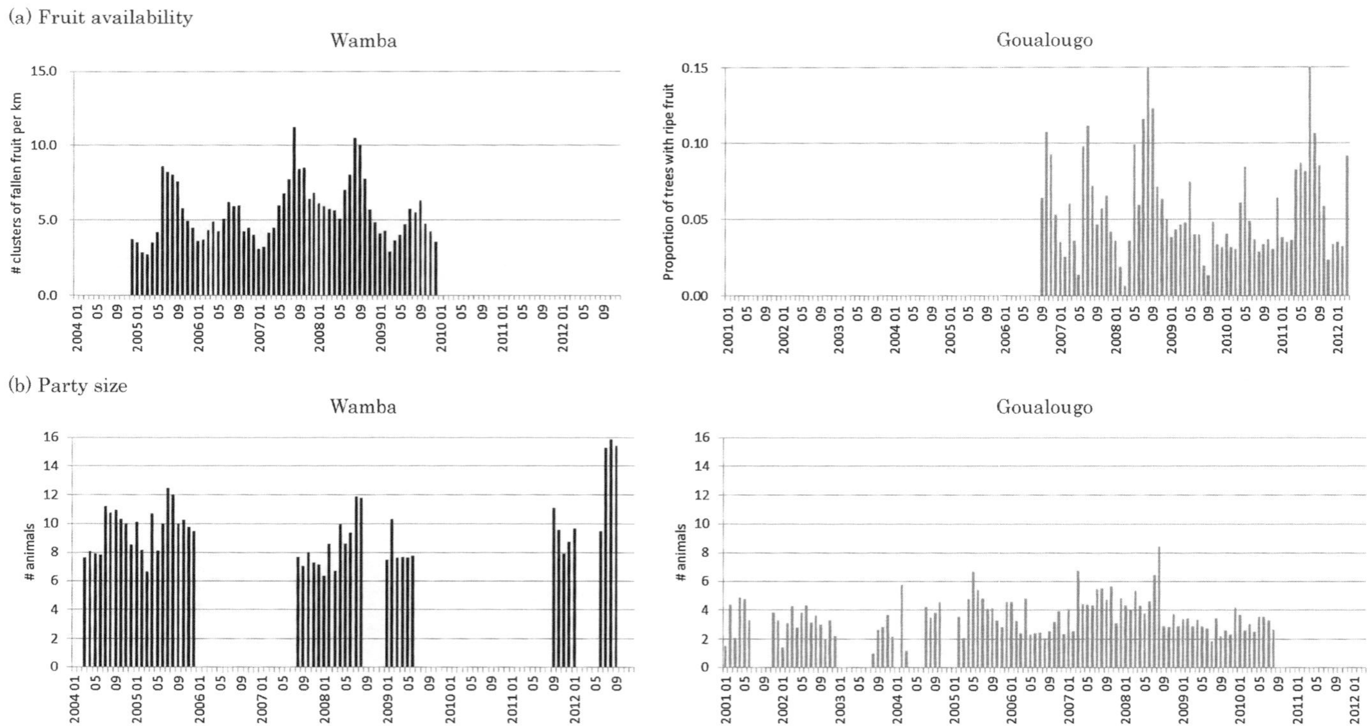

Figure 1. Comparison of ecological variables at Wamba and Goualougo. (a) Seasonal fluctuation in availability of ripe fruit resources at Wamba and Goualougo. Values for Wamba show the number of clusters of ripe fallen fruit per km of transect. Values for Goualougo show proportion of food trees with ripe fruit on the ground. (b) Monthly mean of daily party sizes for the E1 group of bonobos at Wamba and chimpanzees at Goualougo. Party size included adults, adolescents, and juveniles, but excluded infants.

at Wamba for termite fishing, ant dipping, ant fishing, and honey gathering tool use, which is similar to research findings at Lui Kotale (McGrew et al., 2007). We also confirmed the presence of oil palms (*Elaeis guineensis*) at Wamba which are the target of pestle pounding and nut cracking by chimpanzees at Bossou and other sites (Humle & Matsuzawa, 2004).

3.6. Opportunities for social transmission of technical information

As shown in Figure 1b, bonobos at Wamba consistently gathered in larger parties than chimpanzees in Goualougo (monthly mean party size at Wamba: 9.4 ± 2.2 animals, CV $= 0.23$, $N = 50$; Goualougo: 3.5 ± 1.3 animals, CV $= 0.36$, $N = 97$). There was a significant difference in party size if we compared those in the periods for which data was available for both sites (matched-pair t-test, $t = -15.1$, df $= 38$, $p < 0.0001$). This difference in party size was further corroborated by mean nest group size of 9.3 ± 4.9 ($N = 215$, range 1, 24) bonobo nests in Wamba (Mulavwa et al., 2010) versus 2.75 ± 1.88 ($N = 375$, range: 1, 12) chimpanzee nests in Goualougo (Morgan et al., 2006).

3.7. Correlations among examined factors

We examined potential relationships among factors that may influence tool use behaviours: fruit abundance and party size in each site. For Wamba, there was no significant correlation between these factors ($r = 0.31$, n.s.). On the other hand for Goualougo, significant correlation was found between these factors ($r = 0.40$, $n = 48$, $p < 0.01$).

4. Discussion

The aim of this study was to review differences in chimpanzee and bonobo tool use and examine whether these differences could be attributed to current hypotheses outlining various ecological and social factors suggested to be responsible for the emergence and maintenance of tool traditions (Table 1). Our comparison of tool behaviours between bonobos and chimpanzees revealed that the main difference between the two species exists in their repertoires of tool use for feeding, which is large and diverse in chimpanzees and nearly absent in bonobos. On the other hand, our comparisons of ecological and social conditions between Wamba and Goualougo, and comparisons of habitat conditions across various *Pan* study sites, suggested that it is difficult to explain the differences in tool use repertoires between the two species based on current ecological or social conditions. Our review of current evidence for *Pan* tool use led us to conclude that understanding the differences in the expression of tool use between chimpanzees and bonobos will require alternative ecological, behavioural, or social explanations.

In the reported difference in tool use between wild bonobos and chimpanzees, we may have overemphasized the contrast or false dichotomy between the two *Pan* species (Stanford, 1998; Fruth et al., 1999). The size of the Wamba tool repertoire was smaller than that of chimpanzees in Goualougo, but still within the range of the number of tools used by each chimpanzee (Sanz & Morgan, 2007). Although we did not quantify the frequency of tool use in this comparison, our impression is that tool use was less frequently exhibited by bonobos than chimpanzees. This species difference is largely due to the fact that wild bonobos do not use tools in feeding, except for an observation of tool use for drinking water by bonobos in Lomako (Hohmann & Fruth, 2003a). In Goualougo and in other chimpanzee populations, tool-assisted feeding occurs on a regular basis and can occupy a significant portion of the daily activity budget (Pandolfi et al., 2003; Bogart & Pruetz, 2011). Another difference is the absence of investigatory probing in bonobos which is a universal behaviour among chimpanzee populations (Whiten et al., 2001). Many of the behavioural elements involved in investigatory probing can be generalized to tool-assisted feeding situations that involve probes or dipping implements, such as termite fishing, ant dipping, or fluid dipping (Sanz & Morgan, 2010). It has been well-documented that bonobos have the physical and cognitive abilities to exhibit such tool behaviours (Gruber et al., 2010; Boose et al., 2013), and so there may exist other factors which elicit these tool use behaviours in wild chimpanzees, but not in bonobos.

For the Necessity Hypothesis to explain the difference in bonobo and chimpanzee tool use, we would expect to find that bonobos have a more stable food resource base than chimpanzees. This would effectively alleviate the need for tool-assisted fallback strategies. In support of this, chimpanzee tool use was negatively correlated with abundance of ripe fruit at Bossou in Guinea (Yamakoshi, 1998). In environments with few typically preferred foods, tool use may also be a strategy to harvest staple food items. This may be the case for some types of chimpanzee tool use at savannah sites, such as the termite gathering of chimpanzees at Fongoli in Senegal (Bogart &

Pruetz, 2011) and possibly the tuber harvesting by chimpanzees at Ugalla in Tanzania (Hernandez-Aguilar et al., 2007).

However, a recent review of studies that have explored the ecological dimension of the presence or absence of technology clearly shows that necessity may play a less prominent role in prompting and promoting tool use than previously suggested (Sanz & Morgan, 2013b). The Necessity Hypothesis was explicitly tested and a lack of support was cited for nut cracking tool use by capuchins at Boa Vista in Brazil (Spagnoletti et al., 2012), tool use to extract insects from tree holes and extract seeds from Neesia fruit by orangutans across Borneo and Sumatra (Fox et al., 2004), and various forms of chimpanzee tool use at Seringbara in Guinea (Koops et al., 2013). The Goualougo chimpanzee population also does not seem to compensate for the lack of fruit resources by increasing their frequency of tool use for social insects or honey (Sanz & Morgan, 2013b). Rather, opportunities to gather termites, ants, and honey were available throughout the year to this chimpanzee population and enhanced by the use of tool sets (Sanz et al., 2004, 2010; Sanz & Morgan, 2009). Profiles of tool use at the savannah sites of Assirik in Senegal and Ugalla in Tanzania did not fit the traditional predictions of the Necessity Hypothesis in that tool use did not increase during periods of food scarcity (Hernandez-Aguilar et al., 2007; Bogart & Pruetz, 2011), but rather may be a necessary response to the lower overall abundance of preferred resources in arid habitats.

Our comparison between Wamba and Goualougo revealed that availability of ripe fruit, the main food of bonobos and chimpanzees, showed similar patterns of seasonal fluctuation, and therefore did not support the Necessity Hypothesis to explain the *Pan* difference in tool repertoires for feeding. The extent of fluctuation was somewhat greater in Goualougo, suggesting that Goualougo chimpanzees might experience more severe periods of fruit scarcity. However, Sanz & Morgan (2013b) reported that frequency of tool use for feeding by chimpanzees in Goualougo was not related to the proportion of fruiting trees, suggesting that this degree of difference in seasonal fluctuation may not sufficiently explain the marked difference in tool use for feeding between bonobos at Wamba and chimpanzees at Goualougo. During times of fruit scarcity, both apes increased their consumption of terrestrial herbaceous vegetation and leaves which fit the profile of traditional fallback foods (Marshall & Wrangham, 2007). Recent research also showed that there is no substantial difference in the kind and use of fallback food between chimpanzees and bonobos (Harrison & Marshall, 2011). Furthermore, unpublished data by KK and DM showed that the density of terrestrial herbs was not higher in Wamba than in Goualougo (Wamba: 230 plots of 2 × 2 m, 1.51 ± 2.10 stems of Marantaceae and Zingiberaceae/m^2; Goualougo: 7 plots of 5 × 50 m, 2.08 ± 0.74 stems of herbs/m^2). Hence, it is unlikely that the availability of fallback foods during fruit scarcity explains the lack of tool use for feeding in bonobos.

According to the Opportunity Hypothesis, tool use is related to the frequency of encounters with particular tool targets (termites, ants, honey, etc.) or tool materials that may vary seasonally or between sites. For example, a higher abundance of arboreal insects provided increased opportunities for orangutans to invent tool use at Suaq Balimbing compared to other sites (Fox et al., 2004). Termite gathering by chimpanzees at Gombe is also thought to be opportunistic, as it occurs during the rainy season when termites are more accessible (McGrew et al., 1979; McGrew & Collins, 1985). In contrast, termite fishing has been documented throughout the year at several sites within central Africa (Sabater Pi, 1974, 1979; McGrew et al., 1979; Suzuki et al., 1995; Deblauwe, 2009). Termite mounds (*Macrotermes*) are rare and peripheral to the chimpanzee range at Seringbara in Guinea and no evidence of tool use in termite predation has been detected within this population (Koops et al., 2013). However, chimpanzees at Seringbara use tools to harvest army ants, which are both abundant and widespread across the area (Koops et al., 2013).

Our study revealed that habitat type, vegetation, and seasonal variation in climate at Wamba and Goualougo are fairly similar, and we did not identify any obvious differences between the sites that would preclude the possibility for bonobo tool use for feeding. At the bonobo study site Lui Kotal, assessment of opportunities for insectivory revealed that the same tool use opportunities were present and in some cases exceeded abundances reported from chimpanzee sites (McGrew et al., 2007). A detailed study assessing the availability of army ants, termites, nut producing trees and potential tool materials at Wamba is currently underway. Furthermore, the range of habitat types of chimpanzees and bonobos overlap almost completely from savannah to rain forest. Due to such large within-species variation and between-species overlap, it is difficult for the Opportunity Hypothesis to explain presence of tool use for feeding in all studied chimpanzee populations and almost complete absence of it in wild bonobos.

The Relative Profitability Hypothesis suggests that tools will be used to harvest embedded food items when the energetic benefits outweigh the gains from conventional feeding of more easily accessible food items (Rutz et al., 2010; Rutz & St Clair, 2012). There are several indications that the energetic benefits of tool use exceed those of conventional feeding in birds (Tebbich et al., 2002; Rutz et al., 2010). Nut cracking by chimpanzees has also been shown to be an energetically profitable behaviour, which can yield several thousand calories per day (Gunther & Boesch, 1993). Contradictory to this Relative Profitability Hypothesis, past research has shown that some forms of chimpanzee tool use may not necessarily be the most efficient means of calorie or protein intake. For example, a detailed review of ant-fishing behaviour among chimpanzees at Mahale revealed negligible nutritional gain from ant-fishing which was suggested to be a 'leisure' activity (Nishie, 2011). As mentioned above, bonobo tool use occurred mostly in the social and self-directed contexts, and therefore we had no information to examine the profitability of tool use for feeding in bonobos. A more comprehensive evaluation of bonobo diet and nutritional intake may reveal whether or not conventional feeding is more profitable for this species than employing tool-assisted strategies.

With regard to the Invention Hypothesis, it is predicted that more frequent or complex tool use will occur in settings with enhanced opportunities for social transmission. This hypothesis would explain the absence of tool use in bonobos if bonobos gather in smaller parties or have weaker social relations within their groups compared to chimpanzees. Our comparison indicated, however, that bonobo parties at Wamba were larger than chimpanzee parties at Goualougo. A recent comparison of many study sites of chimpanzees and bonobos indicated that differences in party size between the species are not statistically significant but that female bonobos attend party gatherings much more frequently relative to the very low attendance ratio of female chimpanzees (Furuichi, 2009). This seems like a prime social setting for the spread of technological traditions in bonobos. Especially, since studies have reported a female bias in tool use among wild chimpanzees (McGrew, 1979; Lonsdorf, 2005) and captive bonobos (Boose et al., 2013), which has yet to be documented among wild bonobos. Importantly, it remains to be established whether increased gregariousness in bonobos also reflects an increase in close-range social learning opportunities. Moreover, we need to assess how levels of gregariousness and party composition, and thus social learning opportunities, vary across feeding and non-feeding contexts in both chimpanzees and bonobos. Priority of access to food resources exhibited by bonobo females (Furuichi, 1997; Surbeck & Hohmann, 2013) may reduce the need for food gathering with tools. Furthermore, we may want to question whether the party size is indeed an important factor for transmission of tool behaviours. Studies from several sites, including Gombe and Goualougo (Lonsdorf, 2006; Sanz & Morgan, 2013a), suggested that party sizes while termite fishing were relatively small, and the majority of parties were mothers with dependent offspring. If the transmission of tool behaviours mainly occurred between mother and offspring but not among adults, the party size may have no influence on tool behaviours even if the transmission of technological information is important for the existence of tool use.

Thus, as far as we know from the current information, differences in ecological or social conditions may fail to explain differences in bonobo and chimpanzee technological repertoires. It might be because these behaviours evolved in past ecological and social conditions that differ from contemporary settings. A recent study reported that present-day geographical features failed to explain the genetic structure of bonobos, and that current genetic diversity was formed by paleoenvironmental circumstances during the Pleistocene (Kawamoto et al., 2013). During the Pleistocene Epoch which began 2.5 million years ago, forested areas in Africa were reduced to smaller refugia during glacial periods (Mayr & Ohara, 1986; Plana, 2004). Such environmental changes became more conspicuous after 1 million years ago, which coincides with the divergence of the bonobo and chimpanzee lineages (Won & Hey, 2005). The range of ancestral chimpanzees experienced more extensive drying and fragmentation of forests during the late Pleistocene than the habitat of ancestral bonobos on the left bank of the Congo River where large forest refugia were maintained even during dry periods. Therefore, tool-assisted feeding might have had greater adaptive value for the ancestors of modern chimpanzees who were coping with greater variability in food resource availability, more arid habitats, and potentially increased competition with other species than past bonobo populations. It is also possible that feeding technology evolved independently in the different chimpanzee subspecies as they repeatedly experienced dry periods. In their examination of tool use patterns among Ugandan chimpanzee populations, Gruber et al. (2012) suggested that ancestral chimpanzee populations in the region may have coped with harsher environments during the last Ice Age which may have prompted various behavioural innovations, such as extractive tool use. A major problem exists in this explanation however. If the ability for tool use evolved only in chimpanzees under certain conditions in the past, and therefore chimpanzees currently living in all types of habitats inherently use tools for feeding while bonobos in any type of habitat do not, then why is there no substantial difference in the ability for tool use under experimental conditions? This question needs to be further investigated through comparative studies both in the wild and captivity.

Although further studies are needed, it is possible that the differences in tool using propensities between chimpanzees and bonobos in the wild reflect divergent behavioural predispositions rather than differences in cognitive abilities. Comparative studies of the physical and cognitive abilities of great apes have failed to explain the paucity of tool use by bonobos. Within captive settings, bonobo tool use repertoires are as large and diverse as chimpanzees (Gruber et al., 2010). This is unlike gorillas, who use tools for feeding but acquire the behaviour more slowly and perform it less frequently than chimpanzees, suggesting a species difference in predisposition (Boysen et al., 1999; Lonsdorf et al., 2009). Even in experimental settings where individuals had limited experience with tools, both chimpanzees and bonobos demonstrated understanding of tool functional properties (Herrmann et al., 2008).

The species differences in cognition and behavioural propensities between chimpanzees and bonobos were aptly expressed in a range of cognitive problem solving tasks. Bonobos surpassed chimpanzees in solving tasks related to the social world (i.e., theory of mind, understanding of social causality), whereas chimpanzees showed more skill in tasks related to the physical world (i.e., use of tools, understanding of physical causality) (Herrmann et al., 2010). In a comparison of several captive groups, the only major difference between chimpanzee and bonobo tool use was that bonobos of all age and sex classes used tools in a play context (Gruber et al., 2010). The propensity of adult bonobos to engage in play is thought to reflect their neotenous nature, as well as environmental conditions that afford leisure time and a behavioural preference for certain types of social interaction.

As mentioned above, some chimpanzee tool use does not seem to be the most effective means of increasing energy intake, but may be performed when individuals have leisure time (Nishie, 2011). On the other hand, bonobos spend much of their leisure time in play and play is common among adults (Enomoto, 1990; Palagi, 2006). Such differences in behavioural preference may explain why bonobos are likely to perform like chimpanzees in experimental conditions, but do not regularly perform tool use in natural settings. This potential difference in behavioural predispositions toward tool use could be further investigated by comparing levels of object manipulation and exploratory tendency in chimpanzees and bonobos (Koops et al., data not shown).

It has been proposed that we are currently living in the "Anthropocene Epoch" which is shaped by human influences on the environment. Although apes have persisted through shifting climatic condition, there is no historical analogue for the rate and degree of environmental change caused by anthropogenic disturbances. The long-term survival of wild bonobos and chimpanzees is increasingly endangered by poaching, habitat destruction and conversion, and infectious diseases. The Disturbance Hypothesis suggests that ape cultures are fragile and that anthropogenic disturbances may affect the social mechanisms which maintain these traditions (van Schaik, 2001). Recognizing the conservation value of animal cultures, scientists have suggested specific strategies for conserving and managing animals that learn socially and share cultures (Whitehead et al., 2004; Laiolo & Jovani, 2007; Whitehead, 2010). Such measures must be implemented immediately if we hope to continue advancing our understanding of behavioural diversity of our closest living relatives.

Acknowledgements

We are grateful to the Centre de Recherche en Ecologie et Foresterie and the Ministère de Recherche Scientifique (République Démocratique du Congo) for the opportunity to conduct field studies at Wamba. We especially thank I. Monkengo-mo-Mpenge, M. N. Mulavwa, and Mr. Bokasa of these organizations for their continued support. We also thank J. Yamagiwa, T. Matsuzawa, G. Idani, C. Hashimoto and other researchers of Kyoto University, Japan, for their valuable suggestions and support for our study, and to the tracking team and other staff of the Wamba research station for their dedication. Our study at Wamba was financially supported by the Grant-in-Aid for Scientific Research and Asia–Africa Science Platform Program by Japan Society of Promotion of Science, the Environment Research and Technology Development Fund by the Japan Ministry of the Environment, National

Geographic Research and Exploration, and Toyota Foundation. We are also deeply appreciative of the opportunity to work in the Nouabalé-Ndoki National Park and especially the Goualougo Triangle. This work would not be possible without the continued support of the Ministère de l'Economie Forestière et du Développement Durable (République du Congo), Ministère de la Recherche Scientifique (République du Congo), and the Wildlife Conservation Society's Congo Program. Special thanks are due to J. M. Fay, P. Telfer, P. Elkan, S. Elkan, B. Curran, M. Gately, E. Stokes, T. Breuer, P. Ngouembe and D. Dos Santos. We would also like to recognize the tireless dedication of J. R. Onononga, C. Eyana-Ayina, S. Ndolo, A. Nzeheke, W. Mayoukou, M. Meguessa, I. Singono, and the Goualougo tracking team. Grateful acknowledgment of funding is due to the U.S. Fish and Wildlife Service, National Geographic Society, and Columbus Zoological Park. We also acknowledge financial support from the McDonald Institute for Archaeological Research (Cambridge), the Lucie Burgers Foundation for Comparative Behaviour Research (Arnhem, the Netherlands), Homerton College and Newnham College (Cambridge) to K.K. Finally, the authors are grateful to Stephanie Musgrave and two anonymous reviewers for their helpful comments on an earlier version of this manuscript.

References

Beaune, D., Bollache, L., Fruth, B. & Bretagnolle, F. (2012). Bush pig (*Potamochoerus porcus*) seed predation of bush mango (*Irvingia gabonensis*) and other plant species in Democratic Republic of Congo. — Afr. J. Ecol. 50: 509-512.

Beaune, D., Bretagnolle, F., Bollache, L., Hohmann, G., Surbeck, M., Bourson, C. & Fruth, B. (2013). The bonobo–*Dialium* positive interactions: seed dispersal mutualism. — Am. J. Primatol. 75: 394-403.

Boesch, C. & Boesch-Achermann, H. (2000). The chimpanzees of the Tai forest: behavioral ecology and evolution. — Oxford University Press, New York, NY.

Bogart, S.L. & Pruetz, J.D. (2011). Insectivory of savanna chimpanzees (*Pan troglodytes verus*) at Fongoli, Senegal. — Am. J. Phys. Anthropol. 145: 11-20.

Boose, K.J., White, F.J. & Meinelt, A. (2013). Sex differences in tool use acquisition in bonobos (*Pan paniscus*). — Am. J. Primatol. 75: 917-926.

Boysen, S., Kuhlmeier, V.A., Halliday, P. & Halliday, Y.M. (1999). Tool use in captive gorillas. — In: The mentalities of gorillas and orangutans: comparative perspectives (Parker, S.T., Mitchell, R.W. & Miles, H.L., eds). Cambridge University Press, Cambridge, p. 179-187.

Chapman, C.A., Wrangham, R. & Chapman, L. (1994). Indices of habitat-wide fruit abundance in tropical forests. — Biotropica 26: 160-171.

Deblauwe, I. (2009). Temporal variation in insect-eating by chimpanzees and gorillas in Southeast Cameroon: extension of niche differentiation. — Int. J. Primatol. 30: 229-252.

Devos, C., Sanz, C., Morgan, D., Onononga, J.R., Laporte, N. & Huynen, M.C. (2008). Comparing ape densities and habitats in northern congo: surveys of sympatric gorillas and chimpanzees in the odzala and ndoki regions. — Am. J. Primatol. 70: 439-451.

Enomoto, T. (1990). Social play and sexual-behavior of the bonobo (*Pan paniscus*) with special reference to flexibility. — Primates 31: 469-480.

Fox, E.A., Sitompul, A.F. & van Schaik, C.P. (1999). Intelligent tool use in wild Sumatran orangutans. — In: The mentalities of gorillas and orangutans (Parker, S.T., Mitchell, R.W. & Miles, H.L., eds). Cambridge University Press, Cambridge, p. 99-116.

Fox, E.A., van Schaik, C.P., Sitompul, A. & Wright, D.N. (2004). Intra- and interpopulational differences in orangutan (*Pongo pygmaeus*) activity and diet: implications for the invention of tool use. — Am. J. Phys. Anthropol. 125: 162-174.

Fruth, B., Hohmann, G. & McGrew, W.C. (1999). The *Pan* species. — In: The nonhuman primates (Dolhinow, P. & Fuentes, A., eds). Mayfield Publishing, Mountain View, CA, p. 64-72.

Furuichi, T. (1997). Agonistic interactions and matrifocal dominance rank of wild bonobos (*Pan paniscus*) at Wamba. — Int. J. Primatol. 18: 855-875.

Furuichi, T. (2009). Factors underlying party size differences between chimpanzees and bonobos: a review and hypotheses for future study. — Primates 50: 197-209.

Furuichi, T., Hashimoto, C. & Tashiro, Y. (2001). Fruit availability and habitat use by chimpanzees in the Kalinzu forest, Uganda: examination of fallback foods. — Int. J. Primatol. 22: 929-945.

Furuichi, T., Idani, G., Ihobe, H., Hashimoto, C., Tashiro, Y., Sakamaki, T., Mulavwa, M.N., Yangozene, K. & Kuroda, S. (2012). Long-term studies on wild bonobos at Wamba, Luo Scientific Reserve, D. R. Congo: towards the understanding of female life history in a male-philopatric species. — In: Long-term field studies of primates (Kappeler, P.M. & Watts, D.P., eds). Springer, Berlin, p. 413-433.

Gold, K.C. (2002). Ladder use and clubbing by a bonobo (*Pan paniscus*) in Apenheul Primate Park. — Zoo Biology 21: 607-611.

Gruber, T., Clay, Z. & Zuberbuhler, K. (2010). A comparison of bonobo and chimpanzee tool use: evidence for a female bias in the *Pan* lineage. — Anim. Behav. 80: 1023-1033.

Gruber, T., Potts, K.B., Krupenye, C., Byrne, M.-R., Mackworth-Young, C., McGrew, W.C., Reynolds, V. & Zuberbuhler, K. (2012). The influence of ecology on chimpanzee (*Pan troglodytes*) cultural behavior: a case study of five Ugandan chimpanzee communities. — J. Comp. Psychol. 126: 446-457.

Gunther, M. & Boesch, C. (1993). Energetic costs of nut-cracking behavior in wild chimpanzees. — In: Evolution of hands (Chivers, D. & Preuschoft, H., eds). Gustav Fischer, Stuttgart, p. 109-129.

Harrison, M.E. & Marshall, A.J. (2011). Strategies for the use of fallback foods in apes. — Int. J. Primatol. 32: 531-565.

Hashimoto, C., Furuichi, T. & Tashiro, Y. (2001). What factors affect the size of chimpanzee parties in the Kalinzu Forest, Uganda? Examination of fruit abundance and number of estrous females. — Int. J. Primatol. 22: 947-959.

Hashimoto, C., Furuichi, T., Tashiro, Y. & Kimura, D. (1999). Vegetation of the Kalinzu Forest, Uganda: ordination of forest types using principal component analysis. — Afr. Stud. Monogr. 20: 229-239.

Hashimoto, C., Tashiro, Y., Kimura, D., Enomoto, T., Ingmanson, J., Idani, G. & Furuichi, T. (1998). Habitat use and ranging of wild bonobos (*Pan paniscus*) at Wamba. — Int. J. Primatol. 19: 1045-1060.

Hernandez-Aguilar, R.A. (2009). Chimpanzee nest distribution and site reuse in a dry habitat: implications for early hominin ranging. — J. Hum. Evol. 57: 350-364.

Hernandez-Aguilar, R.A., Moore, J. & Pickering, T.R. (2007). Savanna chimpanzees use tools to harvest the underground storage organs of plants. — Proc. Natl. Acad. Sci. USA 104: 19210-19213.

Herrmann, E., Hare, B., Call, J. & Tomasello, M. (2010). Differences in the cognitive skills of bonobos and chimpanzees. — PLoS ONE 5: e12438.

Herrmann, E., Wobber, V. & Call, J. (2008). Great apes' (*Pan troglodytes*, *Pan paniscus*, *Gorilla gorilla*, *Pongo pygmaeus*) understanding of tool functional properties after limited experience. — J. Comp. Psychol. 122: 220-230.

Hohmann, G. & Fruth, B. (2003a). Culture in bonobos? Between-species and within-species variation in behavior. — Curr. Anthropol. 44: 563-571.

Hohmann, G. & Fruth, B. (2003b). Lui Kotal — a new site for field research on bonobos in the Salonga National Park. — Pan Africa News 10: 25-27.

Howard, P.C. (1991). Nature conservation in Uganda's tropical forest reserves. — IUCN, Gland.

Humle, T. & Matsuzawa, T. (2004). Oil palm use by adjacent communities of chimpanzees at Bossou and Nimba Mountains, West Africa. — Int. J. Primatol. 25: 551-581.

Hunt, K.D. & McGrew, W.C. (2002). Chimpanzees in the dry habitats of Assirik, Senegal and Semliki Wildlife Reserve, Uganda. — In: Behavioural diversity in chimpanzees and bonobos (Boesch, C., Hohmann, G. & Marchant, L.F., eds). Cambridge University Press, New York, NY, p. 35-51.

Idani, G., Kuroda, S., Kano, T. & Asato, R. (1994). Flora and vegetation of Wamba forest, central Zaire with reference to bonobo (*Pan paniscus*) foods. — Tropics 3: 309-332.

Ingmanson, E.J. (1996). Tool-using behaviors in wild *Pan paniscus*: social and ecological considerations. — In: Reaching into thought: the minds of the great apes (Russon, A.E., Bard, K.A. & Parker, S.T., eds). Cambridge University Press, Cambridge, p. 190-210.

Inogwabini, B.I., Bewa, M., Longwango, M., Abokome, M. & Vuvuet, M. (2008). The bonobos of the Lake Tumba — Lake Maindombe Hinterland: threats and opportunities for population conservation. — In: The bonobos: behavior, ecology, and conservation (Furuichi, T. & Thompson, J., eds). Springer, New York, NY, p. 273-290.

Jordan, C. (1982). Object manipulation and tool-use in captive pygmy chimpanzees (*Pan paniscus*). — J. Hum. Evol. 11: 35-39.

Kano, T. (1982a). The social group of pygmy chimpanzees (*Pan paniscus*) of Wamba. — Primates 23: 171-188.

Kano, T. (1982b). The use of leafy twigs for rain cover by the pygmy chimpanzees of Wamba. — Primates 23: 453-457.

Kano, T. (1992). The last ape: pygmy chimpanzee behavior and ecology. — Stanford University Press, Stanford, CA.

Kano, T. (1997). Leaf-dropping sexual display exhibited by a male bonobo at Wamba. — Pan Africa News 4: 3-4.

Kano, T. & Mulavwa, M. (1984). Feeding ecology of the pygmy chimpanzees (*Pan paniscus*) of Wamba. — In: The pygmy chimpanzee (Susman, R.L., ed.). Plenum, New York, NY, p. 233-274.

Kawamoto, Y., Takemoto, H., Higuchi, S., Sakamaki, T., Hart, J.A., Hart, T.B., Tokuyama, N., Reinartz, G.E., Guislain, P., Dupain, J., Cobden, A.K., Mulavwa, M.N., Yangozene, K., Darroze, S., Devos, C. & Furuichi, T. (2013). Genetic structure of wild bonobo populations: diversity of mitochondrial DNA and geographical distribution. — PLoS ONE 8: e59660.

Koops, K. (2011). Chimpanzees in the Seringbara region of the Nimba mountains. — In: The chimpanzees of Bossou and Nimba (Matsuzawa, T., Humle, T. & Sugiyama, Y., eds). Springer, Tokyo, p. 277-287.

Koops, K., McGrew, W.C., de Vries, H. & Matsuzawa, T. (2012). Nest-building by chimpanzees (*Pan troglodytes verus*) at Seringbara, Nimba mountains: antipredation, thermoregulation, and antivector hypotheses. — Int. J. Primatol. 33: 356-380.

Koops, K., McGrew, W.C. & Matsuzawa, T. (2013). Ecology of culture: do environmental factors influence foraging tool use in wild chimpanzees, *Pan troglodytes verus*? — Anim. Behav. 85: 175-185.

Laiolo, P. & Jovani, R. (2007). The emergence of animal culture conservation. — Trends Ecol. Evol. 22: 5.

Lonsdorf, E.V. (2005). Sex differences in the development of termite-fishing skills in the wild chimpanzees, *Pan troglodytes* schweinfurthii, of Gombe National Park, Tanzania. — Anim. Behav. 70: 673-683.

Lonsdorf, E.V. (2006). What is the role of mothers in the acquisition of termite-fishing behaviors in wild chimpanzees (*Pan troglodytes* schweinfurthii)? — Anim. Cogn. 9: 36-46.

Lonsdorf, E.V., Ross, S.R., Linick, S.A., Milstein, M.S. & Melber, T.N. (2009). An experimental, comparative investigation of tool use in chimpanzees and gorillas. — Anim. Behav. 77: 1119-1126.

Marshall, A.J. & Wrangham, R.W. (2007). Evolutionary consequences of fallback foods. — Int. J. Primatol. 28: 1218-1235.

Mayr, E. & Ohara, R.J. (1986). The biolgeographic evidence supporting the pleistocene forest refuge hypothesis. — Evolution 40: 55-67.

McGrew, W.C. (1979). Evolutionary implication of sex differences in chimpanzee predation and tool use. — In: The great apes (Hamburg, DA. & McCown, E.R., eds). Benjamin/Cummings, Menlo Park, CA, p. 41-463.

McGrew, W.C. (1992). Chimpanzee material culture: implications for human evolution. — Cambridge University Press, Cambridge.

McGrew, W.C. & Collins, D.A. (1985). Tool use by wild chimpanzees (*Pan troglodytes*) to obtain termites (*Macrotermes herus*) in the Mahale Mountains, Tanzania. — Am. J. Primatol. 9: 47-62.

McGrew, W.C., Tutin, C.E.G. & Baldwin, P.J. (1979). Chimpanzees, tools and termites: cross-cultural comparisons of Senegal, Tanzania, and Rio Muni. — Man 14: 185-214.

McGrew, W.C., Baldwin, P.J. & Tutin, C.E.G. (1981). Chimpanzees in a hot, dry and open habitat: Mt. Assirik, Senegal, West Africa. — J. Hum. Evol. 10: 237-244.

McGrew, W.C., Marchant, L.F., Beuerlein, M.M., Vrancken, D., Fruth, B. & Hohmann, G. (2007). Prospects for bonobo insectivory: Lui kotal, democratic republic of congo. — Int. J. Primatol. 28: 1237-1252.

Moore, J. (1992). "Savanna" chimpanzees. — In: Topics in Primatology, Vol. 1: human origins (Nishida, T., McGrew, W.C., Marler, P., Pickford, M. & de Waal, F.B.M., eds). University of Tokyo Press, Tokyo, p. 99-118.

Morgan, D.B. (2007). Socio-ecology of Chimpanzees (*Pan troglodytes troglodytes*) in the Goualougo Triangle, Republic of Congo. — PhD thesis, Cambridge University, Cambridge.

Morgan, D. & Sanz, C. (2006). Chimpanzee feeding ecology and comparisons with sympatric gorillas in the Goualougo Triangle, Republic of Congo. — In: Primate feeding ecology in apes and other primates: ecological, physiological, and behavioural aspects (Hohmann, G., Robbins, M. & Boesch, C., eds). Cambridge University Press, Cambridge, p. 97-122.

Morgan, D., Sanz, C., Onononga, J.R. & Strindberg, S. (2006). Ape abundance and habitat use in the Goualougo Triangle, Republic of Congo. — Int. J. Primatol. 27: 147-179.

Mulavwa, M., Furuichi, T., Yangozene, K., Yamba-Yamba, M., Motema-Salo, B., Idani, G., Ihobe, H., Hashimoto, C., Tashiro, Y. & Mwanza, N. (2008). Seasonal changes in fruit production and party size of bonobos at Wamba. — In: The bonobos: ecology, behavior, and conservation (Furuichi, T. & Thompson, J., eds). Springer, New York, NY, p. 121-134.

Mulavwa, M.N., Yangozene, K., Yamba-Yamba, M., Motema-Salo, B., Mwanza, N.N. & Furuichi, T. (2010). Nest groups of wild bonobos at Wamba: selection of vegetation and tree species and relationships between nest group size and party size. — Am. J. Primatol. 72: 575-586.

Myers Thompson, J.A. (2002). Bonobos of the Lukuru Wildlife Research Project. — In: Behavioural diversity in chimpanzees and bonobos (Boesch, C., Hohmann, G. & Marchant, L.F., eds). Cambridge University Press, New York, NY, p. 61-70.

Nishida, T. (1990). A quarter century of research in the Mahale Mountains: an overview. — In: The chimpanzees of the Mahale Mountains: sexual and life history strategies (Nishida, T., ed.). University of Tokyo Press, Tokyo, p. 3-35.

Nishie, H. (2011). Natural history of *Camponotus* ant-fishing by the M group chimpanzees at the Mahale Mountains National Park, Tanzania. — Primates 52: 329-342.

Nishihara, T. (1995). Feeding ecology of western lowland gorillas in the Nouabale-Ndoki National Park, Congo. — Primates 36: 151-168.

Palagi, E. (2006). Social play in bonobos (*Pan paniscus*) and chimpanzees (*Pan troglodytes*): implications for natural social systems and interindividual relationships. — Am. J. Phys. Anthropol. 129: 418-426.

Pandolfi, S.S., van Schaik, C.P. & Pusey, A.E. (2003). Sex differences in termite fishing among Gombe chimpanzees. — In: Animal social complexity: intelligence, culture, and individualized societies (de Waal, F.B.M. & Tyack, P.L., eds). Harvard University Press, Cambridge, MA, p. 414-418.

Plana, V. (2004). Mechanisms and tempo of evolution in the African Guineo-Congolian rainforest. — Phil. Trans. Roy. Soc. Lond. B 359: 1585-1594.

Reynolds, V. (2005). The chimpanzees of the Budongo Forest: ecology, behaviour, and conservation. — Oxford University Press, New York, NY.

Rutz, C., Bluff, L.A., Reed, N., Troscianko, J., Newton, J., Inger, R., Kacelnik, A. & Bearhop, S. (2010). The ecological significance of tool use in New Caledonian crows. — Science 329: 1523-1526.

Rutz, C. & St Clair, J.J.H. (2012). The evolutionary origins and ecological context of tool use in New Caledonian crows. — Behav. Proc. 89: 153-165.

Sabater-Pi, J. (1979). Feeding behaviour and diet of chimpanzees (*Pan troglodytes troglodytes*) in the Okorobiko Mountains of Rio Muni (West Africa). — Z. Tierpsychol. 50: 265-281.

Sabater Pi, J. (1974). An elementary industry of the chimpanzees in the Okorobiko Mountains, Rio Muni (Republic of Equatorial Guinea), West Africa. — Primates 15: 351-364.

Samson, D.R. & Hunt, K.D. (2012). A thermodynamic comparison of arboreal and terrestrial sleeping sites for dry-habitat chimpanzees (*Pan troglodytes* schweinfurthii) at the Toro-Semliki Wildlife Reserve, Uganda. — Am. J. Primatol. 74: 811-818.

Sanz, C. & Morgan, D. (2007). Chimpanzee tool technology in the Goualougo Triangle, republic of Congo. — J. Hum. Evol. 52: 420-433.

Sanz, C.M. & Morgan, D.B. (2009). Flexible and persistent tool-using strategies in honey-gathering by wild chimpanzees. — Int. J. Primatol. 30: 411-427.

Sanz, C. & Morgan, D. (2010). Complexity of chimpanzee tool using behaviors. — In: The mind of the chimpanzee: ecological and experimental perspectives (Lonsdorf, E.V., Ross, S.R. & Matsuzawa, T., eds). University of Chicago Press, Chicago, IL, p. 127-140.

Sanz, C.M. & Morgan, D.B. (2013a). The social context of chimpanzee tool use. — In: Tool use in animals (Sanz, C.M., Call, J. & Boesch, C., eds). Cambridge University Press, Cambridge, p. 161-175.

Sanz, C.M. & Morgan, D.B. (2013b). Ecological and social correlates of chimpanzee tool use. — Phil. Trans. Roy. Soc. B 368: 20120416.

Sanz, C., Morgan, D. & Gulick, S. (2004). New insights into chimpanzees, tools, and termites from the Congo basin. — Am. Nat. 164: 567-581.

Sanz, C.M., Schoning, C. & Morgan, D.B. (2010). Chimpanzees Prey on army ants with specialized tool set. — Am. J. Primatol. 72: 17-24.

Serckx, A., Huynen, M.-C., Bastin, J.-F., Hambuckers, A., Beudels-Jamar, R.C., Vimond, M., Raynaud, E. & Kühl, H.S. (2014). Nest grouping patterns of bonobos (*Pan paniscus*) in relation to fruit availability in a forest-savannah mosaic. — PLoS ONE 9: e93742.

Shumaker, R.W., Walkup, K.R. & Beck, B.B. (2011). Animal tool behavior: the use and manufacture of tools by animals. — Johns Hopkins University Press, Baltimore, MD.

Spagnoletti, N., Visalberghi, E., Ottoni, E., Izar, P. & Fragaszy, D. (2012). Stone tool use in wild bearded capuchin monkeys, *Cebus libidinosus*: is it a strategy to overcome food scarcity? — Anim. Behav. 83: 1285-1294.

Stanford, C.B. (1998). The social behavior of chimpanzees and bonobos — empirical evidence and shifting assumptions. — Curr. Anthropol. 39: 399-420.

Surbeck, M. & Hohmann, G. (2013). Intersexual dominance relationships and the influence of leverage on the outcome of conflicts in wild bonobos (*Pan paniscus*). — Behav. Ecol. Sociobiol.: 1-14.

Suzuki, S., Kuroda, S. & Nishihara, T. (1995). Tool-set for termite-fishing by chimpanzees in the Ndoki forest, Congo. — Behaviour 132: 219-234.

Takasaki, H., Nishida, T., Uehara, S., Norikoshi, K., Kawanaka, K., Takahata, Y., Hiraiwa-Hasegawa, M., Hasegawa, T., Hayaki, H., Masui, K. & Huffman, M.A. (1990). Summary of meteorological data at Mahale research camp, 1973–1988. — In: The chimpanzees of the Mahale Mountains: sexual and life history strategies (Nishida, T., ed.). University of Tokyo Press, Tokyo, p. 291-300.

Takemoto, H. (2004). Seasonal change in terrestriality of chimpanzees in relation to microclimate in the tropical forest. — Am. J. Phys. Anthropol. 124: 81-92.

Tebbich, S., Taborsky, M., Fessl, B. & Dvorak, M. (2002). The ecology of tool-use in the woodpecker finch (*Cactospiza pallida*). — Ecol. Lett. 5: 656-664.

van Schaik, C.P. (2001). Fragility of traditions: the disturbance hypothesis for the loss of local traditions in orangutans. — Int. J. Primatol. 23: 527-538.

Wallis, J. (2002). Seasonal aspects of reproduction and sexual behavior in two chimpanzee populations: a comparison of Gombe (Tanzania) and Budongo (Uganda). — In: Behavioural diversity in chimpanzees and bonobos (Boesch, C., Hohmann, G. & Marchant, L.F., eds). Cambridge University Press, Cambridge, p. 181-191.

Watts, D.P. (2012). Long-term research on chimpanzee behavioral ecology in Kibale national park, Uganda. — In: Long-term field studies of primates (Kappeler, P.M. & Watts, D.P., eds). Springer, Heidelberg, p. 313-338.

White, F. (1986). The vegetation of Africa: a descriptive memoir to accompany the UNESCO/AETFAT/UNSO vegetation map of Africa. — UNESCO, Paris.

White, F.J. (1992). Activity budgets, feeding behavior, and habitat use of pygmy chimpanzees at Lomako, Zaire. — Am. J. Primatol. 26: 215-223.

Whitehead, H. (2010). Conserving and managing animals that learn socially and share cultures. — Learn. Behav. 38: 329-336.

Whitehead, H., Rendell, L., Osborne, R.W. & Wursig, B. (2004). Culture and conservation of non-humans with reference to whales and dolphins: review and new directions. — Biol. Conserv. 120: 427-437.

Whiten, A., Goodall, J., McGrew, W.C., Nishida, T., Reynolds, V., Sugiyama, Y., Tutin, C.E.G., Wrangham, R.W. & Boesch, C. (1999). Cultures in chimpanzees. — Nature 399: 682-685.

Whiten, A., Goodall, J., McGrew, W.C., Nishida, T., Reynolds, V., Sugiyama, Y., Tutin, C.E.G., Wrangham, R.W. & Boesch, C. (2001). Charting cultural variation in chimpanzees. — Behaviour 138: 1481-1516.

Whiten, A., Spiteri, A., Horner, V., Bonnie, K.E., Lambeth, S.P., Schapiro, S.J. & de Waal, F.B.M. (2007). Transmission of multiple traditions within and between chimpanzee groups. — Curr. Biol. 17: 1038-1043.

Wilson, M.L., Hauser, M.D. & Wrangham, R.W. (2001). Does participation in intergroup conflict depend on numerical assessment, range location, or rank for wild chimpanzees? — Anim. Behav. 61: 1203-1216.

Won, Y.J. & Hey, J. (2005). Divergence population genetics of chimpanzees. — Mol. Biol. Evol. 22: 297-307.

Wrangham, R.W., Clark, A.P. & Isabirye-Basuta, G. (1992). Female social relationships and social organization of Kibale Forest chimpanzees. — In: Human origins (Nishida, T., McGrew, W.C., Marler, P., Pickford, M. & de Waal, F.B.M., eds). University of Tokyo Press, Tokyo, p. 81-98.

Yamagiwa, J., Maruhashi, T., Yumoto, T. & Mwanza, N. (1996). Dietary and ranging overlap in sympatric gorillas and chimpanzees in Kahuzi-Biega National Park, Zaire. — In: Great ape societies (McGrew, W.C., Marchant, L.F. & Nishida, T., eds). Cambridge University Press, Cambridge, p. 82-98.

Yamakoshi, G. (1998). Dietary responses to fruit scarcity of wild chimpanzees at Bossou, Guinea: possible implications for ecological importance of tool use. — Am. J. Phys. Anthropol. 106: 283-295.

[When citing this chapter, refer to Behaviour 152 (2015) 461–492]

A comparative assessment of handedness and its potential neuroanatomical correlates in chimpanzees (*Pan troglodytes*) and bonobos (*Pan paniscus*)

William D. Hopkins [a,b,*], **Jennifer Schaeffer** [b], **Jamie L. Russell** [b], **Stephanie L. Bogart** [a], **Adrien Meguerditchian** [c] **and Olivier Coulon** [d]

[a] Neuroscience Institute and Language Research Center, Georgia State University, Atlanta, GA 30302, USA

[b] Division of Developmental and Cognitive Neuroscience, Yerkes National Primate Research Center, 954 Gatewood Road, Atlanta, GA 30029, USA

[c] Laboratory of Cognitive Psychology, UMR 7290, Aix-Marseille University, CNRS, Marseille, France

[d] Aix-Marseille Universite, LSIS, UMR, CNRS 7296 Marseille, France

*Corresponding author's e-mail address: whopkins4@gsu.edu or whopkin@emory.edu

Accepted 31 March 2014; published online 8 May 2014

Abstract

The evolutionary origins of human right-handedness remain poorly understood. Some have hypothesized that tool use served as an important preadaptation for the eventual evolution of population-level right-handedness. In contrast, others have suggested that complex gestural and vocal communication served as prerequisite for the evolution of human right-handedness. In this study, we tested these competing hypotheses by comparing the handedness of bonobos and chimpanzees, two closely related species of *Pan*, on three different measures of hand use including simple reaching, manual gestures and coordinated bimanual actions. Chimpanzees are well known for their tool using abilities whereas bonobos rarely use tools in the wild. In contrast, many have suggested that bonobos have a more flexible gestural and vocal communication system than chimpanzees. The overall results showed that chimpanzees were significantly more right-handed than bonobos for all three measures suggesting that adaptations for tool use rather than communication may have led to the emergence of human right-handedness. We further show that species differences in handedness may be linked to variation in the size and asymmetry of the motor-hand area of the precentral gyrus. The results are discussed within the context of evolutionary theories of handedness, as well as some limitations in the approach to handedness measurement in nonhuman primates.

Keywords

handedness, bonobos, chimpanzees, central sulcus morphology, brain asymmetry, motor skill.

1. Introduction

Chimpanzees (*Pan troglodytes*) and bonobos (*Pan paniscus*) are the closest evolutionary relatives to humans. Although humans split from the genus *Pan* 5–7 million years ago, bonobos and chimpanzees shared a common ancestor as recently as 1–2 million years ago (Fischer et al., 2004; Won & Hey, 2005). The vast majority of cognitive and behavioral studies in captivity have been done with chimpanzees, with relatively little information on the bonobo, likely due to the greater availability of chimpanzees compared to bonobos. However, even with the relative scarcity of bonobo data, both species have shown similar cognitive abilities, including for example: symbolic acquisition (Brakke & Savage-Rumbaugh, 1995, 1996; Lyn, 2007), symbol ordering (Greenfield & Savage-Rumbaugh, 1990; Lyn et al., 2011), representational (pretend) play (Lyn et al., 2006) and mirror self-recognition (Gallup, 1970; Hyatt & Hopkins, 1994). Although the physical resemblance and the recently recorded cognitive similarities are manifest, the two species have markedly different social structures and behavioural patterns (de Waal, 1989; Stanford, 1998).

For instance, studies in captivity and in the wild show that bonobo society is decidedly matriarchal, as opposed to the alpha-male dominated chimpanzee society (Parish, 1996; Parish et al., 2000). In addition there is a greater number of detailed reports of tool manufacture and use in chimpanzees (Whiten et al., 1999, 2001), leading some experts to declare bonobo culture as less 'material' (McGrew, 1992), albeit far fewer studies have been conducted in *Pan paniscus* (Toth et al., 1993; Gruber et al., 2010). In contrast, the reports of communication abilities suggest that while both species are highly communicative, bonobos may have a somewhat more flexible and sophisticated communicative system than chimpanzees (Hopkins & Savage-Rumbaugh, 1991; Taglialatela et al., 2003; Pika et al., 2005; Pollick & de Waal, 2007; Clay & Zuberbuhler, 2011). More generally, it has been suggested that bonobos perform better than chimpanzees on tasks that assess different dimensions of social cognition or cooperation whereas chimpanzees perform better on competitive, physical and spatial cognition (Hare & Tomasello, 2004; Hare, 2007; Hare et al., 2007; Herrmann et al., 2010; Rosati & Hare, 2012).

The purpose of the current study was to examine handedness in chimpanzees and bonobos as a means of assessing potential differences in lateralization of the nervous system as it relates to motor functions. Our particular interest in this question stems from historical and contemporary theories on the evolution of right-handedness in humans in relation to emergence of higher cognitive and motor functions including language and speech, tool use and bipedalism (Corballis, 1992; Bradshaw & Rogers, 1993; Bradshaw, 1997). For example, it has been postulated that human right-handedness is a consequence of selection for increasing motor demand for oro-facial motor control or gestural communication (Annett, 2002; Corballis, 2002). Because it has been reported that bonobos have greater flexibility in vocalizations and gestures than chimpanzees (Savage-Rumbaugh, 1984; Hopkins & Savage-Rumbaugh, 1991; Taglialatela et al., 2003; Pollick & de Waal, 2007), it could be hypothesized that they would express more pronounced asymmetries in manual gestures compared to chimpanzees. In contrast, others have suggested that the cognitive and motor demands of tool-use may have served as a prerequisite to the evolution of complex praxic functions that are primarily under control of the left hemisphere (Frost, 1980; Greenfield, 1991; Gibson & Ingold, 1993; McGrew & Marchant, 1997). As noted above, tool-use is widespread in chimpanzees compared to bonobos and there is evidence of population-level handedness in wild chimpanzees for some forms of tool-use such as leaf sponging, ant-dipping and termite fishing but not nut-cracking (Boesch, 1991; McGrew & Marchant, 1992, 1996, 1997; Biro et al., 2003; von Lonsdorf & Hopkins, 2005; Biro et al., 2006; Marchant & McGrew, 2007; Humle & Matsuzawa, 2009; Bogart et al., 2012b). If this theory is correct, then it might be hypothesized that chimpanzees would show greater expression of manual asymmetries for object directed actions compared to bonobos (see also Forrester et al., 2011, 2012). To test these competing hypotheses, we tested a sample of chimpanzees and bonobos on three measures of handedness including manual gestures, simple reaching, and a task measuring coordinated bimanual actions, referred to as the TUBE task.

Beyond the theoretical predictions for handedness in these two species, there are also behavioral and neurological reasons to predict potential differences in handedness. For instance, Hopkins et al. (2009) recently reported that bonobos showed greater leftward asymmetries in the motor-hand area of the precentral gyrus (or KNOB) as well as striatal structures including the putamen and caudate. In humans and chimpanzees, asymmetries in the KNOB region correlate with handedness with a larger KNOB volume found in the hemisphere contralateral to the subject's preferred hand (Hammond, 2002; Hopkins & Cantalupo, 2004). Based on the report by Hopkins et al.

(2009) of a larger left KNOB region in bonobos, it can be hypothesized that this species should show more pronounced right-handedness compared to the chimpanzees. With respect to their behaviour, several reports of handedness in bonobos also suggest potential species differences. Chapelain and colleagues (Chapelain & Hogervorst, 2009; Chapelain et al., 2011) have examined handedness for the TUBE task, and contrary to chimpanzees (Hopkins et al., 2004; Llorente et al., 2010), they reported no evidence of population-level right handedness in the bonobos (see also Hopkins et al., 2011). There are also reports of a lack of population-level handedness for manual gestures in bonobos by several authors (Hopkins & de Waal, 1995; Shafer, 1997; Harrison & Nystrom, 2008), which also contradicts reports in captive chimpanzees (Hopkins et al., 2005a), though methods of handedness assessment may explain some of these differences (Hopkins et al., 2012). Whether these purported species differences reflect methodological factors or are true differences is not clear because the approaches to the assessment of handedness have varied between studies. Thus, the aim of this study was to compare bonobos and chimpanzees on identical measures of handedness when using the same procedures. In addition to hand preferences, we also measured grasping morphology and performance asymmetries when picking up small food items in the bonobos and chimpanzees. Christel et al. (1998) reported that bonobos show nearly exclusive grasping responses that involve the thumb and index finger, notably the lateral portion. In contrast, Hopkins and colleagues (Hopkins et al., 2002, 2005b) have reported that, though some chimpanzees prefer to use the thumb-index grasping response, many prefer to use the index and middle finger to grasp small food items, a pattern apparently rarely observed in the three bonobos studied by Christel et al. (1998). Morphologically, the hand of the bonobo and chimpanzees are quite similar; thus the reports of potential differences in grasping morphology may reflect underlying differences in neural innervations of the digits or other morphological features that may be involved with reach-and-grasp responses such as the wrist and elbow. To further explore whether species differences are evident in grasping morphology, we compared the bonobos' and chimpanzees' grasping responses when picking up small food items. We also assessed whether performance differences in grasping skill were evident between the chimpanzees and bonobos. Previous studies in chimpanzees have shown that they exhibit a small but significant right hand advantage in grasping skill (Hopkins & Russell, 2004) but, to our knowledge, there are no such data available in bonobos. Christel et al. (1998) assessed response times during reach-and grasp responses in three bonobos and found that latencies to reach were shorter for the right compared to left hand but that the time needed to grasp the objects was slightly faster for the left compared to right hand. Comparing the results by Christel et al. (1998) with the findings by Hopkins & Russell (2004) is difficult because different dimensions of motor skill were measured. Thus, in this study, we assessed both chimpanzees and bonobos on the same grasping task in order to compare lateralization in motor skill.

Finally, we present here some preliminary findings on comparative differences in cortical organization between chimpanzees and bonobos. Despite the intense interest in behavioural and cognitive abilities between chimpanzees and bonobos, there is remarkably little research on how these potential differences might map onto variation in cortical organization. Hopkins et al. (2009) measured the volume and areas of eight brains regions in a sample of seven chimpanzees and bonobos and found that bonobos showed greater leftward asymmetries for the motor-hand area of the precentral gyrus, putamen, caudate and hippocampus. More recently, Rilling et al. (2012) used both voxel-based morphometry (VBM) and diffusion tensor imaging (DTI) to compare cortical connectivity in three chimpanzee and bonobo brains. These authors found that bonobos had more grey matter than chimpanzees in the anterior amygdala and insular cortex compared to chimpanzees. Rilling et al. also found significantly larger cortical pathways connecting the amygdala and anterior cingulate cortex compared to the chimpanzees.

Rather than use region-of-interest or VBM analysis, here we examined species difference in the surface area and cortical folding of the central sulcus (CS) using a software program called BrainVisa. BrainVISA 4.1.1 (BV) is freely distributed software (available online at http://brainvisa.info) that measures cortical folding of the brain (Mangin et al., 2004). We selected the CS because the cortex lying anterior and posterior to the CS are the primary motor and somatonsensory cortex, the cortical region where all motor and sensory regions are represented in the brain, most notably the hand (Bailey et al., 1950). For instance, stimulation of the central region of the CS will result in selective movements of the hand and individual digits, suggesting that this is the cortical representation of the hand. Furthermore, there is an anatomical landmark within the precentral gyrus, referred to as the KNOB (Yousry et al., 1997), and several studies have shown that the volume and shape of the KNOB differs in right- and left-handed individuals (Boling et al., 1999; Hammond, 2002) and similar results have been reported in chimpanzees and capuchin monkeys (Hopkins & Cantalupo, 2004; Phillips & Sherwood, 2005; Dadda et al., 2006). The appearance of the KNOB in the brain has evolved as a consequence of the CS having to fold over and around a buried gyrus, referred to as the pli-de-passage frontoparietal moyen parietale (PPFM), which connects the precentral and postcentral gyrus, and was originally described by Broca (Alkadhi & Kollias, 2004). Comparative studies have shown that the KNOB is only evident in great apes and humans and that this is likely attributable to increased gyrification of the brain due to increasing size and connectivity (Hopkins & Pilcher, 2001). Further, using different parameterization tools within BV, the depth of the CS can be quantified along the entire superior to inferior plane (Coulon et al., 2006; Cykowski et al., 2008; Hopkins et al., 2010). This procedure allows for a means of quantifying the PPFM within the entire surface area of the CS. In this study, we examined the surface area and folding pattern of the entire CS as well as the region within the CS corresponding to the PPFM in a sample of chimpanzees and bonobos to test whether (a) differences are evident in CS morphology and (b) whether these differences might map onto variation in motor skill and/or handedness.

2. Methods

2.1. Subjects

At least one measure of handedness was available in 34 bonobos and 34 chimpanzees that were matched on age, sex and rearing histories (see Table 1). The bonobos were housed at three facilities including the Jacksonville Zoo and Gardens, Milwaukee County Zoo and Great Ape Trust of Iowa. The chimpanzees were all housed at the Yerkes National Primate Research Center of Emory University. Within each species, there were 19 females and 15 males and the apes ranged from 3 to 44 years of age (mean = 20.34 years, SE = 1.96). Within each cohort, there were three wild-caught and 31 captive-born individuals. Within the captive-born cohort, there were 27 mother-reared and 4 human-reared individuals. Human-reared apes were those that were removed from their natal groups before 1 month of age and raised in a human nursery setting until they were old enough to be introduced into conspecific social groups. Mother-reared apes were those that were born and raised by their conspecific mothers and raised in nuclear family settings.

Table 1.
Matched subjects, HI scores and N for each handedness measure.

Subject No.	Age	Rearing history	Bonobos			Subject No.	Chimpanzees		
			Gesture	Reach	Tube		Gesture	Reach	Tube
♀♀									
1	46	M	−0.53 (43)	−0.64 (51)	−0.51 (57)	1	−0.13 (30)	0.26 (60)	0.42 (386)
2	44	M	0.87 (30)	−0.08 (52)	−0.81 (43)	2	0.51 (33)	0.13 (53)	0.68 (366)
3	43	W	0.24 (34)	−0.08 (50)	0.49 (86)	3	0.38 (32)	−0.14 (58)	−0.56 (128)
4	28	M	−0.03 (31)	0.47 (53)	−0.46 (89)	4	0.83 (36)	0.44 (50)	−0.16 (90)
5	28	N	−0.20 (30)	−0.05 (55)	−0.77 (33)	5	−0.13 (30)	0.88 (50)	−0.03 (150)
6	26	M	0.61 (31)	−0.48 (50)	−0.71 (63)	6	0.88 (30)	0.62 (53)	0.74 (38)
7	17	M	−0.18 (68)	−0.05 (40)	0.32 (79)	7	−0.32 (34)	0.39 (59)	0.522 (80)
8	16	N	0.88 (51)	−0.03 (60)	−0.93 (123)	8	−0.73 (30)	−0.53 (51)	−0.38 (80)
9	16	M	0.10 (31)	0.0 (52)	−0.71 (28)	9	0.40 (30)	−0.44 (50)	0.50 (56)
10	14	N	0.47 (30)	−0.24 (50)	0.20 (65)	10	0.19 (32)	−0.18 (51)	0.45 (80)
11	14	M	0.64 (22)	−0.24 (50)	0.89 (53)	11	0.26 (38)	0.04 (50)	0.44 (80)
12	12	M	0.06 (32)	−0.11 (54)	−0.60 (50)	12	0.56 (9)	−0.60 (50)	0.56 (80)
13	11	M	–	0.0 (52)	−0.17 (60)	13	0.73 (30)	0.44 (50)	0.30 (123)
14	10	M	1.0 (30)	0.08 (63)	0.82 (100)	14	0.16 (31)	−0.10 (60)	0.27 (123)
15	10	M	−1.0 (18)	0.41 (51)	0.20 (40)	15	0.47 (30)	0.42 (53)	−0.13 (80)
16	8	M	–	0.52 (50)	−0.12 (50)	16	−0.93 (31)	−0.14 (51)	0.50 (104)
17	5	M	1.0 (18)	−0.26 (35)	−0.37 (30)	17	−0.64 (33)	0.56 (59)	0.15 (178)
18	13	M	–	−0.12 (41)	−0.33 (42)	18	0.0 (10)	0.27 (77)	0.68 (80)
19	9	M	–	−0.52 (42)	0.12 (41)	19	0.53 (30)	−0.43 (53)	0.42 (75)

Table 1.
(Continued.)

Subject No.	Age	Rearing history	Bonobos			Subject No.	Chimpanzees		
			Gesture	Reach	Tube		Gesture	Reach	Tube
♂♂									
1	34	M	0.57 (57)	−0.36 (94)	0.26 (84)	1	−0.31 (35)	−0.41 (51)	0.32 (164)
2	33	M	0.07 (30)	0.35 (52)	−0.68 (76)	2	0.67 (7)	0.12 (52)	0.13 (243)
3	33	W	−0.14 (35)	0.06 (51)	−0.45 (59)	3	0.56 (30)	0.20 (63)	0.50 (209)
4	30	M	−0.94 (32)	−0.13 (51)	0.0 (68)	4	0.94 (33)	0.24 (50)	0.67 (167)
5	24	M	0.74 (39)	0.36 (50)	–	5	0.67 (30)	0.23 (65)	0.38 (26)
6	23	M	−0.71 (34)	−0.36 (50)	–	6	−0.13 (30)	0.22 (51)	−0.29 (241)
7	19	M	0.87 (30)	0.48 (50)	0.0 (50)	7	−0.29 (31)	0.31 (73)	0.34 (64)
8	18	M	–	0.12 (50)	−0.96 (134)	8	0.80 (30)	0.31 (55)	0.40 (89)
9	17	M	0.73 (30)	−0.09 (55)	0.73 (67)	9	0.0 (30)	0.30 (50)	0.39 (220)
10	15	N	−0.82 (11)	0.68 (51)	−0.67 (42)	10	0.44 (25)	−0.28 (50)	0.26 (80)
11	15	M	0.0 (32)	0.12 (50)	−0.19 (37)	11	0.45 (47)	0.36 (50)	0.18 (80)
12	13	M	−0.47 (30)	0.45 (51)	−0.38 (187)	12	0.86 (30)	−0.13 (50)	−0.23 (72)
13	11	M	–	0.10 (51)	0.33 (90)	13	0.74 (23)	−0.57 (51)	0.74 (80)
14	40	W	−0.67 (18)	−0.37 (54)	−0.61 (72)	14	−0.47 (30)	0.09 (55)	−0.25 (200)
15	3	M	–	0.30 (40)	–	15	0.16 (31)	0.43 (84)	0.59 (117)

Numbers in parentheses indicate the total number of left and right hand responses produced by each subject and task. M, mother-reared; N, nursery-reared; W, wild caught; –, no available data for that individual on the task. Age is rounded to the closest year. For wild caught apes, ages are estimated.

2.2. Procedure

Handedness was assessed on three measures of hand use previously described in these subjects including manual gestures, simple reaching, and a task measuring coordinated bimanual actions, referred to as the TUBE task (Hopkins, 1995a; Hopkins et al., 2005a, b, 2007). These three measures were selected because they elicited different dimensions of hand preferences and have been shown to be reliable measures across time (Hopkins, 2007). A brief description of each measure is provided below. The testing was done in the indoor and outdoor enclosures of the apes by three different experimenters. The three experimenters participated in data collection from all individuals between the two species. Thus, no one experimenter solely collected data from a single species. Because hand use and grasping morphology were fairly overt manual actions, we did not calculate interrater reliability in their assessment between raters prior to the onset of data collection.

2.2.1. Manual communicative gestures

At the onset of each trial, an experimenter would approach the subject's home cage and centre themselves in front of the subject at a distance of approx. 1.0–1.5 m. If the subject was not already positioned in front of the experimenter at the onset of the trial, the subject would immediately move towards the front of the cage when the experimenter arrived with the food. The experimenter then spoke the ape's name and offered a piece of food until the subject produced a manual gesture, whereupon they were given the food. Only responses in which the apes' unimanually extended the digit(s) through the cage mesh to request the food were considered a response. Hand use was recorded as right or left for each response. Other possible manual responses such as cage banging or clapping were not counted as a gesture. Two-handed gestures, although rare, were not scored as were gestures that were produced by the subject prior to the experimenter arriving in front of the subject's home cage. A minimum of 30 responses were collected from most subjects (range 7–57) (Hopkins et al., 2005c) and the number of trials administered on a given day varied with the subject's motivation and availability for testing. However, at least 5 min separated the administration of each trial. Subjects were tested in both the indoor and outdoor sections of their home enclosures.

2.2.2. *Coordinated bimanual task (TUBE)*

The second handedness measure was a task requiring bimanual coordinated actions, referred to as the TUBE task (Hopkins, 1995a). For the TUBE task, peanut butter or other foods that had adhesive qualities were smeared on the inside edges of poly-vinyl-chloride (PVC) tubes approx. 15 cm in length and 2.5 cm in diameter. Peanut butter was smeared on both ends of the PVC pipe and was placed far enough down the tube such that the subjects could not lick the contents completely off with their mouths but rather had to use one hand to hold the tube and the other hand to remove the substrate. The PVC tubes were handed to the subjects in their home cages and a focal sampling technique was used to collect individual data from each subject. The hand of the finger used to extract the peanut butter was recorded as either right or left by the experimenter. Each time the subject reached into the tube with their finger, extracted peanut butter and brought it to their mouth, the hand used was recorded as left or right. Each subject was tested on at least two occasions and a minimum of 30 responses was obtained from each subject (range 30–386).

2.2.3. *Simple reaching*

During each trial, a raisin was thrown into the subject's home cage. The raisin was thrown by the experimenter to a location at least three meters from the focal subject so that the apes had to locomote to the raisin, pick up the raisin and bring it to their mouth for consumption. When the subject acquired the raisin, the experimenter recorded the hand used as left or right. One, and only one reaching response, was recorded each trial to assure independence of data points. Thus, raisins were not randomly scattered in home cages but rather an individual raisin was thrown into the cage and the subject retrieved the raisin before another was thrown into the cage. Subjects were required to locomote at least three strides between reaching responses to maintain postural readjustment between trials. A minimum of 35 responses was collected from each subject with the range between 35 and 94 for the sample.

In addition to hand use during the simple reaching task, we also recorded the type of grip used by the apes when grasping the raisins. Grip type was characterized as either thumb-index, middle-index, or single digit. Thumb-index gripping was recorded when the subjects abducted the thumb to the lateral or tip of the index finger to grip the food item. Gripping was recorded as middle-index grip when the subjects grasped the food between the index and middle finger with the hand either in a prone or supine position. Thumb-index and middle-index grip types were by far the most common, but occasionally subjects engaged in what was recorded as single digit responses. Single digit responses were instances in which the chimpanzees used one finger to press down hard enough on the food item so that it stuck while being taken to the mouth.

2.3. Grasping errors

For this measure, small food items (quartered peanuts, sunflower seeds, red-hots or skittles) were thrown into the subject's home cage. The use of different food was based on the individual apes' preferences for these items as well as food restrictions at each facility (i.e. one bonobo from JZG was allergic to peanuts, so no peanuts are allowed in the Great Ape area). Food items were dropped in the cage until exactly twenty successful grasping responses with each hand had been obtained from each subject. Because the emphasis of this particular experiment was on obtaining twenty responses with each hand from each subject, the subjects were not required to locomote to the food items in order for the response to be recorded. Instead, the food items were dropped next to one of the subject's hands, thus encouraging them to use one hand over the other. The experimenter watched as the subject attempted to pick up the food item and noted any errors made in grasping the food. An error was recorded when the subjects attempted to grasp but dropped the food or failed to grasp the food item from the floor upon their attempt. If at any point the subject switched hands or used their mouth to obtain the food, the trial was not included. Data were collected ad libitum and during testing, the foods were thrown to either side of the subject's body in a random order in order to facilitate responses with both hands.

2.4. Grasping data analysis

For each hand preference and performance measure, a handedness index (HI) was derived following the formula ($\text{HI} = (R - L)/(R + L)$) where R and L represent the frequency of right and left hand use for each task, respectively. For the hand preference measures, positive HI values indicated a rightward asymmetry and negative values indicated a leftward asymmetry. Note that for the performance task, a positive HI would reflect better performance by the left compared to right hand, reflecting the fact that they would thereby make more errors with the right hand. For each handedness task, binomial z-scores were calculated for each subject based on the total frequency of left and right hand use. Subjects with z-scores greater than or equal to 1.96 or less than or equal to -1.96 were classified as right-handed and left-handed respectively while all other subjects were classified as ambiguously-handed ($z > -1.96$ and $z < 1.96$). Finally, for the grasping morphology analyses, we computed a percentage of thumb-index responses for each subject and hand by summing the total number of left and right thumb-index grasping responses and dividing these values by the total number of grasping responses for the left and right hands, then multiplying by 100. Percentages varied between 0 and 100 with higher values indicating greater thumb-index grasping responses.

2.5. *Magnetic resonance image collection*

Magnetic resonance images (MRI) were obtained from 13 bonobos and 13 chimpanzees that were matched on age, sex and magnet strength. Four bonobos and 4 chimpanzees were scanned *in vivo* and the remaining 9 chimpanzees and 9 bonobos were scanned post-mortem. To minimize stress on the apes, all *in vivo* bonobo and chimpanzee magnetic resonance images were obtained during one of their annual physical examinations. Magnetic resonance image (MRI) scans followed standard procedures at the YNPRC. Thus, the animals were first sedated with ketamine (10 mg/kg) or telazol (3–5 mg/kg) and were subsequently anaesthetized with propofol (40–60 mg/kg per h). They were then transported to the MRI scanning facility and placed in a supine position in the scanner with their head in a human-head coil. Upon completion of the MRI, the apes were singly-housed for 2–24 h to permit close monitoring and safe recovery from the anesthesia prior to return to the home social group. All procedures were approved by the Institutional Animal Care and Use Committees at YNPRC and also followed the guidelines of the Institute of Medicine on the use of chimpanzees in research. The *in vivo* scans were obtained using a 1.5T G.E. echo-speed Horizon LX MR scanner (GE Medical Systems, Milwaukee, WI, USA). T1-weighted images were collected in the transverse plane using a gradient echo protocol (pulse repetition = 19.0 ms, echo time = 8.5 ms, number of signals averaged = 8, matrix size = 256 × 256, with 0.7 × 0.7 × 1.2 resolution). For the post-mortem scanning, either 4.7 or 7T magnets were used and T2-weighted images were collected in the transverse plane using a gradient echo protocol (pulse repetition = 22.0 s, echo time = 78.0 ms, number of signals averaged = 8–12, and a 256 × 192 matrix reconstructed to 256 × 256).

2.6. MRI processing

Initially, using Analyze 8.1, all chimpanzee scans were skull-stripped, cropped, and reformatted at 0.7 cubic isotropic resolution and subsequently imported into BV. Extracting the sulci from the cortex in the brain scans involves a series of steps in a pipeline process within BV (Mangin et al., 2004) (see Figure 1) and these methods are described in detail elsewhere (Bogart et al., 2012a). Initially, the anterior and posterior commissures were manually specified on the MRI where they intersect with the mid-sagittal slice to align the brain. The first step was to correct for special inhomogeneities in the signal intensity providing a spatially smooth bias field with a stable distribution of tissue intensities (Figure 1b). Next, the analysis of the signal histogram

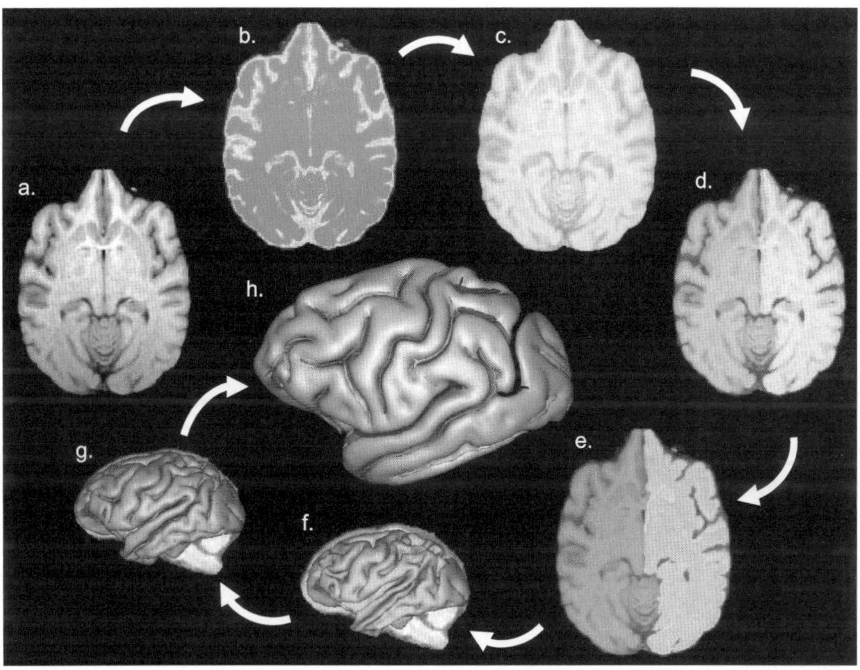

Figure 1. BrainVISA's pipeline processing steps. (a) MR image of a skull-stripped chimpanzee brain, (b) stable tissue intensities after bias field correction, (c) binary mask of the brain, (d) split mask of left and right hemispheres and cerebellum, (e) grey and white interface, (f) negative mould of the white matter, (g) skeletonised mould of cortical folding and (h) cortical fold graph of chimpanzee sulci with the central sulcus in red. This figure is published in colour in the online edition of this journal, which can be accessed via http://booksandjournals.brillonline.com/content/journals/1568539x.

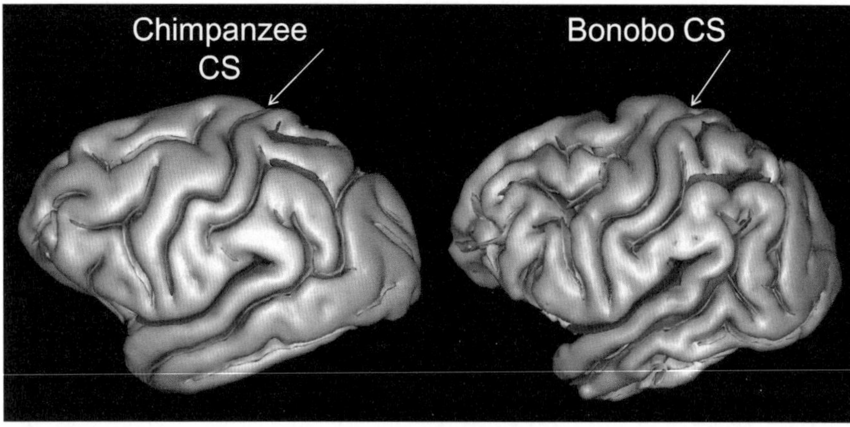

Figure 2. 3-D reconstruction of cortical sulci of the chimpanzee and bonobo brain with the central sulcus (red) indicated in each species. This figure is published in colour in the online edition of this journal, which can be accessed via http://booksandjournals.brillonline.com/content/journals/1568539x.

and mathematical morphology was computed to obtain a binary mask of the brain (Figure 1c). The mask was then split into the left and right hemispheres and the cerebellum (Figure 1d). A negative mould of the white matter was computed from the split-brain mask. The outside boundary of this mould results from a 5 mm morphological closing of the masked hemisphere, filling up the folds. The grey/white interface is the inside boundary of the mould, reconstructed with topography, preserving deformations assuring the spherical topology of the mould (Figure 1e). Finally the mould was skeletonised to detect cortical folding, while topological constraints guarantee the resulting surfaces have no holes (Figure 1e, f). The deepest part of the fold indicates the buried gyrus given the grey/white interface. The final step results in creating a cortical fold graph containing all the extracted sulci (Figure 1h). The central sulcus was manually labelled on the 3-D renderings for each brain (Figure 2).

2.7. *Central sulcus measures*

Central sulcus measures included surface area (mm^2), mean depth (mm) and grey matter thickness (mm) (see Figure 3). The sulcus surface area was computed as the sum of the areas of all the triangles required to mesh the sulcus medial surface. The sulcus mean depth is the average depth computed across all the bottom points of the sulcus along its principal axis of projection

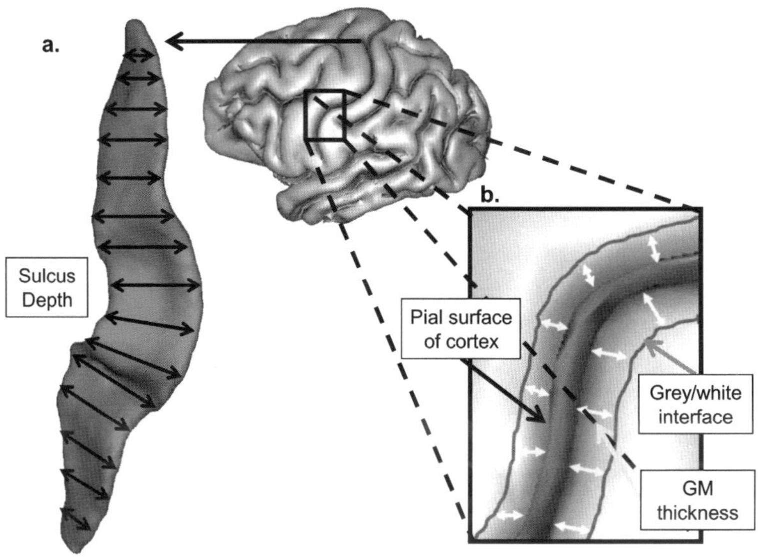

Figure 3. Chimpanzee central sulcus labeled and extracted. (a) The surface area and depth dimensions are shown in the extracted sulcus, as well as the x- and y-coordinates used for computing differences in cortical folding of the CS along the dorsal–ventral axis. (b) Illustration of the landmarks used to quantify grey matter thickness. This figure is published in colour in the online edition of this journal, which can be accessed via http://booksandjournals.brillonline.com/content/journals/1568539x.

(i.e., dorsal–ventral or anterior–posterior). The bottom points are defined from topological properties and correspond to the sulcus edge that is not connected to the cortex hull (see Figure 3a). Hence, this definition can be used even with interrupted sulci. The depth of a bottom point is the length of the shortest path from this point to the cortex hull embedded in the sulcus medial surface (mathematically speaking, the shortest geodesic path). Sulcus grey matter thickness (GM thickness) was quantified by measuring the mean distance between the pial surface and the grey/white interface around the buried sulcus (see Figure 3b).

2.8. Comparative analysis of the motor-hand area

To explore the variation in the central CS region further, we quantified the CS surface area corresponding to the PPFM using previously described methods (Coulon et al., 2006; Cykowski et al., 2008; Hopkins et al., 2010) (see Figure 4).

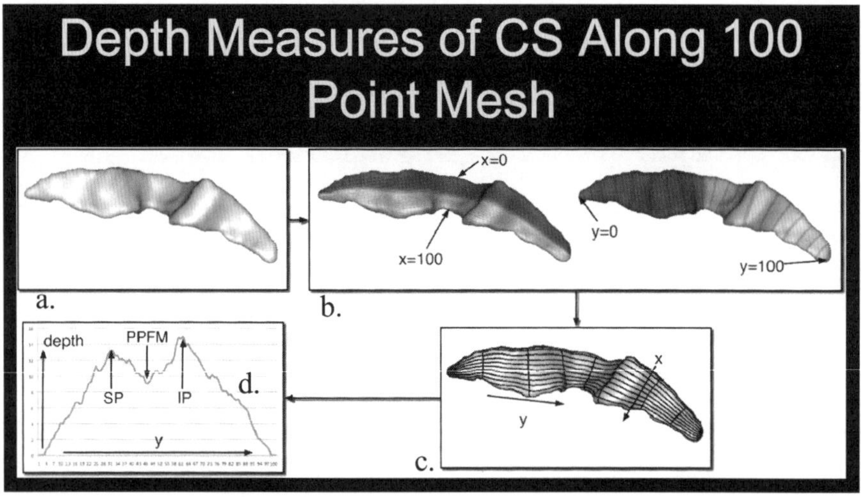

Figure 4. (a) Chimpanzee central sulcus and (b, c) the surface area and depth dimensions are shown in the extracted sulcus, as well as the x- and y-coordinates used for computing differences in cortical folding of the CS along the dorsal–ventral axis (d) outputted data from CS parameterization. Depth of CS is plotted on ordinate and the y coordinate along the abscissa. SP, superior maximum; CS, depth before y coordinate 50; IP, maximum inferior depth after y coordinate 50; PPFM, pli-de-passage moyen parietale, which is the shallowest CS depth measure between the SP and IP y coordinates. This figure is published in colour in the online edition of this journal, which can be accessed via http://booksandjournals.brillonline.com/content/journals/1568539x.

For each subject, the CS was standardized into 100 equally spaced sections along the dorsal–ventral axis and the depth of the sulcus at each point was quantified. The procedure can be briefly summarized as follows: a coordinate system is computed on the sulcus that indicates the position of each point relative to its depth (x coordinate) and its position along the sulcus between the dorsal and ventral extremities (y coordinate), as illustrated in Figure 4b and 4c. At each position, $y = 0$ through 100 along the sulcus length, the depth is defined as the length of the corresponding y iso-coordinate line (see Figures 4d and 5). Various publications provide more technical details about this procedure (Davatzikos & Bryan, 2002; Coulon et al., 2006; Cykowski et al., 2008; Hopkins et al., 2010). From the parameterization, the depth measure of the CS at each of the 100 y-location points are outputted to a text file. From this file, we recorded the greatest depth measure that was superior (SP) and inferior (IP) to the central location of the CS (location 50) (see Figure 5). We then recorded the shallowest CS depth

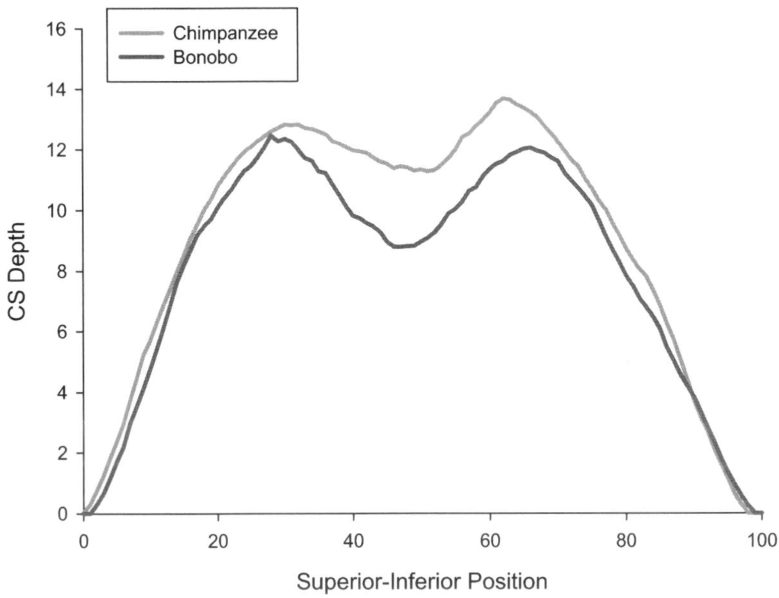

Figure 5. Variation in CS depth along the Superior-Inferior plane in the chimpanzee (purple line) and bonobos (blue line). SP, maximum superior depth; PPFM, shallowest CS depth point between SP and IP; IP, maximum inferior depth. This figure is published in colour in the online edition of this journal, which can be accessed via http://booksandjournals.brillonline.com/content/journals/1568539x.

between the SP and IP locations. The PPFM size was then computed for the left and right hemisphere following the formula (PPFM $= ((\text{depth}_{IP} + \text{depth}_{SP})/2.0) - \text{depth}_{PPFM}$).

2.9. CS data analysis

For all measures (total CS surface area, mean depth, GM thickness), mean scores were computed by averaging the raw values between the left and right hemisphere. Asymmetry quotients (AQ) for all measures were also calculated following the formula: (AQ $= (R - L)/(R + L)$), where R and L represent right and left hemisphere values for each measure, respectively. Positive AQ values reflected a right hemisphere bias whereas negative values reflected a left hemisphere bias. Inferential statistics were used for all analyses with alpha set to $p < 0.05$ unless normality or homogeneity of variance assumptions for the tests could not be met. If these assumptions were not met, then non-parametric tests were used. Post-hoc comparisons on inferential statistics, when used, were conducted using Tukey's Honestly Significant Difference test.

3. Results

3.1. Hand preferences

Individual HI scores for each subject are shown in Table 1. We initially compared the HI values for each task as a function of the species and sex using a mixed model analysis of variance. Handedness measure was the repeated measure while species and sex were the between-group factors. A significant main effect for species was found $F_{1,56} = 8.166$, $p < 0.009$. Overall, the chimpanzees had significantly higher HI values than the bonobos and this was consistent across all three measures. The mean HI score for each species and task are shown in Figure 6. Within each species, we also performed one-sample t-tests to test for population-level asymmetries when considering handedness on a continuous scale of measurement. For all analyses, the estimated population mean for the HI scores was zero and none of the t-tests

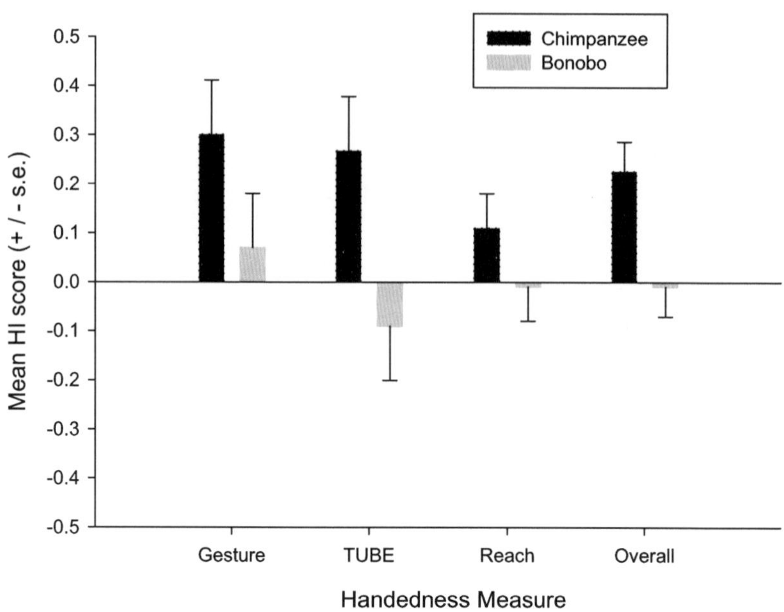

Figure 6. Mean HI scores (±SE) for each handedness task (gesture, TUBE, and reach) and the overall score in bonobos and chimpanzees.

Table 2.
Distribution of handedness based on *z*-scores for each task and species.

Task	Bonobo			Chimpanzee		
	No. R	No. A	No. L	No. R	No. A	No. L
Gestures	12	7	9	21	6	7
TUBE	16	4	11	23	5	6
Reaching	9	15	10	17	9	8

reached statistical significance within the bonobo sample. In contrast, the chimpanzees showed population-level right handedness for manual gestures $t_{33} = 3.27$, $p < 0.01$ and the TUBE task $t_{33} = 4.78$, $p < 0.002$.

The hand preference data based on the classification criteria applied to the *z*-scores largely confirmed analysis of the HI scores. The hand preference distributions based on the *z*-scores are shown in Table 2. Within the bonobo sample, chi-square goodness-of-fit tests failed to reveal significant differences in the number of right- and left-handed individuals for any of the measures. In contrast, for chimpanzees, there were significantly more right- than left-handed individuals for manual gestures $\chi^2(1, N = 28) = 7.00$, $p < 0.02$ and the TUBE task $\chi^2(1, N = 29) = 9.97$, $p < 0.004$.

3.2. *Grip morphology and performance asymmetries*

We next considered whether species and/or sex differences were evident in the grasping morphology and performance asymmetries. We initially calculated the percentage of grasping responses that involved the thumb and index finger for the left and right hands and compared these values between species and sexes using a mixed model analysis of variance. Hand was the repeated measure (left, right) while sex and species served as the between group factors. No significant main effects or interactions were found. The mean percentage of thumb-index grasping responses for bonobos and chimpanzees were 71.6 and 62.9%, respectively. For the grasping error data, overall the chimpanzees (mean HI = −0.078) showed a greater right hand advantage compared to the bonobos (mean HI = +0.013) but the difference in HI values did not reach conventional levels of statistical significance.

3.3. *Central sulcus morphology and asymmetry*

In the first analysis, we compared the bonobos and chimpanzees on the mean surface area, mean depth and mean GM thickness of the overall CS using

Table 3.
Descriptive data on CS and PPFM measurements in chimpanzees and bonobos.

	Bonobo	SD	Chimpanzee	SD
Average				
Surface area	1063.08	193.85	1176.44	229.65
Mean depth	9.367	0.87	9.798	1.23
GM thickness	2.458	0.692	1.984	0.421
PPFM	6.227	2.629	5.167	1.995
Asymmetry				
Surface area	0.042	0.123	−0.048	0.171
Mean depth	0.029	0.054	−0.046	0.111
GM thickness	−0.042	0.092	0.008	0.038
PPFM	−0.201	0.730	0.127	0.641
Handedness	−0.010	0.331	0.215	0.291

Surface area measurements are in mm^2, while mean depth, GM thickness and PPFM measures are in mm.

MANOVA. Species was the between group factors. A significant species difference was found for mean GM thickness $F_{1,24} = 4.463$, $p < 0.05$ with chimpanzees having thinner grey matter (mean = 1.984 mm) compared to bonobos (mean = 2.459 mm) (see Table 3). A second MANOVA using the surface area, mean depth and GM thickness AQ scores revealed significant species differences for the mean depth AQ $F_{1,24} = 4.726$, $p < 0.05$, while a borderline significant difference was found for GM thickness AQ variable $F_{1,24} = 3.199$, $p < 0.09$. The mean AQ scores for each species are shown in Table 3. For comparison, we have also plotted the mean HI scores. As can be seen, bonobos had greater leftward biases than chimpanzees for surface area and mean depth, but greater rightward biases for GM thickness. We next compared bonobos and chimpanzees on the PPFM measures within the CS region. A comparison in the mean PPFM, as well as the PPFM AQ values did not differ between the species; though there was a trend for bonobos to have a greater leftward asymmetry compared to the chimpanzees (see Table 3).

4. Discussion

The results of this study are relatively straightforward. When compared on three measures of hand preference, chimpanzees were significantly more right-handed than the bonobos. No species differences were found in performance asymmetries either in grasping or in their grip preferences. We also found that grey matter thickness within the CS fold was lower in chimpanzees than bonobos. Further, significantly greater leftward asymmetries were found in CS depth in chimpanzees compared to bonobos. In contrast, chimpanzees had greater rightward asymmetries in GM thickness compared to bonobos.

With respect to the hand preference results, the evidence that chimpanzees were significantly more right-handed than the bonobos, particularly for the TUBE task, suggest that previous reports of a lack of population-level handedness in bonobos are not an artefact of the sample size. From an evolutionary perspective, one interpretation of this finding is that selection for increasing motor skill, such as tool-use, may have resulted in increased preferential use of the right hand for coordinated bimanual actions in *Pan troglodytes*. The absence of selection for tool-use in bonobos may explain the less developed hand preferences in this species, at least with regard to the measures of handedness employed in this study. In short, the results of this study favour theories that emphasize the importance of tool-use as a potentially important adaptation for the development of right-handedness in early humans. There are, however, two caveats to this conclusion. First, the best way of directly testing for the potential role of the evolution of tool use on behavioural asymmetries would be to compare data from both bonobos and chimpanzees on tool use tasks. That was not done in this study but these types of data would lead to more definitive conclusions. Second, evidence of hand preference for tool use in wild chimpanzees is not entirely consistent with the theory that increased selection for tool use led to increased right-handedness. When data from wild chimpanzees are combined across field sites, they appear to show right-handedness for wadge dipping and ant -dipping but left hand preferences for termite fishing and no bias for nut-cracking. Of course, it may be the case that tool use per se was not the sole driving force behind the development of human right handedness. For example, though gorillas have been documented to use tools in captivity (Boysen et al., 1999; Parker et al., 1999), evidence of tool use in wild individuals, like bonobos, are fairly limited despite long-term behavioural observations at some field sites. However, of all the great apes and despite early claims by some (McGrew & Marchant, 1993), gorillas appear to show the most consistent evidence of right handedness. For bimanual feeding, there is evidence of right-handedness in wild and captive populations (Byrne & Byrne, 1991; Meguerditchian et al., 2010a; Lambert, 2012; Tablowo &

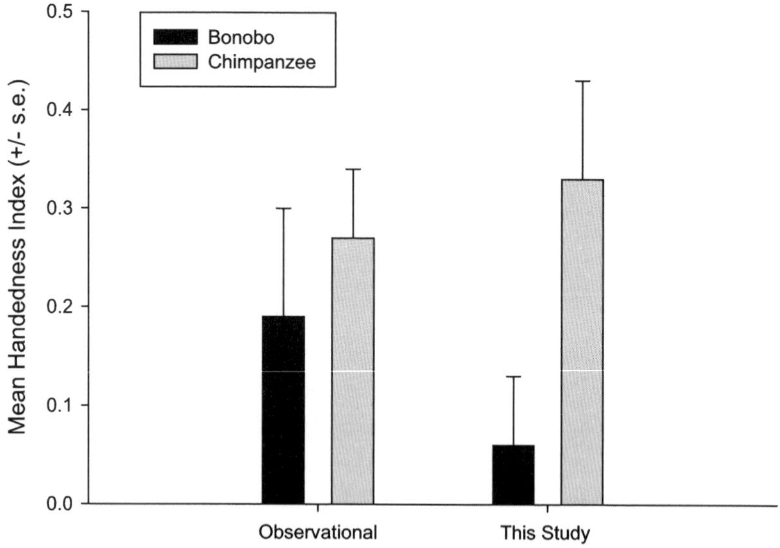

Figure 7. Mean HI scores (±SE) in manual gestures from previous studies compared to the data reported in this study.

Forrester, 2013). Moreover, in two of three captive populations, more than 90% of the gorillas were reported to be right-handed for bimanual feeding. The fact that the chimpanzees were significantly more right-handed for all measures, including manual gestures, was somewhat surprising within the context of existing data on gestural communication in bonobos and chimpanzees. One factor that might explain this result is in the methodology used to assess handedness in chimpanzees and bonobos. Shown in Figure 7 are the mean HI scores for bonobos and chimpanzees for spontaneous communicative gestures pooled from published data in the literature (Hopkins & de Waal, 1995; Shafer, 1997; Hopkins et al., 2005c, 2012; Harrison & Nystrom, 2008; Meguerditchian et al., 2010b). As can be seen, the HI scores are significantly higher in the bonobos during the observational conditions compared to the results reported here, when collected under structured testing conditions. In contrast, the chimpanzee HI scores are relatively comparable between the observational and experimental conditions. Although we cannot rule out the possibility that the differences in HI values between the two bonobo cohorts simply reflects sampling variation, the results suggest that the experimental paradigm used in this study to elicit manual gestures did not induce a population-bias, which appears evident when assessing their natural interspecies communication. This discrepancy was less apparent in the chimpanzees and we would therefore caution against making too strong of a claim regarding the handedness for manual gestures in the bonobos.

The reason for the discrepancy in findings between the two bonobo cohorts is not clear but we would note that the tendency for the bonobos in this study to spontaneously point to foods outside of their home cage was rarely observed. This is quite different from what has been reported in captive living chimpanzees, particularly those housed at the Yerkes National Primate Research Center (Leavens & Hopkins, 1998). In fact, at one zoo, a carestaff member indicated that the bonobos were discouraged from pointing or begging for foods from the humans interacting with them, thereby fundamentally extinguishing this behaviour. To what extent these types of experiences may influence both the propensity and lateralization in gestural communication during these interspecies exchanges is unclear; however, this could prove to be an important variable. Previous studies have shown that hand preferences for reaching at 4 years of age predicts subsequent hand preferences for manual gestures in captive chimpanzees as assessed by the methods employed in this study (Hopkins et al., 2005a). Hopkins et al. (2005a) have previously argued that these early reaching responses become ritualized into a manual gestural response as the subjects grow older. Early in life the young chimpanzees can reach through the cage mesh to retrieve food and they exhibit asymmetries in hand preference for this reaching response (Hopkins, 1995b). As they grow older, their hands become too large to reach through the cage mesh and therefore they develop a more ritualized gestural response to request the food. Thus, their hand preferences for manual gestures largely develop from their early reaching preferences. What may have happened with the bonobos in this study, compared to the chimpanzees, is that these ritualized processes during development were never experienced and therefore they did not come to use gestures functionally the same way as the chimpanzees, or at least those in this study. Though this is speculative, it is plausible and warrants further investigation.

Regarding grasping morphology, the results are largely consistent with the previous report by Christel et al. (1998) with bonobos showing a preponderance of thumb-index grasping responses. Though the findings reported by Christel et al. (1988) in bonobos were certainly suggestive of species differences when compared to previous reports on grasping morphology in chimpanzees (Tonooka & Matsuzawa, 1995; Jones-Engel & Bard, 1996;

Butterworth & Itakura, 1998; Hopkins et al., 2005d), our findings do not confirm the presence of differences in grasping morphology when subject variables such as age, sex and rearing history are controlled for in the species. We also found no significant differences in asymmetries in manual performance as measured by the error rates between the two species nor did we find any population-level biases within each species. Though the performance measures were not significant within and between species, the patterns of results between the hand preference and performance measures were consistent. That is to say, the chimpanzees showed population-level right hand preferences and show overall performance asymmetries which reflected better performance by the right compared to left hand on the grasping task. In contrast, the bonobos showed no population-level hand preferences and performed equally well with the right- and left-hands. The lack of significant population-level asymmetries in grasping performance differs from previous reports but this likely reflects the lower statistical power in this study compared to previous reports. Notably, Hopkins et al. (2002) assessed performance asymmetries in 74 chimpanzees in contrast to the 31 individuals in this group. Hopkins et al. (2002) noted that the effect size for performance asymmetries in grasping were relatively small and required larger samples in order to detect the differences. Thus, with only 31 chimpanzees and bonobos in this study, there may not have been enough statistical power to detect significant differences in grasping performance.

Analysis of the surface area, depth and GM thickness revealed several interesting results. Overall, the chimpanzees had thinner GM within the CS fold compared to bonobos. Previous studies have shown that GM thickness is negatively correlated with white matter volume in chimpanzees (Bogart et al., 2014); therefore, these results indicate that bonobos may have less white matter within the pre- or post-central gyri. Additionally, we found significant differences in asymmetries in mean depth and GM thickness with leftward asymmetries present in chimpanzees compared to bonobos for CS depth and the opposite pattern for GM thickness. These results are also consistent with the suggestion that differences in white matter between bonobos and chimpanzees in the pre- or post-central gyri may underlie the effects. It has been suggested that phylogenetic variation in cortical folding is driven, in part, by increases in white matter (van Essen, 1997; Herculano-Houzel et al., 2010). According to this view, regions that are strongly connected form more connections and results in the surface of the cortex folding inward and thereby forming gyri and sulci. By this definition, the greater CS depth in the left hemisphere of chimpanzees expands the surface area of the fold; the consequence is an increased thinning of the grey matter within the left hemisphere, which is necessary, assuming a finite and consistent amount of grey matter between hemispheres. Because the right hemisphere is less folded, it has grey matter that can be more thickly dispersed along the surface area.

Another important result was the differences in PPFM asymmetry. Though not significant at $p < 0.05$, nonetheless, the data reported here are entirely consistent with the behavioural data on handedness (see Figure 6). Chimpanzees were more right-handed than the bonobos and they also showed much more rightward asymmetries in the PPFM. This may appear paradoxical but it is important to recall that a greater right AQ value for the PPFM actually reflects a larger region in the left hemisphere. Because we are measuring the depth of the fold, which is being impeded by the presence of the buried PPFM gyrus, a smaller PPFM value likely indicates a larger buried gyrus connecting the pre- and post-central gyri. Presumably this would indicate potentially greater intrahemispheric connectivity between the motor and somatosensory functions.

Finally, the robust leftward asymmetry in the PPFM found here in bonobos is consistent with the reported evidence of leftward asymmetries in the volume motor-hand region of the precentral gyrus or KNOB (Hopkins et al., 2009).

In summary, this study found that chimpanzees are significantly more right-handed than bonobos for a series of measures of hand preference. Like other studies on handedness in a different set of closely related species (*Macaque*, *Cebus*) (Spinozzi et al., 1998; Westergaard et al., 2001; Lilak & Phillips, 2007; Meunier & Vauclair, 2007), the results reported here suggest that differences in ecological or sociological factors may play an important role in the development of handedness in closely-related primates. There has historically been a tendency to consider primate handedness in very limited, linear evolutionary terms, with a very anthropocentric view of human uniqueness relative to other apes (Cashmore et al., 2008; Cashmore, 2009; Uomini, 2009). If nothing else, we believe our results suggest that many evolutionary factors influence the development of handedness and that there is need to more broadly consider the comparative study of handedness, particularly among closely-related species with the goal of trying to find common ecological or social factors that might explain between-species similarities and differences in handedness. This approach, we believe, may be more fruitful in identifying the unique factors that have selected for the pronounced expression of handedness seen in humans.

Acknowledgements

This research was supported in part by NIH grants NS-42867, NS-73134, HD-38051, and HD-56232 to WDH. O. Coulon is funded by the French National Research Agency (ANR-09-BLAN-0038-1, BrainMorph). The Yerkes Center is fully accredited by the American Association for Accreditation of Laboratory Animal Care. American Psychological Association guidelines for the ethical treatment of animals were adhered to during all aspects of this study. We are grateful to the helpful assistance of the bonobo veterinarians and keepers at the Milwaukee and Jacksonville Zoos, as well as the carestaff at the Iowa Primate Learning Center.

References

Alkadhi, H. & Kollias, S.S. (2004). Pli de passage fronto-parietal moyen of Broca separates the motor homoculus. — Am. J. Neuroradiol. 25: 809-812.

Annett, M. (2002). Handedness and brain asymmetry: the right shift theory. — Psychology Press, Hove.

Bailey, P., von Bonin, G. & McCulloch, W.S. (1950). The isocortex of the chimpanzee. — University of Illinois Press, Urbana-Champaign, IL.

Biro, D., Inoue-Nakamura, N., Tonooka, R., Yamakoshi, G., Sousa, C. & Matsuzawa, T. (2003). Cultural innovation and transmission of tool use in wild chimpanzees: evidence from field experiments. — Anim. Cogn. 6: 213-223.

Biro, D., Sousa, C. & Matsuzawa, T. (2006). Ontogeny and cultural propagation of tool use by wild chimpanzees at Bossou, Guinea: case studies in nut cracking and leaf folding. — In: Cognitive development of chimpanzees (Matsuzawa, T., Tomonaga, T. & Tanaka, M., eds). Springer, New York, NY, p. 476-507.

Boesch, C. (1991). Handedness in wild chimpanzees. — Int. J. Primatol. 6: 541-558.

Bogart, S.L., Mangin, J.-F., Schapiro, S.J., Reamer, L., Bennett, A.J., Pierre, P.J. & Hopkins, W.D. (2012a). Cortical sulci asymmetries in chimpanzees and macaques: a new look at an old idea. — NeuroImage 61: 533-541.

Bogart, S.L., Pruetz, J.D., Ormiston, L.K., Russell, J.L., Meguerditchian, A. & Hopkins, W.D. (2012b). Termite fishing laterality in the Fongoli savanna chimpanzees (*Pan troglodytes verus*): further evidence of a left hand preference. — Am. J. Phys. Anthropol. 149: 591-598.

Bogart, S.L., Bennett, A.J., Schapiro, S.J., Reamer, L.A. & Hopkins, W.D. (2014). Different early rearing experiences have long term effects on cortical organziation in captive chimpanzees (*Pan troglodytes*). — Dev. Sci. 17: 161-174.

Boling, W., Olivier, A., Bittar, R. & Reutens, D. (1999). Localization of hand motor activation in Broca's *pli de passage moyen*. — J. Neurosurg. 91: 903-910.

Boysen, S.T., Kuhlmeier, K., Halliday, P. & Halliday, Y.M. (1999). Tool use in captive gorillas. — In: The mentality of gorillas and orangutans (Parker, S.T., Mitchell, R.W. & Miles, H.L., eds). Cambridge University Press, Cambridge.

Bradshaw, B. & Rogers, L. (1993). The evolution of lateral asymmetries, language, tool-use and intellect. — Academic Press, San Diego, CA.

Bradshaw, J.L. (1997). Human evolution: a neuropsychological perspective. — Psychology Press, Hove.

Brakke, K.E. & Savage-Rumbaugh, E.S. (1995). The development of language skills in bonobo and chimpanzee — I. Comprehension. — Lang. Commun. 15: 121-148.

Brakke, K.E. & Savage-Rumbaugh, E.S. (1996). The development of language skills in *Pan* — II. Production. — Lang. Commun. 16: 361-380.

Butterworth, G. & Itakura, S. (1998). Development of precision grips in chimpanzees. — Dev. Sci. 1: 39-43.

Byrne, R.W. & Byrne, J.M. (1991). Hand preferences in the skilled gathering tasks of mountain gorillas (*Gorilla gorilla berengei*). — Cortex 27: 521-536.

Cashmore, L. (2009). Can hominin "handedness" be accuartely assessed? — Ann. Hum. Biol. 36: 624-641.

Cashmore, L., Uomini, N. & Chapelain, A. (2008). The evolution of handedness in humans and great apes: a review and current issues. — J. Anthropol. Sci. 86: 7-35.

Chapelain, A. & Hogervorst, E. (2009). Hand preferences for bimanual coordination in 29 bonobos (*Pan paniscus*). — Behav. Brain Res. 196: 15-29.

Chapelain, A., Hogervorst, E., Mbonzo, P. & Hopkins, W.D. (2011). Hand preferences for bimanual coordination in 77 bonobos (*Pan paniscus*): replication and extension. — Int. J. Primatol. 32: 491-510.

Christel, M.I., Kitzel, S. & Niemitz, C. (1998). How precisely do bonobos (*Pan paniscus*) grasp small objects? — Int. J. Primatol. 19: 165-194.

Clay, Z. & Zuberbuhler, K. (2011). Bonobos extract meaning from call sequences. — PLoS One 6: e18786.

Corballis, M.C. (1992). The lopsided brain: Evolution of the generative mind. — Oxford University Press, New York, NY.

Corballis, M.C. (2002). From hand to mouth: the origins of language. — Princeton University Press, Princeton, NJ.

Coulon, O., Clouchoux, C., Operato, G., Dauchot, K., Sirigu, A. & Anton, J.-L. (2006). Cortical localization via surface parameterization: a sulcus-based approach. — NeuroImage 31: S46.

Cykowski, M.D., Coulon, O., Kochunov, P.V., Amunts, K., Lancaster, J.L., Laird, A.R., Glahn, C. & Fox, P.T. (2008). The central sulcus: an observer-independent characterization of sulcal landmarks and depth asymmetry. — Cerebr. Cort. 18: 1999-2009.

Dadda, M., Cantalupo, C. & Hopkins, W.D. (2006). Further evidence of an association between handedness and neuroanatomical asymmetries in the primary cortex of chimpanzees (*Pan troglodytes*). — Neuropsychologia 44: 2482-2486.

Davatzikos, C. & Bryan, R.N. (2002). Morphometric analysis of cortical sulci using parametric ribbons: a study of the central sulcus. — J. Comput. Assis. Tomogr. 26: 298-307.

de Waal, F.B.M. (1989). Behavioral contrasts between bonobo and chimpanzee. — In: Understanding chimpanzees (Heltne, P.G. & Marquardt, L.A., eds). Harvard University Press, Cambridge, MA, p. 154-175.

Fischer, A., Wiebe, V., Pääbo, S. & Przeworski, M. (2004). Evidence for a complex demographic history of chimpanzees. — Mol. Biol. Evol. 21: 799-808.

Forrester, G.S., Leavens, D.A., Quaresmini, C. & Vallortigara, G. (2011). Target animacy influences gorilla handedness. — Anim. Cogn. 14: 903-907.

Forrester, G.S., Quaresmini, C., Leavens, D.A., Spiezio, C. & Vallortigara, G. (2012). Target animacy influences chimpanzee handedness. — Anim. Cogn. 15: 1121-1127.

Frost, G.T. (1980). Tool behavior and the origin of laterality. — J. Hum. Evol. 9: 447-459.

Gallup, G.G. (1970). Chimpanzees: self-recognition. — Science 167: 86-87.

Gibson, K.R. & Ingold, T. (1993). Tools, language and cognition in human evolution. — Cambridge University Press, Cambridge.

Greenfield, P.M. (1991). Language, tools, and brain: the ontogeny and phylogeny of hierarchically organized sequential behavior. — Behav. Brain Sci. 14: 531-550.

Greenfield, P.M. & Savage-Rumbaugh, E.S. (1990). Grammatical combination in *Pan paniscus*: processes of learning and invention in the evolution and development of language. — In: "Language" and intelligence in monkeys and apes: comparative developmental perspectives (Parker, S.T. & Gibson, K.R., eds). Cambridge University Press, New York, NY, p. 540-578.

Gruber, T., Clay, Z. & Zuberbuhler, K. (2010). A comparison of bonobo and chimpanzee tool use: evidence for a femle bias in the Pan lineage. — Anim. Behav. 80: 1023-1033.

Hammond, G. (2002). Correlates of human handedness in primary motor cortex: a review and hypothesis. — Neurosci. Biobehav. Rev. 26: 285-292.

Hare, B. (2007). From nonhuman to human mind. what changed and why? — Curr. Direct. Psychol. Sci. 16: 60-64.

Hare, B. & Tomasello, M. (2004). Chimpanzees are more skillful in competitive than in cooperative cognitive tasks. — Anim. Behav. 68: 571-581.

Hare, B., Melis, A.P., Woods, V., Hastings, S. & Wrangham, R. (2007). Tolerance allows bonobos to outperform chimpanzees on a cooperative task. — Curr. Biol. 17: 619-623.

Harrison, R.M. & Nystrom, P. (2008). Handedness in captive bonobos (*Pan paniscus*). — Folia Primatologica 79: 253-268.

Herculano-Houzel, S., Mota, B., Wong, P. & Kaas, J.H. (2010). Connectivity-driven white matter scaling and folding in primate cerebral cortex. — Proc. Natl. Acad. Sci. USA 107: 19008-19013.

Herrmann, E., Hare, B., Call, J. & Tomasello, M. (2010). Differences in the cognitive skills of bonobos and chimpanzees. — PLoS One 5: e12438.

Hopkins, W.D. (1995a). Hand preferences for a coordinated bimanual task in 110 chimpanzees: cross-sectional analysis. — J. Comp. Psychol. 109: 291-297.

Hopkins, W.D. (1995b). Hand preferences in juvenile chimpanzees: continuity in development. — Dev. Psychol. 31: 619-625.

Hopkins, W.D. (2007). Hemispheric specialization in chimpanzees: evolution of hand and brain. — In: Evolutionary cognitive neuroscience (Shackelford, T., Keenan, J.P. & Platek, S.M., eds). MIT Press, Cambridge, MA, p. 99-120.

Hopkins, W.D. & Cantalupo, C. (2004). Handedness in chimpanzees is associated with asymmetries in the primary motor but not with homologous language areas. — Behav. Neurosci. 118: 1176-1183.

Hopkins, W.D. & de Waal, F.B.M. (1995). Behavioral laterality in captive bonobos (*Pan paniscus*): replication and extension. — Int. J. Primatol. 16: 261-276.

Hopkins, W.D. & Pilcher, D.L. (2001). Neuroanatomical localization of the motor hand area with magnetic resonance imaging: the left hemisphere is larger in Great Apes. — Behav. Neurosci. 115: 1159-1164.

Hopkins, W.D. & Russell, J.L. (2004). Further evidence of a right hand advantage in motor skill by chimpanzees (*Pan troglodytes*). — Neuropsychologia 42: 990-996.

Hopkins, W.D. & Savage-Rumbaugh, E.S. (1991). Vocal communication as a function of differential rearing experiences in *Pan paniscus*: a preliminary report. — Int. J. Primatol. 12: 559-583.

Hopkins, W.D., Cantalupo, C., Wesley, M.J., Hostetter, A.B. & Pilcher, D. (2002). Grip morphology and hand use in chimpanzees (*Pan troglodytes*): evidence of a left hemisphere specialization in motor skill. — J. Exp. Psychol. 131: 412-423.

Hopkins, W.D., Wesley, M.J., Izard, M.K., Hook, M. & Schapiro, S.J. (2004). Chimpanzees are predominantly right-handed: replication in three colonies of apes. — Behav. Neurosci. 118: 659-663.

Hopkins, W.D., Russell, J., Freeman, H., Buehler, N., Reynolds, E. & Schapiro, S. (2005a). The distribution and development of handedness for manual gestures in captive chimpanzees (*Pan troglodytes*). — Psych. Sci. 16: 487.

Hopkins, W.D., Russell, J., Hook, M., Braccini, S. & Schapiro, S. (2005b). Simple reaching is not so simple: association between hand use and grip preferences in captive chimpanzees. — Int. J. Primatol. 26: 259-277.

Hopkins, W.D., Russell, J.L., Freeman, H., Buehler, N., Reynolds, E. & Schapiro, S.J. (2005c). The distribution and development of handedness for manual gestures in captive chimpanzees (*Pan troglodytes*). — Psychol. Sci. 16: 487-493.

Hopkins, W.D., Russell, J.L., Hook, M., Braccini, S. & Schapiro, S.J. (2005d). Simple reaching is not so simple: association between hand use and grip preferences in captive chimpanzees. — Int. J. Primatol. 26: 259-277.

Hopkins, W.D., Russell, J.L. & Cantalupo, C. (2007). Neuroanatomical correlates of handedness for tool use in chimpanzees (*Pan troglodytes*): implication for theories on the evolution of language. — Psychol. Sci. 18: 971-977.

Hopkins, W.D., Lyn, H. & Cantalupo, C. (2009). Volumetric and lateralized differences in selected brain regions of chimpanzees (*Pan troglodytes*) and bonobos (*Pan paniscus*). — Am. J. Primatol. 71: 988-997.

Hopkins, W.D., Coulon, O. & Mangin, J.F. (2010). Observer-independent characterization of sulcal landmarks and depth asymmetry in the central sulcus of the chimpanzee brain. — Neuroscience 171: 544-551.

Hopkins, W.D., Phillips, K.A., Bania, A., Calcutt, S.E., Gardner, M., Russell, J.L., Schaeffer, J.A., von Lonsdorf, E., Ross, S. & Schapiro, S.J. (2011). Hand preferences for coordinated bimanual actions in 777 great apes: implications for the evolution of handedness in Hominins. — J. Hum. Evol. 60: 605-611.

Hopkins, W.D., Pika, S., Liebal, K., Bania, A., Meguerditchian, A., Gardner, M. & Schapiro, S.J. (2012). Handedness for manual gestures in great apes: a meta-analysis. — In: Current developments in non-human primate gesture research (Pika, S. & Liebal, K., eds). John Benjamins Publishing, Amsterdam.

Humle, T. & Matsuzawa, T. (2009). Laterality in hand use across four tool use behaviors among the wild chimpanzees of Bossou, Guinea, West Africa. — Am. J. Primatol. 71: 40-48.

Hyatt, C.W. & Hopkins, W.D. (1994). Self-awareness in bonobos and chimpanzees: a comparative perspective. — In: Self-awareness in animals and humans: developmental perspectives (Parker, S.T., Mitchell, R.W. & Boccia, M.L., eds). Cambridge University Press, New York, NY, p. 248-253.

Jones-Engel, L.E. & Bard, K.A. (1996). Precision grips in young chimpanzees. — Am. J. Primatol. 39: 1-15.

Lambert, M. (2012). Hand preference for bimanual and unimanual feeding in captive gorillas: extension in a second colony of apes. — Am. J. Phys. Anthropol. 148: 641-647.

Leavens, D.A. & Hopkins, W.D. (1998). Intentional communication by chimpanzee (*Pan troglodytes*): a cross-sectional study of the use of referential gestures. — Dev. Psychol. 34: 813-822.

Lilak, A.L. & Phillips, K.A. (2007). Consistency in hand preference across low-level and high-level tasks in capuchin monkeys (*Cebus apella*). — Am. J. Primatol. 69: 1-12.

Llorente, M., Palou, L., Carrasco, L., Riba, D., Mosquera, M., Colell, M., Fabre, M. & Feliu, O. (2010). Population-level right handedness for a coordinated bimanual task in naturalistic housed chimpanzees: replication and extension in 114 animals from Zambia and Spain. — Am. J. Primatol. 71: 1-10.

Lyn, H. (2007). Mental representation of symbols as revealed by vocabulary errors in two bonobos (*Pan paniscus*). — Anim. Cogn. 10: 461-475.

Lyn, H., Greenfield, P.M. & Savage-Rumbaugh, E.S. (2006). The development of representational play in chimpanzees and bonobos: evolutionary implications, pretense, and the role of interspecies communication. — Cogn. Dev. 21: 199-213.

Lyn, H., Greenfield, P.M. & Savage-Rumbaugh, E.S. (2011). Semiotic combinations in Pan: a cross-species comparison of communication in a chimpanzee and a bonobo. — First Lang. 31: 300-325.

Mangin, J.F., Riviere, D., Cachia, A., Duchesnay, E., Cointepas, Y., Papadopoulos-Orfanos, D., Collins, D.L., Evans, A.C. & Regis, J. (2004). Object-based morphometry of the cerebral cortex. — Med. Imag. 23: 968-982.

Marchant, L.F. & McGrew, W.C. (2007). Ant fishing by wild chimpanzees is not lateralised. — Primates 48: 22-26.

McGrew, W.C. (1992). Chimpanzee material culture: implications for human evolution. — Cambridge University Press, Cambridge.

McGrew, W.C. & Marchant, L.F. (1992). Chimpanzees, tools, and termites: hand preference or handedness? — Curr. Anthropol. 33: 114-119.

McGrew, W.C. & Marchant, L.F. (1993). Are gorillas right-handed or not? — Hum. Evol. 8: 17-23.

McGrew, W.C. & Marchant, L.F. (1996). On which side of the apes? — In: Great ape societies (McGrew, W.C., Marchant, L.F. & Nishida, T., eds). Cambridge University Press, Cambridge, p. 255-272.

McGrew, W.C. & Marchant, L.F. (1997). Using the tools at hand: manual laterality and elementary technology in *Cebus* spp. and *Pan* spp. — Int. J. Primatol. 18: 787-810.

Meguerditchian, A., Calcutt, S.E., von Lonsdorf, E., Ross, S.R. & Hopkins, W.D. (2010a). Captive gorillas are right-handed for bimanual feeding. — Am. J. Phys. Anthropol. 141: 638-645.

Meguerditchian, A., Vauclair, J. & Hopkins, W.D. (2010b). Captive chimpanzees use their right hand to communicate with each other: implications for the origins of hemispheric specialization for language. — Cortex 46: 40-48.

Meunier, H. & Vauclair, J. (2007). Hand preferences on unimanual and bimanual tasks in white-face capuchins. — Am. J. Primatol. 69: 1064-1069.

Parish, A.R. (1996). Female relationships in bonobos (*Pan paniscus*) — evidence for bonding, cooperation, and female dominance in a male-philopatric species. — Hum. Nat. Interdisc. Biosoc. Perspect. 7: 61-96.

Parish, A.R., de Waal, F.B.M. & Haig, D. (2000). The other "closest living relative": how bonobos (*Pan paniscus*) challenge traditional assumptions about females, dominance, intra- and intersexual interactions, and hominid evolution. — Ann. NY Acad. Sci. 907: 97-113.

Parker, S., Kerr, M., Markowitz, H. & Gould, J. (1999). A survey of tool use in zoo gorillas. — In: The mentalities of gorillas and orangutans: comparative perspectives (Parker, S.T. & Mitchell, R.W., eds). Cambridge University Press, Cambridge.

Phillips, K.A. & Sherwood, C.C. (2005). Primary motor cortex asymmetry is correlated with handedness in capuchin monkeys (*Cebus apella*). — Behav. Neurosci. 119: 1701-1704.

Pika, S., Liebal, K. & Tomasello, M. (2005). The gestural repertoire of bonobos (*Pan paniscus*): flexibility and use. — Am. J. Primatol. 65: 39-61.

Pollick, A.S. & de Waal, F.B.M. (2007). Ape gestures and language evolution. — Proc. Natl. Acad. Sci. USA 104: 8184-8189.

Rilling, J.K., Scholz, J., Preuss, T.M., Glasser, M.F., Errangi, B.V. & Behrens, T.E.J. (2012). Differences between chimpanzees and bonobos in neural systems supporting social cognition. — Soc. Cogn. Affect. Neurosci. 7: 369-379.

Rosati, A.G. & Hare, B. (2012). Chimpanzees and bonobos exhibit divergent spatial memory development. — Dev. Sci. 15: 840-853.

Savage-Rumbaugh, E.S. (1984). *Pan paniscus* and *Pan troglodytes*: contrast in preverbal communicative competence. — In: The pygmy chimpanzee: evolutionary biology and behavior (Susman, R.L., ed.). Plenum Press, New York, NY, p. 395-413.

Shafer, D.D. (1997). Hand preference behaviors shared by two groups of captive bonobos. — Primates 38: 303-313.

Spinozzi, G., Castornina, M.G. & Truppa, V. (1998). Hand preferences for unimanual and coordinated-bimanual tasks in tufted capuchin monkeys (*Cebus apella*). — J. Comp. Psychol. 112: 183-191.

Stanford, C.B. (1998). The social behavior of chimpanzees and bonobos: empirical evidence and shifting assumptions. — Curr. Anthropol. 39: 399-420.

Tablowo, E. & Forrester, G.S. (2013). Stuctured bimanual actions and hand transfers reveal population-level right handedness in captive gorillas. — Anim. Behav. 86: 1049-1057.

Taglialatela, J.P., Savage-Rumbaugh, E.S. & Baker, L.A. (2003). Vocal production by a language-competent *Pan paniscus*. — Int. J. Primatol. 24: 1-17.

Tonooka, R. & Matsuzawa, T. (1995). Hand preferences in captive chimpanzees (*Pan troglodytes*) in simple reaching for food. — Int. J. Primatol. 16: 17-34.

Toth, N., Schick, K.D., Savage-Rumbaugh, E.S., Sevcik, R.A. & Rumbaugh, D.M. (1993). Pan the tool-maker: investigations into the stone tool-making and tool-using capabilities of a bonobo (*Pan paniscus*). — J. Archaeol. Sci. 20: 81-91.

Uomini, N.T. (2009). The prehistory of handedness: archeological data and comparative ethology. — J. Hum. Evol. 57: 411-419.

van Essen, D.C. (1997). A tension-based theory of morphogenesis and compact wiring in the central nervous system. — Nature 385: 313-318.

von Lonsdorf, E. & Hopkins, W.D. (2005). Wild chimpanzees show population level handedness for tool use. — Proc. Natl. Acad. Sci. USA 102: 12634-12638.

Westergaard, G.C., Lussier, I.D. & Higley, J.D. (2001). Between-species variation in the development of hand preference among macaques. — Neuropsychologia 39: 1373-1378.

Whiten, A., Goodall, J., McGrew, W., Nishida, T., Reynolds, V., Sugiyama, Y., Tutin, C., Wrangham, R. & Boesch, C. (2001). Charting cultural variation in chimpanzees. — Behaviour 138: 1489-1525.

Whiten, A., Goodall, J., McGrew, W.C., Nishida, T., Reynolds, V., Sugiyama, Y., Tutin, C.E.G., Wrangham, R.W. & Boesch, C. (1999). Cultures in chimpanzees. — Nature 399: 682-685.

Won, Y.-J. & Hey, J. (2005). Divergence population genetics of chimpanzees. — Mol. Biol. Evol. 22: 297-307.

Yousry, T.A., Schmid, U.D., Alkadhi, H., Schmidt, D., Peraud, A., Buettner, A. & Winkler, P. (1997). Localization of the motor hand area to a knob on the precentral gyrus. A new landmark. — Brain 120: 141-157.

[When citing this chapter, refer to Behaviour 152 (2015) 493–520]

Bonobos and chimpanzees exploit helpful but not prohibitive gestures

Evan L. MacLean [a,*] **and Brian Hare** [a,b]

[a] Department of Evolutionary Anthropology, Duke University, Box 90383 Biological Sciences, Durham, NC 27705, USA

[b] Center for Cognitive Neuroscience, Duke University, Durham, NC 27705, USA

[*]Corresponding author's e-mail address: maclean@duke.edu

Accepted 28 March 2014; published online 7 May 2014

Abstract

Previous research has shown that chimpanzees exploit the behavior of humans and conspecifics more readily in a competitive than a cooperative context. However, it is unknown whether bonobos, who outperform chimpanzees in some cooperative tasks, also show greater cognitive flexibility in competitive contexts. Here we tested the cooperative-competitive hypothesis further by comparing bonobos and chimpanzees in a series of tasks where a human gesture indicated the correct (cooperative) or incorrect (competitive) choice. A human either pointed cooperatively to the object a subject should choose, or competitively to the object subjects should avoid choosing. In contrast to previous research, subjects were most skilled at choosing the correct location when the communicator was cooperative and there were no major differences between bonobos and chimpanzees. Analysis of gaze direction revealed that in some cases subjects visually followed the direction of the experimenter's gesture despite choosing incorrectly, dissociating gesture following from gesture comprehension. This supports the hypothesis that, unlike human children, nonhuman apes respond to the direction of social gestures more readily than they understand the communicative intentions underlying them. We evaluate these findings in regard to previous studies comparing the cooperative and communicative skills of bonobos and chimpanzees.

Keywords

cognition, chimpanzee, bonobo, cooperation, competition.

1. Introduction

A range of animals are able to communicate in flexible ways depending on their audience, the context, and the state of the receiver (Searcy & Nowicki, 2010). However, even the most adaptable animal communicators do not approach the flexibility observed in humans (Tomasello, 2008). Human infants are thought to begin developing skills for intentional communication as early as 9 months as evidenced first through their comprehension, and then production of pointing gestures (Carpenter et al., 1998; Behne et al., 2005). It has been proposed that the early expression of these pointing gestures is fundamental to the development of human cultural cognition and language (Carpenter et al., 1998; Butterworth, 2003) and the emergence of pointing in infants has become a central marker for the development of uniquely human social cognition (Tomasello, 2006).

The skills of other great apes have been examined to test whether the pointing abilities observed in human infants are unique to our species development (Hare, 2011). While nonhuman apes do regularly use manual gestures to flexibly communicate with one another in their natural interactions (Nishida, 1980; Goodall, 1986), evidence from observational and experimental studies suggest the same cognitive mechanisms are not always responsible for gestures in human and nonhuman apes (Call & Tomasello, 2007). Longitudinal studies examining the natural gestures used by chimpanzees show that the production of these gestures develops relatively slowly through a process of social ritualization (Tomasello et al., 1994; Tomasello & Call, 1997). Nonhuman apes have largely only been observed to produce imperative gestures, and rarely if ever make declarative or informative gestures (Melis et al., 2006, 2009; Tomasello, 2006; Lyn et al., 2011). A series of studies also have suggested that in the most basic communicative contexts nonhuman apes show little flexibility in comprehending human gestures. Unlike human infants, mother-reared chimpanzees do not spontaneously use cooperative social gestures when searching for hidden objects or food (Call et al., 1998, 2000; Itakura et al., 1999; Barth et al., 2005). They do not show spontaneous skill even when the target of the gesture is made more obvious by placing the hiding locations as much as 2 m apart (Itakura et al., 1999; Herrmann & Tomasello, 2006; but see Mulcahy & Call, 2009). Moreover, once chimpanzees learn to use one human gesture to find food, they do not readily generalize this skill to a slightly different gesture. For example, when chimpanzees are trained to search in the location toward which a human extends their finger within a few centimeters, they no longer reliably follow the pointing gesture if the experimenter simply stands one meter away from the location to which he points (Povinelli et al., 1997, 1999). Finally, in a direct longitudinal comparison of human, bonobo and chimpanzee infants, only human infants showed spontaneous skill at reading human gestures (Wobber et al., 2013). Taken together, this work supports the hypothesis that the sudden and early emerging gestural flexibility seen in human infants is a completely derived trait in our lineage (Herrmann et al., 2007).

This conclusion has been challenged more recently. Some apes with extensive human socialization have shown skills similar to human infants in basic tests of their comprehension of human gestures (Call et al., 2000; Leavens et al., 2005; Lyn et al., 2010). Some researchers have suggested that wild as well as human socialized bonobos occasionally make declarative gestures (Veà & Sabater-Pi, 1998; Lyn et al., 2011). Nonhuman great apes also adjust to their audience by making visual gestures more often when visible to the recipient, using tactile or vocal signals when not visible (Kaminski et al., 2004; Call & Tomasello, 2007; Tempelmann et al., 2011), and signaling danger more often when the receiver is ignorant than knowledgeable of an imminent threat (Crockford et al., 2012). Finally, it has been suggested that apes only struggle to spontaneously read gestures in cooperative contexts but are skilled when interpreting the same gestures in a competitive context (Hare & Tomasello, 2004). Almost universally nonhuman apes have been tested for their comprehension of human gestures in experimental contexts that require them to understand the cooperative-communicative intentions of a human experimenter. Yet primates rarely share information about the location of monopolizable food (Dittus, 1984; Coussi-Korbel, 1994; Hauser, 1996; Hirata & Matsuzawa, 2001). This suggests that more flexible social cognitive skills will be observed in apes when they are competing (Hare, 2001). In support of this hypothesis chimpanzees were more skilled at locating hidden food using gestures when competing against the signaler than when cooperating with him (Hare & Tomasello, 2004). Bonobos and chimpanzees were also more skilled at understanding the intentions behind a gesture meant to communicate which location they were prohibited from searching than when a similar gesture was made in a cooperative manner (Herrmann & Tomasello, 2006). However, this study included only two bonobo subjects, limiting the ability for a direct comparison of the two *Pan* species.

These findings present the possibility that nonhuman apes can understand the communicative intentions behind a human gesture when searching for food in a competitive context; it is only that they do not understand the cooperative-communicative intentions of a human when the signaler is trying to help them locate monopolizable food. Moreover, it is not that apes fail to understand the meaning of human gestures because they cannot follow the directionality of these signals, because there is overwhelming evidence that apes follow the gaze direction of others to external targets (Tomasello et al., 1998, 1999, 2001, 2007b; MacLean & Hare, 2012), but do not use these same cues to locate food in cooperative tasks (Call et al., 1998). This leads to the prediction that in cooperative contexts nonhuman apes may initially follow the direction of a human gesture, but do not reliably search in this location because they simply do not understand the informative motive behind the gesture in this context (Tomasello et al., 2007a). In the current series of studies we examine the ability of a group of bonobos and chimpanzees to use the pointing gestures of a human experimenter. We chose to investigate subjects' visual orienting and choice behavior in response to a pointing cue because many previous studies document that apes struggle to interpret this gesture in cooperative-communicative contexts (Call et al., 1998, 2000; Itakura et al., 1999; Barth et al., 2005). Thus, using a pointing gesture in both a cooperative and competitive context allows direct comparison to the previous literature, and also allows us to test the hypothesis that apes follow the direction of these cues faithfully, without understanding their communicative intent (Tomasello et al., 2007a).

Similar to Herrmann & Tomasello (2006) we either cooperatively indicated where subjects should choose or competitively indicated where subjects were 'prohibited' from choosing. In order to test the orienting hypothesis we also varied whether the food being chosen was visible or hidden and we coded from video the first orienting response of subjects after the experimenter pointed. We predicted that subjects would choose correctly when the experimenter's gesture was prohibitive, but not cooperative, but would first orient toward the cued location in both contexts. This would support the idea that nonhuman apes follow the direction of pointing gestures, but do not understand the cooperative-communicative intentions underlying them. Lastly we directly compared the performance of bonobos and chimpanzees to assess whether bonobos, who outperform chimpanzees in some cooperative tasks (Hare et al., 2007; Tan & Hare, 2013), and are affected by human competitors (Rosati & Hare, 2012), also show greater proficiency in the cooperative-communicative object choice paradigm.

2. General methods

All chimpanzee subjects live, and were tested at the Tchimpounga Sanctary, Pointe Noire, Republic of Congo. Bonobo subjects live and were tested at Lola ya Bonobo, Kinshasa, Democratic Republic of Congo. For a more detailed description of these field sites see Wobber & Hare (2011). Subjects were tested in a familiar dormitory room with metal grid walls. Subject demographics and experiment participation are shown in Table 1. The majority of subjects for both species (19 bonobos, 13 chimpanzees) had some previous experience in object choice tasks using cooperative-communicative gestures (see Table 1). All subjects were relatively unfamiliar with the experimenter (E1) prior to the experiment (E1 had briefly participated in research with some subjects approximately 1 year beforehand). Food was hidden under small plastic bowls (17.5 × 5.5 cm) positioned on a table with a sliding platform (50 × 81 cm) used to present subjects with the choices on each trial. On trials when food was hidden in containers a large plastic occluder was used to hide the baiting process from the subject. The sliding platform was marked in two places to delineate three 27 cm areas (left, center and right) used in scoring choices (see below). All trials were recorded from two video cameras. The first camera was positioned behind and to the side of E1 and recorded E1, the subject and procedure. The second camera was mounted to the rear of the table where choices were located and recorded the subject's face during trials (Figure 1). This angle was used to score looking behavior.

Data were analyzed using nonparametric statistics in IMB SPSS Version 20. We compared performance to chance expectation using one-sample Wilcoxon signed-rank tests. We compared performance between conditions and species using Mann–Whitney U-tests, and assessed learning between sessions using Related-Samples Wilcoxon signed-rank tests. We report the standardized test statistic (T) from each of these tests throughout.

3. Experiment 1

In Experiment 1 subjects were introduced to the experimenter either as a cooperator, who shared food, or a competitor who stole food from the subject. Apes then participated in a series of object-choice trials in which the experimenter either helpfully pointed to the (visible) food item that subjects should choose (cooperative condition), or prohibitively pointed to the item that subjects should not choose (competitive condition).

Table 1.
Subject information, individual performance and previous experience.

Species	Subject	Sex	Age	Experiment 1		Experiment 2		Experiment 3		Experiment 4		Previous experience (Reference)
				Condition	% Correct	Condition	% Correct	Condition	% Correct	Condition	% Correct	
Bonobo	Api	M	8	H	65	H	90	–	–	–	–	1
	Boyoma	M	4	H	55	H	*85*	H	45	H	53	2
	Dilolo	M	8	P	50	H	70	–	–	–	–	1
	Kalena	F	11	P	50	H	70	–	–	–	–	1
	Kasongo	M	7	P	50	H	*95*	P	50	P	50	1, 2
	Kikongo	M	8	P	56	H	41	–	–	–	–	1
	Kikwit	M	11	P	50	H	*74*	H	50	H	60	1, 2
	Luozi	M	6	H	58	H	*75*	–	–	–	–	2
	Mabali	M	8	P	40	H	70	–	–	–	–	1, 2
	Maniema	M	8	H	*78*	H	*85*	H	58	H	56	1, 2
	Manono	M	15	H	*94*	H	*90*	P	47	P	25	1, 2
	Matadi	M	8	H	70	H	65	P	45	P	47	1
	Maya	F	16	H	*85*	H	60	–	–	–	–	–
	Mbandaka	M	8	P	20	H	65	–	–	–	–	1
	Muanda	F	6	H	*74*	H	*79*	H	53	H	53	2
	Nioki	F	10	H	50	H	65	H	40	H	55	1, 2
	Opala	F	14	P	42	H	70	P	55	P	45	1
	Tchilengue	F	6	P	50	H	70	P	50	P	55	1, 2
	Vanga	M	5	P	40	H	70	H	40	H	50	2
	Yolo	M	6	H	50	H	65	P	35	P	40	1, 2

Table 1.
(Continued.)

Species	Subject	Sex	Age	Experiment 1		Experiment 2		Experiment 3		Experiment 4		Previous experience (Reference)
				Condition	% Correct	Condition	% Correct	Condition	% Correct	Condition	% Correct	
Chimpanzee	Bob	M	17	P	45	H	55	P	45	P	50	1, 2
	Boumbo	M	19	H	65	H	70	P	45	P	45	1, 2
	Christophe	M	19	H	*85*	H	*95*	H	50	H	*75*	1, 2
	Elikia	M	21	P	45	H	*90*	H	45	H	50	1, 2
	Kefan	M	9	H	65	H	60	H	50	H	*75*	1
	Kimenga	M	6	H	65	H	*85*	P	42	P	45	1, 2
	Marcel	F	7	H	*80*	H	*95*	H	50	H	60	2
	Maya	M	15	H	*75*	H	*85*	H	50	H	45	1, 2
	Mvouti	F	7	P	50	H	50	P	55	P	55	2
	Ngoro	F	5	P	25	H	*75*	H	40	H	55	2
	Oulengue	M	8	H	*80*	H	*100*	P	40	P	32	2
	Petit Pere	F	13	P	35	H	*75*	P	55	P	40	–
	Ramsay	M	10	P	45	H	70	H	50	H	60	1
	Tabonga	M	9	H	65	H	*85*	H	55	H	65	–
	Talianne	M	12	P	47	H	50	H	50	H	55	–
	Tchibanga	M	8	H	60	H	70	P	70	P	45	–
	Tiki	M	11	H	55	H	60	P	37	P	45	–
	Timi	M	5	P	25	H	60	P	35	P	55	–
	Vitika	F	9	P	40	H	*74*	H	50	H	50	1
	Yoko	M	13	P	35	H	*75*	P	40	P	40	–

Scores in italics are significantly above chance (binomial test). H, Helpful; P, Prohibitive.
[1] From Herrmann et al. (2010).
[2] From Wobber et al. (2013).

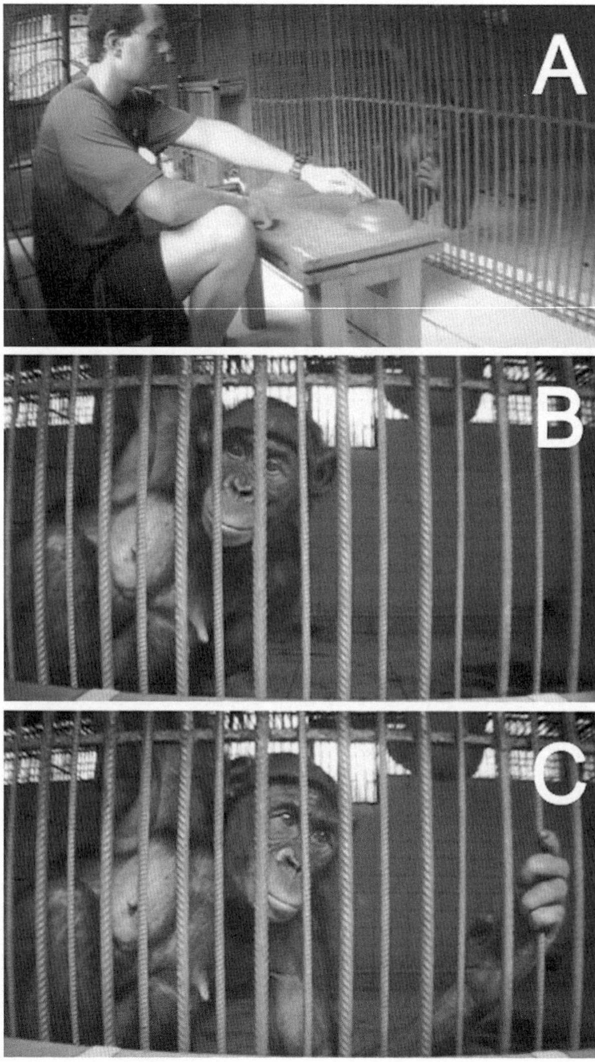

Figure 1. (A) The pointing gesture used in Experiments 1 and 2. Panels B and C show images from the coding camera corresponding to a subject looking at choices to the left (B) or right (C) of the blue lines defining the choice areas. This figure is published in colour in the online edition of this journal, which can be accessed via http://booksandjournals.brillonline.com/content/journals/1568539x.

3.1. Method

3.1.1. Subjects

We tested 20 chimpanzees (Table 1; 15 males, 5 females, mean age = 11 years) and 20 bonobos (Table 1; 14 males, 6 females, mean age = 9 years).

3.1.2. Procedure

3.1.2.1. Familiarization. We conducted 4 familiarization trials at the beginning of each session to introduce subjects to E1's cooperative or competitive motive. In each of these trials E1 placed a banana piece in the center of the table and slid the platform forward allowing subjects to reach for the food. Choices were defined as subjects touching or positioning their hand over the marked center area of the table where the food was positioned. In the cooperative condition E1 allowed subjects to retrieve the food. In the competitive condition E1 pulled the sliding platform away from subjects as they reached for the food, preventing them from obtaining it. We conducted these familiarization trials immediately before each of the two test sessions.

3.1.2.2. Test. In test trials E1 placed a piece of banana on each side of the table. He then called the subject's name and pointed to one of the banana pieces. The pointing gesture was performed with the contralateral arm, whole hand outstretched, fingers approximately 5 cm from the banana. E1 held this gesture for approx. 1 s before sliding the table forward for subjects to choose, maintaining the gesture until subjects made a choice or the trial timed out (see Design and analysis). Choices were defined as subjects touching or positioning their hand over the left or right marked areas of the table. Touching or positioning the hand over the center area was disregarded and E1 waited for subjects to choose either the left or right option. If a subject used both hands attempting to choose two options simultaneously, E1 retracted the table and repeated the trial. In the cooperative condition, if subjects chose the food E1 pointed to they were allowed to consume the reward, whereas choices to the non-indicated food ended the trial with no reward. In contrast, in the competitive condition if subjects chose the food E1 pointed to, E1 withdrew the table ending the trial, whereas subjects were allowed to consume the reward if they chose the non-indicated food. Therefore, in both conditions E1 provided subjects with the same social cue, but the meaning of this cue differed between the cooperative and competitive conditions.

3.1.2.3. Design and analysis. We conducted two 10-trial sessions with each subject. These sessions were separated by a 5-min break. Half of the subjects within each species were tested in the cooperative condition while the other half were tested in the competitive condition. The location (left or right) that E1 pointed to was counterbalanced within each session. Subjects were allowed 30 s to choose on each trial and if no choice was made during this time the trial was scored as 'no choice' and the next trial was conducted. We used the percentage of correct choices (including only trials in which subjects made a choice) as the dependent measure for analysis of choice behavior. From video, we recorded the direction of subjects looking behavior (Figure 1) on each trial during the period beginning when E1 gave the cue, and ending when he pushed the platform forward for subjects to choose. Only the direction of the first look (left or right) was recorded and if subjects did not look to either location in this period the trial was scored as 'no look'. If the subject's face was out of view during the coding window, the trial was scored as 'off camera' and excluded from analysis. Because the lighting conditions and enclosure mesh varied between testing rooms, only a subset of videos could be coded for looking behavior (bonobo $N = 11$, chimpanzee $N = 8$). A second individual blind to the condition and hypotheses scored looking behavior and subjects' choices in 20% of trials to assess inter-observer reliability. Inter-observer reliability was good for looking direction (kappa = 0.73; Altman, 1991), very good for choice location (kappa = 0.93), and scores from the primary coder were retained in cases of disagreement.

3.2. Results

3.2.1. Choice accuracy

Subjects in the cooperative condition chose the food that E1 pointed to significantly more often than expected by chance (mean = 69 ± 3% correct, $T = 3.73$, $N = 20$, $p < 0.01$) and showed no effect of learning between the first and second sessions ($T = -0.17$, $N = 20$, $p = 0.87$). Eight subjects (4 bonobos, 4 chimpanzees) performed above chance expectation as individuals in the cooperative condition (binomial tests, $p < 0.05$). Subjects in the competitive condition also tended to choose the food that E1 pointed to, leading to overall scores significantly worse than chance expectation (mean = 42 ± 2% correct, $T = -2.99$, $N = 20$, $p < 0.01$). However, subjects in the competitive condition improved between the first and second test sessions, decreasing their tendency to choose the food that E1 pointed to

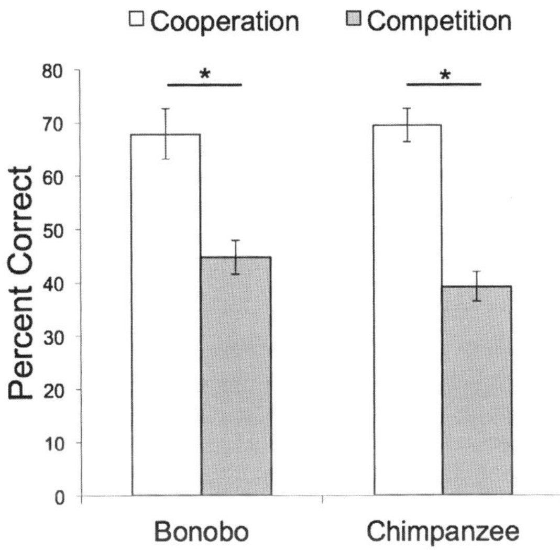

Figure 2. The results from Experiment 1. Both bonobos and chimpanzees performed better when E1's gesture cooperatively indicated the correct choice than when his gesture prohibitively indicated the incorrect choice.

over time (session 1 mean $= 36 \pm 3\%$ correct; session 2 mean $= 48 \pm 3\%$ correct; $T = 2.54$, $N = 20$, $p = 0.01$). At the individual level, no subjects performed above chance expectation in the competitive condition (binomial tests, $p > 0.05$). Overall performance was significantly higher in the cooperative than the competitive condition in both species (Figure 2; chimpanzees: $T = 3.80$, $N = 20$, $p < 0.01$; bonobos: $T = 3.24$, $N = 20$, $p < 0.01$).

There were no species differences in overall performance in either the cooperative or competitive conditions (cooperation: $T = 0.45$, $N = 20$, $p = 0.68$; competition: $T = -1.65$, $N = 20$, $p = 0.11$). However, in the competitive condition chimpanzees improved between the two test sessions ($T = 2.46$, $N = 10$, $p = 0.01$), but bonobos did not ($T = 0.67$, $N = 10$, $p = 0.50$).

3.2.2. Looking measures

For the subset of individuals for whom we were able to code looking behavior, we analyzed the pattern of looking behavior following E1's gesture, and the relationship between looking direction and subsequent choice accuracy. Subjects in the cooperation condition looked at one of the two items during the coding window in $95 \pm 2\%$ of trials. Of these trials, their first look was to the item that E1 pointed to 91% of the time. Thus subjects visually followed

E1's gesture to the correct item significantly more often than they chose this item ($T = 2.52$, $N = 8$, $p = 0.01$). After an initial look to the correct location these subjects chose with $73 \pm 5\%$ accuracy, while choices following an initial look to the incorrect location tended to be less accurate ($50 \pm 22\%$ correct; $T = -0.81$, $N = 5$, $p = 0.42$). Subjects in the competitive condition looked to one of the two items during the coding window in $88 \pm 3\%$ of trials. Of these trials, their first look was to the item that E1 pointed to (the incorrect choice) 88% of the time. After initially looking at this item subjects chose at a very low accuracy (mean $= 36 \pm 3\%$), whereas choices following a look to the item E1 did *not* point to were significantly more accurate (mean $= 76 \pm 11\%$; $T = -2.29$, $N = 10$, $p = 0.02$).

3.3. Discussion

The main finding from Experiment 1 was that apes of both species performed significantly better when the experimenter's gesture was cooperative, and indicated the object they should choose, than when the gesture was competitive, and indicated the object they should not choose. This result contrasts with previous studies in which chimpanzees used prohibitive gestures at higher levels than cooperative-communicative gestures (Herrmann & Tomasello, 2006). However, several features of our design may explain this discrepancy. First, the prohibitive gesture in our study indicated the object that subjects should not choose, whereas the gesture used by Hermann & Tomasello (2006) directed the subjects' attention to the object they should choose (though it was intended to prohibit subjects from doing so). Thus it may be that the gesture directed subjects' attention to the indicated object, strongly biasing them to choose this object regardless of whether the experimenter's gesture was helpful or prohibitive. Our analysis of looking data supports this hypothesis as subjects in both the cooperative and competitive condition initially oriented toward (and tended to choose) the indicated object on the majority of trials. Second, in this task food was visible to subjects at both the correct and incorrect locations, and the experimenter's gesture served to indicate which item the subject should select rather than informing the (ignorant) subject of where a single piece of food was hidden. Consequently subjects were not reliant on the experimenter's gesture to infer the location of the reward, and instead tended to select the first visible piece of food that they oriented toward. In Experiment 2 we address the impact of whether the food is hidden or visible by testing subjects using the same gesture from Experiment 1 in a context where the food is hidden under one of two containers.

4. Experiment 2

In Experiment 2 we tested subjects in a traditional object choice task where food was hidden in one of two containers and the experimenter provided a helpful pointing gesture to indicate the food's location.

4.1. Method

4.1.1. Subjects

We tested the same subjects from Experiment 1 (approximately 1 week later).

4.1.2. Procedure

4.1.2.1. Familiarization. We conducted 4 familiarization trials at the beginning of each session to introduce subjects to finding food underneath the containers used in test trials. On each trial E1 visibly placed a piece of banana under one of the two containers positioned at the right and left sides of the table and slid the platform forward for the subject to make a choice. Choices were defined as subjects touching or positioning their hand over the marked areas at the left or right sides of the table. If a subject chose the baited container, E1 lifted this container allowing the subject to eat the reward. If a subject chose the un-baited container, E1 lifted this container to show the subject that it was empty, retracted the table and visibly removed the banana from underneath the correct container. Subjects were required to correctly choose the baited container on at least 3/4 familiarization trials to advance to the test. All subjects met this criterion.

In test trials E1 blocked the subject's view of the table with the plastic occluder and baited or sham baited each of the containers. E1 then removed the occluder, performed the pointing gesture (identical to Experiment 1) to indicate the baited container, and pushed the platform forward for subjects to make a choice. If a subject chose the baited container, E1 lifted this container allowing the subject to eat the reward. If a subject chose the unbaited container, E1 lifted this container to show the subject that it was empty, retracted the table and visibly removed the banana from underneath the baited container.

4.1.3. Design and analysis

We conducted two 10-trial sessions with each subject. These sessions were separated by a 5-min break. The location that the experimenter pointed to was counterbalanced within each session. Subjects were allowed 30 s to choose on each trial and if no choice was made during this time the trial was scored as 'no choice' and the next trial was conducted. We used the percentage of correct choices (including only trials where subjects made a choice) as the dependent measure for analysis of choice behavior. From video, we recorded the direction of subjects looking behavior using the same procedures (and subjects) from Experiment 1 (bonobo $N = 11$, chimpanzee $N = 8$). A second individual blind to the condition and hypotheses scored looking behavior and subjects' choices for 20% of trials to assess inter-observer reliability. Inter-observer reliability was very good for both measures (looking: kappa = 0.85; choice: kappa = 0.95; Altman, 1991) and scores from the primary coder were retained in cases of disagreement.

4.2. Results

4.2.1. Choice accuracy

Overall subjects chose the container that E1 pointed to significantly more often than chance expectation (mean = 73 ± 2% correct, $T = 5.35$, $N = 40$, $p < 0.01$). Nineteen subjects (8 bonobos, 11 chimpanzees) performed above chance at the individual level (binomial tests, $p < 0.05$). Of these subjects, 12 individuals had previously participated in the cooperative condition in Experiment 1, whereas 7 subjects had previously participated in the competitive condition. Averaging data across both test sessions, subjects who had previously participated in the cooperative condition in Experiment 1 did not differ from subjects who had previously participated in the competitive condition ($T = 1.89$, $N = 40$, $p = 0.06$). However, these groups did differ significantly in the first test session, with subjects who had previously participated in the cooperative condition performing better than subjects who had previously participated in the competitive condition (Figure 3; previously cooperative: 79 ± 4% correct; previously competitive: 65 ± 4% correct; $T = 2.23$, $N = 40$, $p = 0.03$). Nonetheless, both groups performed above chance in the first test session (previously cooperative: $T = 3.55$, $N = 20$, $p < 0.01$; previously competitive: $T = 2.66$, $N = 20$, $p = 0.01$). There was no significant difference between these groups in the second test session ($T = 1.25$, $N = 40$, $p = 0.21$).

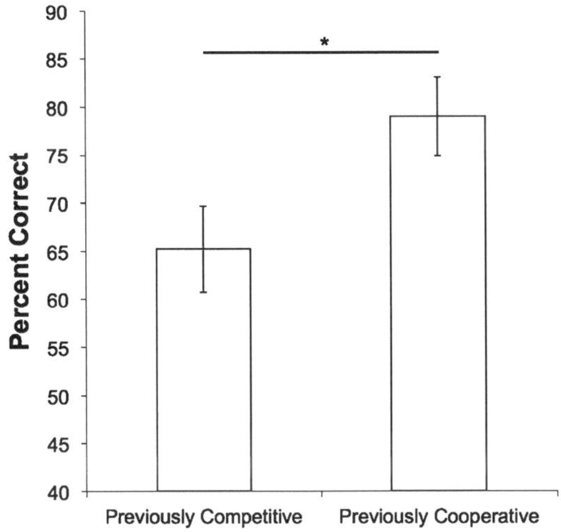

Figure 3. Performance in the first test session of Experiment 2 as a function of whether subjects previously participated in the cooperative or competitive condition in Experiment 1.

There were no differences between species regardless of whether subjects had previously participated in the cooperative ($T = 0.77$, $N = 20$, $p = 0.48$) or competitive ($T = 0.27$, $N = 20$, $p = 0.80$) condition in Experiment 1. However, among the subjects that participated in the competitive condition of Experiment 1, chimpanzees performed above chance expectation in the first test session of this study (mean = 68 ± 6% correct, $T = 2.33$, $N = 10$, $p = 0.02$) whereas bonobos did not (mean = 62 ± 7% correct, $T = 1.36$, $N = 10$, $p = 0.17$).

4.2.2. Looking measures

For the subset of individuals for whom we were able to code looking behavior, we analyzed the pattern of looking behavior following E1's gesture, and the relationship between looking direction and subsequent choice accuracy. Overall subjects looked toward one of the two containers during the coding window on 95 ± 1% of trials. Of these trials, they first looked to the location that E1 pointed to 89% of the time. Following an initial look to the correct location subjects responded correctly on 78 ± 3% of trials. In contrast, when subjects first looked toward the incorrect option the chose correctly only 48 ± 10% of the time. The difference in accuracy following initial looks toward the correct and incorrect containers was significant ($T = 2.33$, $N = 16$, $p = 0.02$). Lastly, we compared whether subjects who participated in the cooperative condition of Experiment 1 were more likely to choose the location that they initially oriented toward more often when the food was visible (Experiment 1) than when the food was hidden (Experiment 2). Subjects chose the object they initially oriented toward at similar frequencies between these conditions (food visible: mean $= 72 \pm 4\%$; food hidden: mean $= 70 \pm 6\%$; $T = -0.28$, $p = 0.78$).

4.3. Discussion

Overall, subjects of both species used the cooperative-communicative gesture at high levels, and nearly half of our subjects performed above chance expectation at the individual level. These unusually high levels of performance may be due to the saliency of our pointing gesture, which was proximal to the indicated container (approx. 5 cm), and was sustained until subjects had made a choice. Indeed, previous research indicates that cues conferring local enhancement generally facilitate performance in this type of task (Itakura et al., 1999; Miklósi & Soproni, 2006; Schmitt et al., 2014). Moreover, the similarity in performance between Experiment 1 (cooperative condition) and Experiment 2 indicates that it was most likely the salience of the pointing cue, not simply that food was visible, that was responsible for subjects' high accuracy with the cooperative cue in Experiment 1.

Interestingly, in the first session of Experiment 2, subjects who had previously interacted with the experimenter as a competitor used the experimenters' cooperative gestures at lower levels than subjects who had previously interacted with the experimenter as a cooperator. Thus, even though this experiment was conducted approximately 1 week after Experiment 1, subjects' previous experience constrained or facilitated their initial ability to use cooperative-communicative cues in this study. However, after minimal experience interacting with the experimenter as a cooperator (10 trials), the difference between these groups abated. Thus, previous experience interacting with humans (both immediate and long term), is likely to influence apes' behavior in cooperative-communicative tasks (Lyn et al., 2010).

Finally, analysis of looking behavior in this study largely echoed the findings from Experiment 1. When subjects initially looked toward the correct container, they chose with very high accuracy whereas choices following an initial look to the incorrect location tended to be inaccurate. However, on ~20% of trials, subjects did initially look to the correct location, but subsequently chose incorrectly. Thus we found some support for the hypothesis that apes follow the direction of gestures in object choice tasks more readily than they understand the communicative intentions underlying them (Tomasello et al., 2007a). Our comparison of gaze direction following the gesture and the subject's subsequent choice in the hidden food and visible food manipulations of the first two experiments suggest that hiding the food has little effect on subject's performance. This argues against the previous suggestion that subjects perform poorly in the object choice task because they interpret the human's gesture as indicating the hiding location itself and not the food hidden within (Call et al., 1998).

5. Experiment 3

In Experiments 1 and 2, E1 gestured by extending his hand approx. 5 cm from the cued location. Opposite to our prediction that subjects would perform better when E1's gesture was prohibitive, both species performed best when E1 pointed cooperatively to the location they should choose, but poorly when E1 pointed to the location they were prohibited from choosing. However, the difference in performance between the cooperative and competitive conditions in Experiment 1 may simply reflect an effect of local enhancement in which the physical proximity of E1's hand to one of the choices biased subjects to attend to and select this option. Thus, even if apes possess an advantage with competitive communication, this effect may have been overshadowed in Experiment 1 by strong local enhancement cues. Therefore in Experiment 3 we introduced a more subtle gesture in which E1's arm was equidistant from the two options, reducing the potential confound of local enhancement. As in Experiment 1, this gesture was used either cooperatively to indicate the object that subjects should choose, or prohibitively to indicate the object that subjects should not choose.

5.1. Method

5.1.1. Subjects

The subjects were the same as Experiments 1 and 2, with the exception that only 12 bonobos (8 males, 4 females, mean age = 8.3 years) were tested in this study. Experiment 3 was conducted approximately 1 week after Experiment 2.

5.1.2. Procedure

We conducted 4 familiarization trials, identical to those from Experiment 1, at the beginning of each session to introduce subjects to E1's cooperative

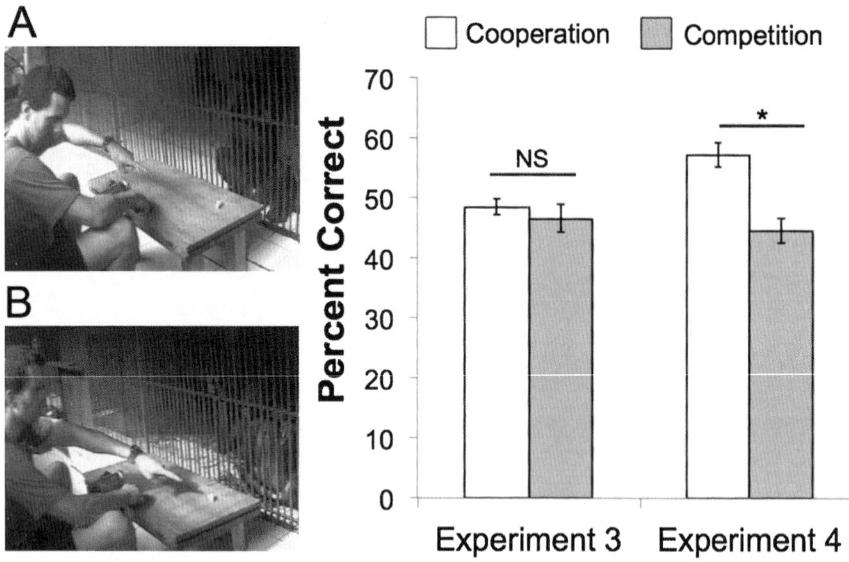

Figure 4. The pointing gesture and results from Experiments 3 and 4. In Experiment 3 subjects did not use a subtle pointing gesture (A) either when this gesture cooperatively indicated the correct location, or competitively indicated the incorrect location. In Experiment 4 E1 provided a more salient gesture (B) and subjects used this cue at higher levels in the cooperative than the competitive condition. This figure is published in colour in the online edition of this journal, which can be accessed via http://booksandjournals.brillonline.com/content/journals/1568539x.

or competitive motive. Test trials were identical to Experiment 1 with the exception that E1 provided subjects with a more subtle pointing gesture. Specifically, E1 held his hand centered in front of his chest, with his index finger extended toward one of the visible pieces of food (Figure 4A). While holding this gesture E1 said the subject's name and alternated his gaze between the subject and the location where he pointed two times. Following this cue E1 pushed the platform forward for subjects to choose. This cue was similar to that used by Hermann et al. (2007) with the exception that E1 stopped pointing before pushing the platform forward for subjects to choose. As in Experiment 1, half of the subjects within each species were tested in the cooperative condition, in which E1's gesture indicated the food that subjects should choose. For the other half of subjects (competitive condition), E1's gesture indicated the piece of food that subjects should not choose.

5.1.3. Design and analysis

The design and analysis were identical to Experiment 1. As in Experiments 1–2 we attempted to score looking behavior between the time when E1 provided the cue and subjects' ultimate choice. However, the direction of subjects' looks was much less obvious in response to the subtle gesture used in this study, and inter-rater agreement (blind to the hypotheses and condition) was only moderate for looking direction (kappa: 0.56; Altman, 1991). Therefore we could not reliably code looking direction in this experiment and did not analyze these data. Inter-rater agreement for subjects' choices was very good (kappa: 0.96).

5.2. Results

Subjects in the cooperative condition did not choose the item that E1 pointed to more often than expected by chance (mean $= 49 \pm 1\%$ correct; $T = -1.27$, $N = 16$, $p = 0.20$) and showed no learning between the first and second test session ($T = -1.71$, $N = 16$, $p = 0.09$). Similarly, subjects in the competitive condition struggled to use E1's pointing gesture and also performed at chance levels (mean $= 47 \pm 2\%$ correct; $T = -1.58$, $N = 16$, $p = 0.11$) with no learning between sessions ($T = -1.23$, $N = 16$, $p = 0.22$). No subjects performed above chance expectation at the individual level in either condition (binomial tests, $p > 0.05$). A comparison of overall accuracy between the cooperative and competitive conditions revealed no significant difference (Figure 4; $T = 1.06$, $N = 32$, $p = 0.31$).

There were no species differences in either the cooperative ($T = 0.35$, $N = 16$, $p = 0.79$) or competitive conditions ($T = -0.60$, $N = 16$, $p = 0.56$) and neither species performed above chance expectation in either the cooperative condition (chimpanzee: $T = -0.82$, $N = 10$, $p = 0.41$; bonobo: $T = -0.95$, $N = 6$, $p = 0.34$) or the competitive condition (chimpanzee: $T = -1.28$, $N = 10$, $p = 0.20$; bonobo: $T = -0.92$, $N = 6$, $p = 0.36$). Finally, performance did not differ between the cooperative and competitive conditions within either species (chimpanzee: $T = 1.24$, $N = 20$, $p = 0.25$; bonobo: $T = 0.08$, $N = 12$, $p = 1.0$).

5.3. Discussion

The main finding from Experiment 3 was that neither chimpanzees nor bonobos used a subtle pointing gesture (pointed finger centered in front of the experimenter's chest) either when this cue cooperatively indicated the correct choice, or prohibitively indicated the incorrect choice. While this cue eliminated the element of local enhancement that may have supported performance in Experiments 1 and 2, it also greatly reduced the salience of the cue leading to chance performance in both conditions. In contrast to Experiments 1 and 2, subjects' looking behavior could not be scored reliably, and it was not apparent that subjects visually followed the direction of the gestures.

6. Experiment 4

In Experiment 4 we tested apes in the same task from Experiment 3, but used a pointing gesture intermediate to that from Experiments 1 and 2, and Experiment 3.

6.1. Method

6.1.1. Subjects

We tested the same subjects from Experiment 3 (approx. 4 days later). Subjects were tested in the same condition (cooperative or competitive) that they had participated in during Experiment 3.

6.1.2. Procedure

The procedure was identical to Experiment 3 with the exception of the pointing gesture that E1 performed. In Experiment 4 E1 pointed toward one of the food items with his index finger extended ∼25 cm from the item he pointed to (Figure 4B). This cue was maintained until subjects made a choice or the trial timed out. Thus the gesture used in this experiment was intermediate to that used in Experiment 1 (proximal pointing) and Experiment 3 (distal pointing).

6.1.3. Design and analysis

The design and analysis were identical to Experiment 3. Again we attempted to code the direction that subjects looked following E1's gesture. However, as in Experiment 3 the direction that subjects looked could not be judged reliably (kappa = 0.61; Altman, 1991) and therefore was not included in analysis. Reliability for subjects' choices was very good (kappa = 0.93). One bonobo subject in the competitive condition became reluctant to choose, making choices on only 4/10 test trials in the first test session, and 0/10 trials in the second test session. Therefore we excluded this subject from analysis.

6.2. Results

Overall, subjects in the cooperative condition chose the item that E1 pointed to significantly more often than expected by chance (mean = 57 ± 2% correct; $T = 2.88$, $N = 16$, $p < 0.01$) and showed no learning between the first and second test sessions ($T = -1.24$, $N = 16$, $p = 0.22$). Two subjects (both chimpanzees) performed above chance at the individual level in the cooperative condition (binomial tests, $p < 0.05$). Subjects in the competitive condition also tended to choose the item that E1 pointed to, leading to performance significantly worse than chance expectation (mean = 46 ± 2%, $T = -2.08$, $N = 15$, $p = 0.04$). Subjects in the competitive condition showed no improvement between the first and second test sessions ($T = 0.45$, $N = 15$, $p = 0.66$), and no subject performed above chance expectation at the individual level (binomial tests, $p > 0.05$). Overall subjects in the cooperative condition performed significantly better than subjects in the competitive condition (Figure 4; $T = 3.55$, $N = 31$, $p < 0.01$).

There were no species differences in either the cooperative ($T = 0.77$, $N = 16$, $p = 0.49$) or the competitive condition ($T = -0.69$, $N = 15$, $p = 0.51$). Consistent with the results across species, both chimpanzees and bonobos performed above chance in the cooperative condition (chimpanzees: $T = 2.26$, $N = 10$, $p = 0.02$; bonobos: $T = 2.03$, $N = 6$, $p = 0.04$), and tended to score below chance in the competitive condition (chimpanzees: $T = -1.90$, $N = 10$, $p = 0.06$; bonobos: $T = -0.92$, $N = 5$, $p = 0.36$). Similarly, performance was better in the cooperative than the competitive condition within both species (chimpanzee: $T = 2.91$, $N = 20$, $p < 0.01$; bonobos: $T = 2.02$, $N = 11$, $p = 0.05$).

Lastly, for subjects that participated in the same condition (cooperative or competitive) across Experiments 1, 3 and 4, we conducted a Friedman's analysis of variance by ranks to determine whether performance differed significantly across the three types of pointing gesture. For subjects in the cooperative condition performance varied significantly across cue types ($\chi^2 = 9.94$, $N = 9$, df = 2, $p < 0.01$) and post-hoc tests revealed that performance in Experiment 1 was better than performance in Experiment 3, but that no other pairs differed. For subjects in the competitive condition performance also varied significantly across cue types ($\chi^2 = 9.33$, $N = 8$, df = 2, $p = 0.01$), and post hoc tests revealed that performance was significantly worse in Experiment 1 than in Experiment 4, but that no other pairs differed.

6.3. Discussion

The results from Experiment 4 are consistent with those from Experiment 1. Subjects of both species used the cooperative cue more readily than the competitive cue to guide their choice behavior. The main difference from Experiment 1 was that here, subjects were less skilled with the pointing gesture, most likely because E1's finger was less proximal to the indicated object. As in Experiment 3, subjects' initial orienting responses were not easily detected, and could not be scored reliably.

7. General discussion

Our findings provide no support for the competitive communicator hypothesis. In the three studies where subjects were tested in either a cooperative or competitive scenario, subjects in the competitive condition never responded above chance expectation, and were significantly worse than chance expectation in several cases. In contrast, many subjects in the cooperative condition responded above chance expectation in multiple contexts.

Interestingly we found some support for the hypothesis that apes respond to the direction of social gestures more readily than they understand the communicative intentions underlying them. In Experiments 1–2, subjects tended to follow E1's gesture to the correct referent, but only subsequently chose this object approx. 75% of the time. Thus apes followed the direction of the gesture considerably more often than they used the gesture to guide their choice. This tendency to look toward the correct choice more often than choosing it occurred both when the food was visible, or hidden from the subjects during the procedure. This argues against the possibility that apes struggle with the object choice task because they interpret the gesture as referring to the hiding container itself (which has no value to the subject), rather than the hidden food within it (Call et al., 1998, 2000). However, in Experiments 3 and 4 subjects' looking direction was less salient and could not be coded reliably. Given that looking direction was less clear in these contexts, it is possible that apes struggled to use the pointing gesture in part due to a failure to link the gesture to its referent. Therefore, determining whether apes follow the directionality of more subtle pointing gestures (such as those from Experiments 3 and 4) remains an important question for future research, and will likely be aided by improved technology for noninvasively monitoring eye movements (e.g., Hattori et al., 2010; Kano & Tomonaga, 2011).

Our results raise the question of why we did not replicate the findings of previous research on this topic (e.g., Hare & Tomasello, 2004; Herrmann & Tomasello, 2006). One possible explanation could be subtle methodological differences between the different experiments. For example, in Hare & Tomasello (2004) procedures were designed to maximize the competitive nature of the interaction. The human experimenter indicated displeasure when a subject chose the correct hiding location and when the subject chose incorrectly, the experimenter ate the food reward to clearly indicate that he was actively competing. It may be that these types of social cues are necessary to indicate that a human — who typically is neutral or helpful — is actively competing. Like previous studies (Hare & Tomasello, 2004; Herrmann & Tomasello, 2006), our experiments did not use behavioral criteria to verify that a cooperative or competitive relationship had been established prior to the test. Thus an important design consideration for future studies will be to determine the specific procedures required to effectively establish a cooperative or competitive relationship, as well as behavioral measures to assess the impact of these manipulations.

A second plausible explanation is that the competitive gestures used by Hare & Tomasello (2004) and Herrmann & Tomasello (2006), still directed subjects' attention to the correct response (e.g., a demonstrator reached toward the object unsuccessfully, or forbid subjects from selecting the object), whereas our competitive gesture directed subjects' attention to the object they should not choose. Consequently, in our studies subjects may have struggled to avoid inhibiting the incorrect response after the gesture directed their attention to this object. Indeed, in Experiments 1 and 2 subjects did visually orient toward the indicated location on the majority of trials, regardless of whether the experimenter's gesture was cooperative or competitive in nature. Therefore, the salience of this gesture may have inadvertently introduced a problem of inhibitory control in which subjects in the competitive condition were required to choose the location opposite to where their attention was originally directed. Given that inhibitory control actively constrains primate problem solving in many contexts (Santos et al., 1999; Amici et al., 2008; Vlamings et al., 2009; MacLean et al., in press), it may also partially explain the generally poor performance in the competitive conditions of these tasks. Therefore, it may be that the competitive advantage is only detectable when both the cooperative and competitive gestures initially direct attention toward the correct response, and there are few constraints imposed by inhibitory control.

Another important difference between our study and previous publications testing the competitive communicator hypothesis is differences in the rearing and experimental histories of the subjects. It is possible that the difference in results are in part a result of the daily human contact that the majority of our subjects receive from human surrogate mothers for the first several years they live at the sanctuary; they receive helpful pointing gestures from humans in a way that mother-reared zoo-born apes typically do not. Additionally, the majority of our subjects had some previous experience in cooperative-communicative tasks (Table 1) perhaps facilitating their performance in these studies. As documented in other populations of apes, rearing history and previous experience can influence performance in these types of tasks in some cases (Tomasello & Call, 2004; Lyn et al., 2010). Thus an important question for future research will be whether these findings generalize to novel gestures that subjects have no experience with prior to the test (e.g., foot pointing).

Regardless of which of these explanations best accounts for our data, it is clear that the competitive communication hypothesis has little power to explain the results of these studies. Therefore future research on the methodological variables that elicit a competitive response, and the role of other task demands (e.g., inhibiting a response when attention is directed toward the incorrect option) will be important for further evaluation of the competitive communicator hypothesis.

Lastly, we found no differences between bonobos and chimpanzees in either the cooperative or competitive conditions. Bonobos, who are characterized by higher social tolerance than chimpanzees (Hare et al., 2012), have been shown to outperform chimpanzees in some cooperative problem-solving tasks (Hare et al., 2007; Tan & Hare, 2013; but see MacLean & Hare, 2013). In addition, bonobos are sensitive to human competitors as their preferences for risk in a foraging task shifted in similar ways to chimpanzees when both species competed against a human (Rosati & Hare, 2013). However, the explanation for the species differences in cooperative abilities implicates differences in social tolerance, not differences in an understanding of cooperative or communicative intentions (Hare et al., 2007). Our results lend further support for this interpretation, as the two *Pan* species appear not to differ in their understanding of cooperative or competitive gestures.

Acknowledgements

We thank R. Atencia, L. Pharoah, D. Cox, K. Brown and the chimpanzee caretakers for facilitating research at Tchimpounga Chimpanzee Sanctuary, C. Andre, V. Dhanani, D. Morel, P. Mbonzo, F. Mehl and the bonobo caretakers for their assitance at Lola ya Bonobo, C. Boukindi, G. Shorland and S. Kwetuenda for their assistance with data collection, and K. Leimberger, S. Laderman, and K. Almon for data coding. We also thank the Ministry of Scientific Research and Technological Inovation, Republic of Congo, and the Ministry of Scientific Research and Technology, Democratic Republic of Congo (permit: MIN.RS/SG/004/2009), for the permits under which this research was conducted. We thank the North Carolina Zoo, and chimpanzee zookeepers for supporting pilot studies of these experiments. We thank all members of the Hare group and Department of Evolutionary Anthropology at Duke University for their helpful comments on these studies. This work was supported in part by European Research Commission Advanced Grant Agreement 233297, National Institute of Health Grant 5 R03 HD070649-02, National Science Foundation grants NSF-BCS-08-27552 and NSF-BCS-10-25172 to B.H.

References

Altman, D.G. (1991). Practical statistics for medical research. — Chapman and Hall, London.

Amici, F., Aureli, F. & Call, J. (2008). Fission–fusion dynamics, behavioral flexibility, and inhibitory control in primates. — Curr. Biol. 18: 1415-1419.

Barth, J., Reaux, J.E. & Povinelli, D.J. (2005). Chimpanzees' (*Pan troglodytes*) use of gaze cues in object-choice tasks: different methods yield different results. — Anim. Cogn. 8: 84-92.

Behne, T., Carpenter, M. & Tomasello, M. (2005). One-year-olds comprehend the communicative intentions behind gestures in a hiding game. — Dev. Sci. 8: 492-499.

Butterworth, G. (2003). Pointing is the royal road to language for babies. — In: Pointing: where language, culture, and cognition meet (Kita, S., ed.). Lawrence Erlbaum Associates, Mahwah, NJ, p. 9-33.

Call, J. & Tomasello, M. (2007). The gestural communication of apes and monkeys. — Lawrence Erlbaum Associates, Publishers, Mahwah, N.J.

Call, J., Hare, B. & Tomasello, M. (1998). Chimpanzee gaze following in an object-choice task. — Anim. Cogn. 1: 89-99.

Call, J., Agnetta, B. & Tomasello, M. (2000). Cues that chimpanzees do and do not use to find hidden objects. — Anim. Cogn. 3: 23-24.

Carpenter, M., Nagell, K. & Tomasello, M. (1998). Social cognition, joint attention, and communicative competence from 9 to 15 months of age. — Monogr. Soc. Res. Child Dev. 63: 1-143.

Coussi-Korbel, S. (1994). Learning to outwit a competitor in mangabeys (*Cercocebus torquatus torquatus*). — J. Comp. Psychol. 108: 164.

Crockford, C., Wittig, R.M., Mundry, R. & Zuberbühler, K. (2012). Wild chimpanzees inform ignorant group members of danger. — Curr. Biol. 22: 142-146.

Dittus, W.P. (1984). Toque macaque food calls: semantic communication concerning food distribution in the environment. — Anim. Behav. 32: 470-477.

Goodall, J. (1986). The chimpanzees of gombe: patterns of behavior. — Belknap Press of Harvard University Press, Cambridge, MA.

Hare, B. (2001). Can competitive paradigms increase the validity of experiments on primate social cognition? — Anim. Cogn. 4: 269-280.

Hare, B. (2011). From hominoid to hominid mind: what changed and why? — Ann. Rev. Anthropol. 40: 293-309.

Hare, B. & Tomasello, M. (2004). Chimpanzees are more skilful in competitive than in cooperative cognitive tasks. — Anim. Behav. 68: 571-581.

Hare, B., Melis, A.P., Woods, V., Hastings, S. & Wrangham, R. (2007). Tolerance allows bonobos to outperform chimpanzees on a cooperative task. — Curr. Biol. 17: 619-623.

Hattori, Y., Kano, F. & Tomonaga, M. (2010). Differential sensitivity to conspecific and allospecific cues in chimpanzees and humans: a comparative eye-tracking study. — Biol. Lett. 6: 610-613.

Hauser, M.D. (1996). The evolution of communication. — The MIT Press.

Herrmann, E. & Tomasello, M. (2006). Apes' and children's understanding of cooperative and competitive motives in a communicative situation. — Dev. Sci. 9: 518-529.

Herrmann, E., Call, J., Hernandez-Lloreda, M.V., Hare, B. & Tomasello, M. (2007). Humans have evolved specialized skills of social cognition: the cultural intelligence hypothesis. — Science 317: 1360-1366.

Hirata, S. & Matsuzawa, T. (2001). Tactics to obtain a hidden food item in chimpanzee pairs (*Pan troglodytes*). — Anim. Cogn. 4: 285-295.

Itakura, S., Agnetta, B., Hare, B. & Tomasello, M. (1999). Chimpanzee use of human and conspecific social cues to locate hidden food. — Dev. Sci. 2: 448.

Kaminski, J., Call, J. & Tomasello, M. (2004). Body orientation and face orientation: two factors controlling apes' begging behavior from humans. — Anim. Cogn. 7: 216-223.

Kano, F. & Tomonaga, M. (2011). Species difference in the timing of gaze movement between chimpanzees and humans. — Anim. Cogn. 14: 879-892.

Leavens, D.A., Hopkins, W.D. & Bard, K.A. (2005). Understanding the point of chimpanzee pointing — epigenesis and ecological validity. — Curr. Dir. Psychol. Sci. 14: 185-189.

Lyn, H., Greenfield, P.M., Savage-Rumbaugh, S., Gillespie-Lynch, K. & Hopkins, W.D. (2011). Nonhuman primates do declare! A comparison of declarative symbol and gesture use in two children, two bonobos, and a chimpanzee. — Lang. Commun. 31: 63-74.

Lyn, H., Russell, J.L. & Hopkins, W.D. (2010). The impact of environment on the comprehension of declarative communication in apes. — Psychol. Sci. 21: 360-365.

MacLean, E.L. & Hare, B. (2012). Bonobos and chimpanzees infer the target of another's attention. — Anim. Behav. 83: 345-353.

MacLean, E. & Hare, B. (2013). Spontaneous triadic engagement in bonobos (*Pan paniscus*) and chimpanzees (*Pan troglodytes*). — J. Comp. Psychol. 127: 245-255.

MacLean, E.L., Hare, B., Nunn, C.L., Addessi, E., Amici, F., Anderson, R.C., et al. (in press). The evolution of self-control. — Proc. Natl. Acad. Sci. USA 111.

Melis, A.P., Hare, B. & Tomasello, M. (2006). Chimpanzees recruit the best collaborators. — Science 311: 1297-1300.

Melis, A.P., Hare, B. & Tomasello, M. (2009). Chimpanzees coordinate in a negotiation game. — Evol. Hum. Behav. 30: 381-392.

Miklósi, A. & Soproni, K. (2006). A comparative analysis of animals' understanding of the human pointing gesture. — Anim. Cogn. 9: 81-93.

Mulcahy, N.J. & Call, J. (2009). The performance of bonobos (*Pan paniscus*), chimpanzees (*Pan troglodytes*), and orangutans (*Pongo pygmaeus*) in two versions of an object-choice task. — J. Comp. Psychol. 123: 304-309.

Nishida, T. (1980). The leaf-clipping display: a newly-discovered expressive gesture in wild chimpanzees. — J. Hum. Evol. 9: 117-128.

Povinelli, D.J., Bierschwale, D.T. & Cech, C.G. (1999). Comprehension of seeing as a referential act in young children, but not juvenile chimpanzees. — Br. J. Dev. Psychol. 17: 37-60.

Povinelli, D.J., Reaux, J.E., Bierschwale, D.T., Allain, A.D. & Simon, B.B. (1997). Exploitation of pointing as a referential gesture in young children, but not adolescent chimpanzees. — Cogn. Dev. 12: 327-365.

Rosati, A.G. & Hare, B. (2012). Decision making across social contexts: competition increases preferences for risk in chimpanzees and bonobos. — Anim. Behav. 84: 869-879.

Santos, L.R., Ericson, B.N. & Hauser, M.D. (1999). Constraints on problem solving and inhibition: object retrieval in cotton-top tamarins (*Saguinus oedipus oedipus*). — J. Comp. Psychol. 113: 186-193.

Schmitt, V., Schloegl, C. & Fischer, J. (2014). Seeing the experimenter influences the response to pointing cues in long-tailed macaques. — Plos One 9: e91348.

Searcy, W.A. & Nowicki, S. (2010). The evolution of animal communication: reliability and deception in signaling systems. — Princeton University Press, Princeton, NJ.

Tan, J. & Hare, B. (2013). Bonobos share with strangers. — Plos One 8: e51922.

Tempelmann, S., Kaminski, J. & Liebal, K. (2011). Focus on the essential: all great apes know when others are being attentive. — Anim. Cogn. 14: 433-439.

Tomasello, M. (2006). Why don't apes point? — In: Roots of human sociality: culture, cognition and interaction (Enfield, N.J. & Levinson, S.C., eds). Berg, New York, NY, p. 506-524.

Tomasello, M. (2008). Origins of human communication. — MIT Press, Cambridge, MA.

Tomasello, M. & Call, J. (1997). Primate cognition. — Oxford University Press, New York, NY.

Tomasello, M. & Call, J. (2004). The role of humans in the cognitive development of apes revisited. — Anim. Cogn. 7: 213-215.

Tomasello, M., Call, J., Nagell, K., Olguin, R. & Carpenter, M. (1994). The learning and use of gestural signals by young chimpanzees: a trans-generational study. — Primates 35: 137-154.

Tomasello, M., Call, J. & Hare, B. (1998). Five primate species follow the visual gaze of conspecifics. — Anim. Behav. 55: 1063-1069.

Tomasello, M., Hare, B. & Agnetta, B. (1999). Chimpanzees, *Pan troglodytes*, follow gaze direction geometrically. — Anim. Behav. 58: 769-777.

Tomasello, M., Hare, B. & Fogleman, T. (2001). The ontogeny of gaze following in chimpanzees, *Pan troglodytes*, and Rhesus macaques, *Macaca mulatta*. — Anim. Behav. 61: 335-343.

Tomasello, M., Carpenter, M. & Liszkowski, U. (2007a). A new look at infant pointing. — Child Dev. 78: 705-722.

Tomasello, M., Hare, B., Lehmann, H. & Call, J. (2007b). Reliance on head versus eyes in the gaze following of great apes and human infants: the cooperative eye hypothesis. — J. Hum. Evol. 52: 314-320.

Veà, J. & Sabater-Pi, J. (1998). Spontaneous pointing behaviour in the wild pygmy chimpanzee (*Pan paniscus*). — Folia Primatol. 69: 289-290.

Vlamings, P.H., Hare, B. & Call, J. (2009). Reaching around barriers: the performance of the great apes and 3–5-year-old children. — Anim. Cogn. 13: 273-285.

Wobber, V., Herrmann, E., Hare, B., Wrangham, R. & Tomasello, M. (2013). Differences in the early cognitive development of children and great apes. — Dev. Psychobiol. 56: 547-573.

Wobber, V.T. & Hare, B. (2011). Psychological health of orphan bonobos and chimpanzees in african sanctuaries. — Plos One 6: e17147.

[When citing this chapter, refer to Behaviour 152 (2015) 521–544]

Preference or paradigm? Bonobos show no evidence of other-regard in the standard prosocial choice task

Jingzhi Tan [a,*], **Suzy Kwetuenda** [b] **and Brian Hare** [a,c]

[a] Department of Evolutionary Anthropology, Duke University, Durham, NC, USA

[b] Lola ya Bonobo, Kinshasa, Democratic Republic of Congo

[c] Center for Cognitive Neuroscience, Duke University, Durham, NC, USA

[*]Corresponding author's e-mail address: jingzhi.tan@duke.edu

Accepted 4 August 2014; published online 19 September 2014

Abstract

Bonobos are the only ape species, other than humans, that have demonstrated prosocial behaviors toward groupmates and strangers. However, bonobos have not been tested in the most frequently used test of prosociality in animals. The current study tested the other-regarding preferences of bonobos in two experiments using the prosocial choice task. In the first experiment subjects preferred a food option that would benefit both themselves and another bonobo. This preference was likely the result of a location bias developed in the pretest since they showed the same preference in the non-social control condition within test sessions. A second experiment was designed to help subjects overcome this bias that might interfere with their social choices. Bonobos again did not prefer to choose the prosocial option. However, results suggest constraints of this paradigm in revealing social preferences. In discussing our results we consider why bonobos show robust prosocial preferences in other paradigms but not here. While others have suggested that such contradictory results might suggest interesting motivational or cognitive differences between humans and non-humans, we propose that the current 'standard' paradigm has failed validation due to three methodological constraints. Across the dozens of studies completed few have demonstrated that non-human subjects understand the causal properties of the apparatus, non-social biases quickly develop in inadequately counterbalanced pretests that typically explain subjects' choices in the test, and even human children found this choice task too cognitively demanding to consistently show prosocial preferences. We suggest it is time to consider switching to a variety of more powerful and valid measures.

Keywords

bonobo, *Pan paniscus*, prosocial behavior, other-regarding preferences.

1. Introduction

A central question in animal cognition is the origin of human sharing and cooperation. Nonhuman animals and humans have been compared for their prosocial behaviors and the underlying motivations. Currently there is tension in the literature regarding the extent to which non-human primates share prosocial preferences with humans (Cronin, 2012). Some studies suggest that non-human primates are willing to help and share without immediate, tangible rewards (i.e., they are showing other-regarding preferences, Burkart et al., 2007; Warneken et al., 2007; de Waal et al., 2008; Lakshminarayanan & Santos, 2008; Cronin et al., 2010; Massen et al., 2010; Horner et al., 2011; Melis et al., 2011; Yamamoto et al., 2012). Other studies report negative results suggesting that non-human primates are not motivated to help or share — even when there is no cost to the actor (Silk et al., 2005; Jensen et al., 2006; Vonk et al., 2008; Cronin et al., 2009; Stevens, 2010; Silk & House, 2011; Skerry et al., 2011; Liebal et al., 2014). Resolving this disagreement will be required in order to understand the evolution of human other-regarding preferences.

Bonobos are one of our species' two closest relatives. Their close phylogenetic relationship to humans makes them especially important to understand when testing hypotheses regarding human cognitive evolution. In the case of prosocial behaviors, bonobos, like humans and chimpanzees, are prosocial toward groupmates in various forms. They spontaneously console victims of aggression (Clay & de Waal, 2013a, b). This consolation reduces the anxiety level of the victims and protects them from future aggression (Palagi & Norscia, 2013). In an experimental context, bonobos voluntarily disengage from playing to help groupmates obtain out-of-reach food, even when there are no immediate, tangible rewards (Tan & Hare, 2013, 2014).

Bonobos are especially important in understanding human sharing since in this regard they are very different from chimpanzees (Hare et al., 2012). Several differences between the two *Pan* species make bonobos a strong candidate for human-like other-regarding preferences: First, relative to chimpanzees bonobos are highly food tolerant. Bonobos co-feed on the same food resource that quickly becomes monopolizable between chimpanzees (Hare et al., 2007; Wobber et al., 2010a; but see Jaeggi et al., 2010a), obtain equal food shares after collaboration (Hare et al., 2007), and become anxious compared to chimpanzees in anticipation of food being monopolized (Wobber et al., 2010b). Second, relative to the xenophobia observed in chimpanzees, bonobos are highly tolerant toward strangers (Wrangham, 1999; Tan & Hare, 2013, 2014). In the wild intergroup encounters of bonobos can be affiliative with strangers and engage in grooming, genital rubbing, copulation and co-feeding (Furiuichi, 2011). Transferring bonobos between groups is relatively free of aggression compared to chimpanzee transfer (Idani, 1991; Pfalzer & Ehret, 1995; Gold, 2001). Third, bonobos are prosocial toward strangers without immediate, tangible rewards. In two recent studies, bonobos voluntarily shared monopolizable food with a stranger and preferred to share with a stranger over a groupmate (Hare & Kwetuenda, 2010; Tan & Hare, 2013). In a helping task, bonobos are also willing to pay an energetic cost to help a stranger obtain out-of-reach food (Tan & Hare, 2013; see also Tan & Hare, 2014). These findings illustrate the unique character of the bonobo social system and raise the possibility that bonobos show high levels of other-regard previously thought to be unique to humans.

While bonobos have shown choices consistent with other-regard toward groupmates and strangers in multiple paradigms, they have yet to be tested using what has become the 'standard' non-verbal method for measuring prosociality. First developed by Silk et al. (2005) and Jensen et al. (2006), the standard approach to examining prosocial preferences requires subjects to make a direct choice between a prosocial option, which delivers a food reward to both the actor and a recipient in an adjacent room (1/1), and an asocial option, which delivers the same amount of food to the actor but none to the recipient (1/0). Because the payoff to the subjects is identical in either option, subjects should show a preference for the prosocial option that benefits the recipient if they have other-regarding preferences. As shown in Table 1, variations of the prosocial choice task have been administered in six non-human primate species representing New World monkeys, Old World monkeys and great apes (and even non-primates, e.g., Di Lascio et al., 2013). Although these studies yield inconsistent results within and between species (Cronin, 2012), many have taken them as core evidence to support human uniqueness in sharing (e.g., Jaeggi et al., 2010b; Silk & House, 2011, 2012; Jensen, 2012).

Here we tested whether bonobos, whose prosociality is well-established in other contexts, would be prosocial in what many think of as the conventional task (Jensen, 2012; Silk & House, 2012). Bonobos should prefer to aid the recipient given their high level of social tolerance and observed other-regarding preferences in other contexts. In experiment 1 benefiting the

Table 1.
Summary of the design features of non-human primate experiments using the prosocial choice task.

Species	Experiment	Design	Self-regard pretest?	Alternative to self-regard pretest	How to counterbalance?
Bonobo	Current study, experiment 1	Lateral	Yes*		Between-subject
	Current study, experiment 2	Lateral	Yes		Between-subject
Chimpanzee	Jensen et al. (2006), experiment 1	Lateral	Yes*		No
	Jensen et al. (2006), experiment 2	Lateral	Yes		No
	Jensen et al. (2006), experiment 3	Lateral	Yes		No
	Yamamoto & Tanaka (2010), experiment 1	Token	Yes*		No
	Yamamoto & Tanaka (2010), experiment 2	Token	No	Experience trials × 12	No
	Horner et al. (2011)	Token	No	Experience trials × 10	Between-subject
	Brosnan et al. (2009)	Vertical	No	Experience trials × ?	Within-subject
	Silk et al. (2005), population 2	Vertical	No	No alternative	Within-subject
	Silk et al. (2005), population 1	Opposite	No	Experience trials × 2	Within-subject
Long-tailed macaque	Massen et al. (2010)	Lateral	No	Experience trials × ?	No
	Massen et al. (2011), dyadic condition	Lateral	No	Unspecified	No
Capuchin	Lakshminarayana & Santos (2008)	Vertical	Yes		Within-subject
	de Waal et al. (2008)	Token	No	Experience trials × 5	Between-subject
	Suchak & de Waal (2012)	Token	No	Experience trials × 30	Unspecified
	Fletcher (2008)	Opposite	No	No alternative	Within-subject
	Takimoto et al. (2010), experiment 1	Opposite	No	Experience trials	Within-subject
	Takimoto et al. (2010), experiment 2	Opposite	No	Experience trials	Within-subject
	Takimoto et al. (2010), experiment 3	Opposite	No	Experience trials	Within-subject
Common marmoset	Burkart et al. (2007)	Vertical	Yes		Within-subject
Cottontop tamarin	Cronin et al. (2009), experiment 1	Vertical	Yes		Within-subject
	Cronin et al. (2009), experiment 2	Vertical	Yes		Within-subject
	Stevens (2010)	Lateral	Yes		Within-subject

Only 9 of 22 experiments (40.9%) have conducted the necessary self-regard pretest. Some studies with similar apparatus are not included because they are not a choice task (e.g., Vonk et al., 2008; Cronin et al., 2010).

* Subjects developed a baseline bias.

recipient (a groupmate or a stranger) was cost-free because subjects received equal rewards if they chose either a selfish or other-regarding option; in experiment 2 subjects had to pay a cost to benefit the recipient, making their prosocial preference costlier.

2. Experiment 1

In this experiment subjects were presented with a prosocial option and an asocial option (see Figure 1a). The prosocial option (the 1/1 option) had two slices of food with one delivered to the subject and the other to a recipient. The asocial option (the 1/0 option) also had two slices with one delivered to the subject and the other discarded by experimenters. To increase the intuitiveness of the task, we have two human experimenters each holding a plastic tray to present and distribute the options. We chose this social presentation method since this approach has been used successfully in a variety of tasks (in some cases leading to improved performance over non-social versions, e.g., Wobber & Hare, 2009; Wobber et al., 2010a).

One important methodological feature here is that we included a pretest to show that our subjects could take advantage of their understanding of this paradigm to maximize their own benefits before testing whether they provided benefits to others (i.e., they showed self-regards before being tested for other-regards). To show self-regard understanding, we tested each subject alone, allowed them free access to the recipient room, and presented them with exactly the same options they would receive in the test phase (i.e., 1/1 and 1/0). They would receive one extra unit of food by choosing 1/1 because they could travel to the recipient room. Only if they prefer the 1/1 option in this context will they demonstrate an understanding of the consequences to themselves and to the recipient room (Jensen et al., 2006; Melis et al., 2010; Hare & Tan, 2011). We assume that subjects had the basic understanding that conspecifics would eat food. Once subjects knew which option would deliver food to the recipient room, they would therefore understand that the food would be consumed if another bonobo were in the room. Without such a demonstration of self-regards, one cannot determine whether subjects lose track of where the food pieces would go or perceive the task as food competition against the recipient (see Table 1 for a summary of published prosocial choice experiments).

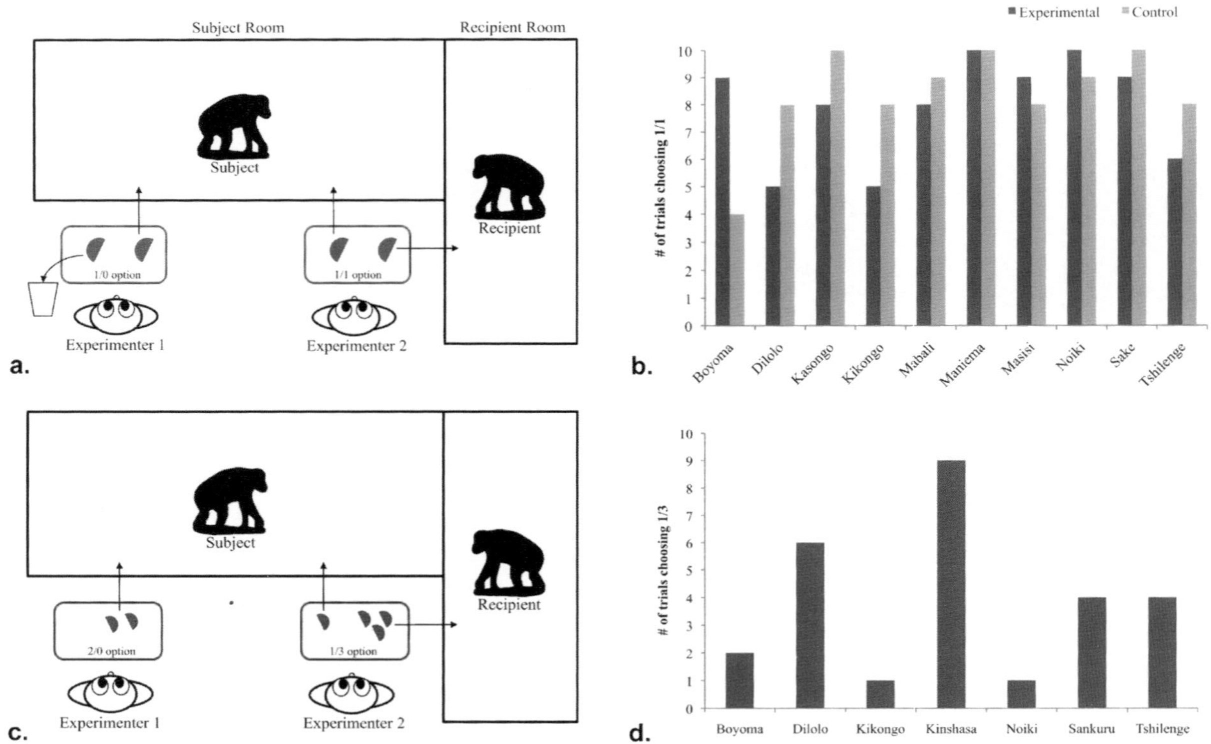

Figure 1. (a) The setup of the experimental condition of experiment 1; (b) the results of experiment 1; (c) the setup of the experimental condition of experiment 2; (d) the results of experiment 2. Three subjects in this experiment (Kinshasa, Kikongo and Noiki) demonstrated an individual preference ($p < 0.05$, binomial test).

2.1. Methods

2.1.1. Subjects

Ten bonobos (4M:6F) from Lola ya Bonobo sanctuary were tested in this experiment (mean age 8.20 ± 0.59, Table 2). All bonobos lived in one of three social groups with access to large forested enclosures in the day and indoor sleeping enclosures at night. Although they were orphans from bushmeat trade, it is evident that the cognition of orphan bonobos was not different from that of mother-reared bonobos (Wobber & Hare, 2011). Subjects could quit at any time by refusing to eat and/or sitting next to the exit of the testing rooms. Ten other bonobos participated but did not finish the pretest phases because three were uncomfortable being separated (Fizi, Lisala and Opala) and seven could not pass the pretests by the end of our field trip (Api, Kalina, Katako, Luozi, Sankuru, Tembo and Waka). Subjects were never food-/water-deprived. To maximize subjects' motivation, the experiment was conducted prior to their morning or afternoon meal.

Each subject was paired with a recipient in the experimental trials of the test phase (see Procedure). Nine bonobos (5M:4F) played the role of the recipient and three of them were also subjects (Boyoma, Kikongo, and Noiki). In five pairs, the recipient was a familiar groupmate; in the other five pairs, the recipient was a stranger from another social group. We maximized possible combinations by including as many individuals as possible to play

Table 2.
Participant information of experiment 1.

Subject			Recipient				
Name	Sex	Age*	Name	Sex	Age*	Familiarity	Location
Boyoma	M	6	Noiki	F	11	Ingroup	Left
Dilolo	M	10	Lukuru	F	6	Outgroup	Left
Kasongo	M	9	Chibombo	M	6	Outgroup	Right
Kikongo	M	9	Api	M	9	Ingroup	Right
Mabali	M	10	Kodoro	M	4	Outgroup	Right
Maniema	M	8	Kikongo	M	9	Ingroup	Left
Masisi	F	6	Chibombo	M	6	Outgroup	Left
Noiki	F	11	Boyoma	M	6	Ingroup	Right
Sake	F	6	Katako	F	7	Outgroup	Right
Tshilenge	F	7	Muanda	F	6	Ingroup	Left

* Based on estimates from Wobber et al. (2010b).

the role of recipient. To exclude potential reciprocity during the experiment, the subjects and the recipients did not reverse their roles except in the pairs of Noiki and Boyoma (this was due to the availability of other participants, e.g., they were sick or in the forest). In total, we created 10 dyads comprised of 16 unique individuals.

2.1.2. Setup

The general setup consisted of two adjacent testing rooms, the subject room and the recipient room (Figure 1a). The experiment was conducted in two different buildings. In one building, both rooms were 15 m^2; in the other building, the recipient room was 8.6 m^2, and the subject room was 15 m^2. The two rooms were separated by mesh and were connected by a sliding door. Two experimenters each held a blue, plastic tray (30×20 cm) in front of the subject room to present food. The trays were two meters apart and were controlled by coordinated behaviors of the experimenters so that subjects could never reach both simultaneously (see Procedure). Therefore, subjects could only choose one tray. Fruit slices (bananas or apples) were used as reward.

2.1.3. Procedure

The experiment consisted of four phases: number pretest, experience pretest, self-regard pretest and test. Each phase followed the general procedure with two experimenters (E1, E2) holding and controlling the food trays as well as a third experimenter (E3) centering the subject. A trial began as E3 lured the subject to the middle point between E1 and E2 with milk or peanuts. E1 and E2 simultaneously showed the contents of the food trays by holding them in front of the subject and calling its name three times. They then quickly pushed the trays aside to two predetermined locations that were 2 m apart and against the mesh. The subject had 60 s to make a choice. The holder of the chosen tray distributed the food according to the rules of specific phases; meanwhile the other experimenter immediately retracted the other tray. A trial was aborted and re-run immediately if the subject did not make a choice within 60 s or if it moved to either side before the trays were pushed forward. Each phase was usually finished in one testing day. If subjects lost motivation, they continued on a subsequent day.

2.1.3.1. Number pretest. This phase was designed to test whether the subjects could discriminate quantities and choose between the contents of the food trays. The subjects were individually tested in the subject room. In each trial, one tray contained one slice of food while the other had two slices. The quantities of food on the two trays were switched every trial, and the locations of E1 and E2 were switched every other trial. The subject was directly handed all the food on the chosen tray. The subject could proceed to the next phase if they made correct choices in 8 out of 10 consecutive trials within one day. For those who succeeded in passing the criterion, the average number of trials needed was 18.10 ± 7.20 (10–28 trials).

2.1.3.2. Experience pretest. This was a four-trial session designed to introduce the 1/0 option to the subjects. This phase was similar to the number pretest with three major exceptions. First, an opaque bucket would be placed next to the experimenter who was at the distal side of the recipient room. The bucket was out of the subject’s reach and was shown to the subject that it was empty at the beginning of this pretest. Therefore, one tray was proximal to the recipient room (a.k.a. the recipient tray) and the other was proximal to the bucket (a.k.a. the bucket tray). Second, the bucket tray represented the 1/0 option. It always contained two slices of food but only delivered one slice to the subject if chosen. The holder picked up the first slice, showed it to the subject, called its name twice to attract attention, dropped that slice into the bucket, and then gave the second slice to the subject. The recipient tray was always empty to make sure that the subject would pay attention to the bucket tray. Third, the locations of the two trays (and the two rooms) did not change within subjects but were counterbalanced between subjects. E1 and E2 switched locations every trial. One subject (Dilolo) did not participate in this pretest because we directly administered the trials of the next phase by mistake, but he did experience the 1/0 option at least four times during the next phase (see below).

2.1.3.3. Self-regard pretest. This phase was designed to demonstrate that the subject understood the consequences of each option. In this pretest, the door connecting the subject room and the recipient room was open, allowing the subject to freely travel between both rooms. Like the experience pretest, the bucket tray contained two slices of food and only one slice would be given to the subject (1/0). Here the recipient tray also contained two fruit slices and represented the 1/1 option. If chosen, the experimenter showed the first slice to the subject, called its name twice to attract attention, dropped that slice in the recipient room, and directly handed the second slice to the subject. Therefore, the recipient tray gave the subject both slices, one in the subject room and the other in the recipient room. The locations of the two trays were consistent with the experience pretest and counterbalanced between subjects. E1 and E2 switched locations every trial. The subject could proceed to the next phase if they chose the recipient tray in 8 of 10 consecutive trials. The average number of trials needed was 12.40 ± 5.95 (8–25 trials).

2.1.3.4. Test. This 20-trial session consisted of two types of trials. The first and the last five trials were control trials, while the 10 trials in the middle were experimental trials. The contents and locations of the food trays were similar to the self-regard pretest, but the door connecting the subject room and the recipient room was closed. In the experimental trials, a recipient was in the recipient room (see Figure 1a); in the control trials, the recipient room was empty. Because the subject could not enter the recipient room, it received only one slice no matter which tray it chose. However, in the experimental trials choosing the recipient tray would donate one slice to the recipient (i.e., the prosocial option) while choosing the bucket tray would leave the recipient with nothing (i.e., the asocial option).

2.1.4. Coding and analysis

A choice was coded if the subject stuck out fingers, lips or straws toward one tray through the mesh, or if they sat down in front of one tray with their body touching the mesh. We did not code gestures of the recipient due to the narrow mesh of some testing rooms that blocked gestural attempts. A second coder coded all the trials and inter-coder reliability was perfect (Cohen's kappa = 1). All statistics were two-tailed and non-parametric due to limited sample size.

2.2. Results

Figure 1b demonstrates the main results of experiment 1. The subjects preferred the 1/1 option to the 1/0 option in both the experimental and the control conditions (experimental: mean $= 7.90 \pm 1.77$, $N = 10$, $p = 0.011$; control: mean $= 8.40 \pm 1.76$, $N = 10$, $p = 0.006$, one-sample Wilcoxon signed rank test, two-tailed). However, their preferences for the 1/1 option did not differ between the two conditions throughout the test phase ($N = 10$, $Z = 1.018$, $p = 0.309$, one tie, Wilcoxon signed rank test, two-tailed). Neither stranger pairs nor groupmate pairs showed a preference (stranger: $N = 5$, $Z = -1.511$, $p = 0.131$, no tie; groupmate: $N = 5$, $Z = 0$, $p = 1.0$, one tie; all Wilcoxon signed rank test, two-tailed). In either condition, the subjects' preferences did not differ when the first 5 and the last 5 trials were compared (experimental: $N = 10$, $Z = 0.333$, $p = 0.739$, four ties; control: $N = 10$, $Z = 0.966$, $p = 0.334$, five ties, Wilcoxon test, two-tailed). At the individual level, in each condition there were five subjects choosing the 1/1 option for 9 or 10 times (see Figure 1b). We scored the difference in the frequency of choosing the 1/1 option between the experimental and the control condition as an index of each subject's prosociality. The recipient's group membership had no effect on this difference score ($N = 10$, $U = 9.5$, $p = 0.526$; Mann–Whitney U-test, two-tailed).

2.3. Discussion

Although the subjects preferred the prosocial option to the asocial option when the recipient was present, they showed the same preference when the recipient was absent. If we consider only the pairs of strangers or only those of groupmates, the results remain the same. The individual analysis is consistent with this pattern. Half of the subjects preferred the 1/1 option in the control condition. This result suggests that subjects have developed a bias toward the content or the location of the 1/1 option.

It is unlikely that the subjects were simply attracted by the content of the 1/1 option because both options always contained two pieces of food. Alternatively, the subjects might have preferred the 1/1 option in an attempt to retrieve the fruit slice delivered into the recipient room even though they had no access. However, they were tested in their sleeping building and thus understood the separation caused by the sliding door.

The subject's apparent location bias was likely introduced during the self-regard pretest because the subjects were required to choose the 1/1 option 8 of 10 times before the test phase. In the test phase the two options conferred equal payoff to the subjects, so there was no cost to them if they kept choosing the same location. Our results are consistent with two chimpanzee experiments with a similar self-regard pretest (Jensen et al., 2006, experiment 1; Yamamoto & Tanaka, 2010, experiment 1). Experiment 1 of Jensen et al. (2006) had a similar lateral setup of the two options, and their subjects developed a location bias toward the 1/1 option; in experiment 1 of Yamamoto & Tanaka (2010), their subjects were trained to associate each option with a button in a fixed color and developed a location bias toward one particular color.

Some proposed to avoid this problem by not conducting the self-regard pretest (Horner et al., 2011). We disagree with this proposal because it will be unclear whether subjects understand the consequences of their choices. Placing this self-regard phase after the test phase is also problematic since it is possible that subjects enter the test phase without any understanding of the setup but gradually develop such understanding during the test. Others suggested that this location bias could be avoided by making the location of the prosocial option contrary to the biased location so that subjects would clearly have overcome the pre-existing bias if they show a prosocial preference (Massen et al., 2010). Similarly in a token-exchange version of this task, Horner et al. (2011) designated the least preferred token in the baseline as the prosocial token in the test phase.

Following this logic we kept the self-regard pretest and adjusted the reward distributions in the second experiment. The prosocial option delivered one fruit slice to the subjects and three slices to the recipient room (1/3). The other option consisted of two pieces and all went to the subjects (2/0). In the self-regard pretest subjects would have to prefer the 2/0 option to understand they would obtain less food if the 1/3 option were chosen because the access door between rooms was closed; in the test they would need to choose the 1/3 option to benefit the recipient (and thus overcome the location bias toward the 2/0 option). As a result, we also created a costly sharing situation where the prosocial act could be regarded as altruistic.

3. Experiment 2

This experiment presented subjects with an asocial option and a prosocial option (Figure 1c). The asocial option (the 2/0 option) had two slices of food that would all be given to the subjects; while the prosocial option (the 1/3 option) had four slices with one delivered to the subjects and three to the recipient.

3.1. Methods

3.1.1. Subjects

Seven bonobos (3M:4F) participated in this experiment. Five of them were subjects in experiment 1 (Table 3). Four other bonobos were dropped out of the experiment because they were not able to pass the pretests by the end of our field trip (Mabali, Maniema, Muanda and Tembo). Each subject

Table 3.
Participant information of experiment 2.

Subject			Recipient			
Name	Sex	Age*	Name	Sex	Age*	Location
Boyoma	M	6	Noiki	F	11	Right
Dilolo	M	10	Mabali	M	10	Right
Kikongo	M	9	Tchilenge	F	7	Left
Kinshasa	F	6	Muanda	F	6	Left
Noiki	F	11	Kikongo	M	9	Left
Sankuru	F	11	Tchilenge	F	7	Right
Tshilenge	F	7	Kikongo	M	9	Left

* Based on estimates from Wobber et al. (2010b).

was paired with a groupmate recipient since in experiment 1 there was no difference in prosociality between stranger pairs and groupmate pairs. Five bonobos played the role of recipient and three of them were also subjects (Kikongo, Noiki and Tchilenge). We followed the same rules as in experiment 1 to determine the pairings. There was no role-reversal except in the pairs of Kikongo and Tchilenge due to the availability of other participants. In total, we created seven dyads comprised of nine unique individuals.

3.1.2. Setup and procedure

Figure 1c shows the setup of the current experiment. Its setup and general procedure were identical to experiment 1 with two food trays and three experimenters. This experiment consisted of four phases: number pretest I, number pretest II, self-regard pretest and test. Each phase was usually finished in one testing day.

3.1.2.1. Number pretest I. This phase was designed to test whether subjects could distinguish different contents on the food trays. It was identical to the number pretest in experiment 1 (i.e., one piece versus two pieces). The subject could proceed to the next phase if they chose the option with two food slices in 8 of 10 consecutive trials. The average number of trials needed was 16.14 ± 6.41 (10–26 trials).

3.1.2.2. Number pretest II. This phase was designed to test whether the subjects could prefer four fruit slices to two. It was identical to the number pretest I except that one tray contained two slices and the other contained four. Subjects needed 10.14 ± 1.25 (9–13) trials to pass this phase.

3.1.2.3. Self-regard pretest. This phase was designed to introduce the 1/3 and the 2/0 options and demonstrate that the subjects understood the consequences of each option. In this phase, the recipient room was empty and was separated from the subject room by open mesh. The recipient tray was the 1/3 option and always contained four fruit slices. If the subjects chose this tray, the holder picked up three slices, showed them to the subject by calling its name twice, dropped them into the recipient room and directly handed the remaining slice to the subject at the end. The subject could see but could never retrieve the food in the recipient room (i.e., a 1/3 payoff). The other tray contained two slices. The subject would receive all of them if it chose it (i.e., a 2/0 payoff). The location of the recipient room (and the recipient tray) did not change within subjects but was counterbalanced between subjects. The two tray-holders switched locations every trial. To pass this phase, the subject had to choose the 2/0 option in 8 of 10 consecutive trials. The average number of trials needed was 28.43 ± 10.72 (16–46 trials). Because we needed to establish a preference for the 2/0 option here, we could not open the door to the recipient room as we did in experiment 1. However, since subjects have already demonstrated that they paid attention to the tray with four pieces in the number pretest II, their preference for 2/0 to 1/3 in the current phase suggests that they have now understood the payoffs of each option to self and to the recipient room.

3.1.2.4. Test. This session has 10 trials and was identical to the self-regard pretest except that there was a recipient in the recipient room. The test was conducted no more than one day after the self-regard pretest. The locations of the two options were the same as the self-regard pretest. Therefore, choosing the 1/3 option was altruistic because it benefited the recipient at a cost to the subject. Moreover, an altruistic choice would require the actor to overcome any location bias toward the 2/0 option established in the self-regard pretest. It is important to note that here we compared subjects' choices to chance level, instead of comparing a social condition to a non-social condition. This was because if we had administered a non-social condition, it would be equivalent to the self-regard pretest and subjects would have more experience in this hypothetical non-social condition than in the social condition.

3.1.3. Coding and analysis

All coding followed the same criteria as in experiment 1. A second coder coded all trials and inter-coder was perfect (Cohen's kappa = 1). Due to limited sample size, all statistics were non-parametric and two-tailed.

3.2. Results

As shown in Figure 1d, subjects as a group did not prefer any option (mean = 3.86 ± 2.91, $N = 7$, $p = 0.303$, one-sample Wilcoxon signed rank test, two-tailed). They did so in the first and the second half of the test (first half: mean = 2.14 ± 1.46, $N = 7$, $p = 0.391$; second half: mean = 1.71 ± 1.49, $N = 7$, $p = 0.2$, one-sample Wilcoxon signed rank test, two-tailed). At the individual level, three subjects showed a preference of one option in 9 of 10 trials: Noiki and Kikongo preferred the 2/0 option, and Kinshasa preferred the 1/3 option.

3.3. Discussion

The bonobos did not prefer the 2/0 option collectively, although two subjects did show such an individual preference. They did not prefer the 1/3 option as a group either, but one individual preferred the 1/3 option. It is plausible that the subjects at least demonstrated some degree of prosociality because they did not overwhelmingly choose the 2/0 option (Massen et al., 2010). A more parsimonious interpretation is that subjects' choices might have been affected by local enhancement, i.e., they were attracted by the presence/proximity of the recipient so that they did not exclusively choose the 2/0 option during the test (Jaeggi et al., 2010b). Finally similar to Jensen et al. (2006) the current results show considerable individual variability with two subjects sticking to the selfish option and one subject preferring the altruistic option. Future research should adjust the payoffs of the two options to examine if subjects are sensitive to the cost of the prosocial option. For example, when the selfish option becomes 4/0, if subjects choose 4/0 less often when the prosocial option is 3/1 than when it is 1/3, it would suggest that subjects' choices were not simply a result of local enhancement.

4. General discussion

Overall we found little evidence for other-regarding preferences in this prosocial choice task with bonobos. In experiment 1 the subjects could aid a stranger or a groupmate recipient at no cost to themselves, but they showed a preference for the prosocial option regardless of the presence of the recipient. Because both options in experiment 1 had the same amount of food, the observed preference was likely due to a bias toward the location of the prosocial option developed during the self-regard pretest. In experiment 2 prosocial choices could not be explained by a location bias since subjects had to choose the opposite location chosen in the pretest. In addition, they would obtain less food by doing so. As a group the bonobos showed no preference for any option or location with two subjects preferring the selfish option and one preferring the altruistic option, which could be explained parsimoniously by local enhancement.

One interpretation of our results is that bonobos, like chimpanzees, show little evidence of other-regard in this paradigm, which supports the argument that humans have unique propensity to share among great apes (Warneken & Tomasello, 2009; Jaeggi et al., 2010b; Silk & House, 2011, 2012; Jensen, 2012). The genus *Pan* may have psychological constraints preventing the expression of other-regard in the prosocial choice task and, by extrapolation, the context of food provisioning more generally (Warneken & Tomasello, 2009; Silk & House, 2011). Although the current study could not rule out this food inhibition hypothesis, it has been repeatedly challenged by findings that bonobos and chimpanzees demonstrate other-regard in other tasks that directly provision the recipient with food (Greenberg et al., 2010; Melis et al., 2011; Tan & Hare, 2013). In fact, chimpanzees, which are highly competitive over food, show other-regard in some other tasks using visible food, invisible food and non-food (Warneken et al., 2007; Melis et al., 2011), whereas reducing the visibility of food does not necessarily lead to positive results in the prosocial choice task (Vonk et al., 2008). If the relatively food-intolerant chimpanzees appear unaffected by the presence or the visibility of food, these factors should not constrain the prosociality of the relatively food-tolerant bonobos (see also Cronin, 2012).

Alternatively our results may tell us more about methodological constraints intrinsic to this standard paradigm for prosociality. Given this task requires the use of an apparatus, subjects must first show they understand the physical setup by demonstrating self-regard (Jensen et al., 2006; Melis et al., 2010; Hare & Tan, 2011). However, this introduction seems to create a location bias that then carries over into the test and likely interferes with any preference the subjects might naturally express. A location bias could explain all three 1/1-vs-1/0 experiments on great apes in which a self-regard pretest was conducted (Jensen et al., 2006, experiment 1; Yamamoto & Tanaka, 2010, experiment 1; current study, experiment 1). In fact, while Jensen et al., 2006 is normally cited as evidence against other-regards in chimpanzees, the subjects in experiment 1 show a strong bias *for* the prosocial option in the test trial just as the bonobos do here. However, like our bonobos, the chimpanzees showed the same preference in the self-regard pretests and controls. This pattern does not suggest an interpretable failure on the part of the subjects.

One approach to avoid the development of a location bias is to make the prosocial choice in test trials at the opposite location from where the same food distribution was presented in the self-regard pretest. In this way subjects have to overcome the location bias to demonstrate prosocial preferences (e.g., Massen et al., 2010; Horner et al., 2011). While we have designed experiment 2 in this way, this design feature only protects against a false positive due to a location bias — but remains open to alternate interpretations such as local enhancement. Subjects could choose the prosocial option because they were simply attracted by the presence of the recipient.

Other approaches to avoid a location bias focus on removing or replacing the self-regard pretest. For example, some propose to omit the self-regard pretest or conduct it after the test phase (e.g., Horner et al., 2011; Yamamoto & Tanaka, 2010, experiment 2). However, without a self-regard pretest it is almost impossible to know if the subjects understood the consequences of their choices in the actual test. That means any results from no-pretest experiments will be equally difficult to interpret since they might be false negative (e.g., subjects pay no attention to the other end of the apparatus) or false positive (e.g., subjects attempt to compete for the extra food piece). Table 1 shows that over 50% of published experiments omitted the self-regard pretest.

A common alternative to the self-regard pretest is experience trials (i.e., subjects are given 2–30 forced-choice trials with each option). This approach is problematic for two reasons. First, the number of experience trials given is arbitrary. No study has ever validated that amount of experience leads to an understanding of the consequences of each option. For example, the current study shows that there was tremendous variability of how much experience each subject needed to understand the consequences of each option (the number of trials needed to pass the self-regard pretest: experiment 1: 8–25; experiment 2: 16–46). Second, experience trials alone cannot show subjects understand what would happen to the recipient room. Because subjects cannot enter the recipient room, they only experience the reward of each option for themselves. This approach seems to be based on an assumption: within an arbitrary number of experience trials, their subjects will somehow pay attention and understand what happens on the other end of the apparatus through observation alone. This assumption is likely unwarranted unless an additional pretest is conducted to show that subjects indeed pay attention to the reward that would eventually go into the recipient room (e.g., as in the number pretest II in experiment 2 of the current paper).

The last approach is proposed by Cronin et al. (2010) to measure the change of subjects' choices over time in the test phase. For example, if their choices of the prosocial option decrease in the non-social control but not in the experimental condition, that should be evidence of prosocial preferences. However, given the Cronin study itself included a rigorous self-regard pretest, it remains to be further tested if measuring the temporal change of subjects' choices alone is sufficient. A combination of this approach and the self-regard pretest might provide the strongest evidence of subjects' understanding of the setup (as in Cronin et al., 2010).

We suggest a self-regard pretest is crucial for interpreting the results from this paradigm but that counterbalancing the location of the prosocial option is critical. As shown in Table 1, when considering experiments with a self-regard pretest, a location bias only develops in those studies using a between-subjects design in regards to the location of the prosocial choice (or no counterbalancing at all, Jensen et al., 2006; Yamamoto & Tanaka, 2010; the current study). This design leads to a location bias since the 1/1 option, in both the pretest and the test, always leads to maximal payoff and it stays in the same location. To corroborate, studies that are able to use a design that counterbalances the location of the 1/1 option *within* each subject did not generate a location bias (Burkart et al., 2007; Lakshminarayana & Santos, 2008; Cronin et al., 2009; Stevens, 2010). Therefore, it is not that all pretests create a location bias as previously suggested (Horner et al., 2011), but that, in particular, a between-subjects design does.

Not all designs are equal in terms of their ability to allow for a counterbalanced pretest. Table 4 presents a qualitative comparison of the four major setups previously used based on the difficulty of running self-regard pretest and that of counterbalancing the choice options within subjects. For example, when subjects face each other across opposite rooms no experiments have conducted a self-regard pretest (Silk et al., 2005; Fletcher, 2008, population 1; Takimoto et al., 2010). This is likely because the necessary configuration of rooms that would allow for a subject to travel in a separate tunnel to the recipient room for the pretest typically does not exist. In addition, while the 'opposite' and 'vertical' designs allow for easy counterbalancing, the 'lateral' and 'token' designs do not and likely encourage competing biases to develop in the pretest. One notable exception is the lateral design used by Stevens (2010), in which within-subject counterbalancing occurred

Table 4.
Summary of the four designs of the prosocial choice task.

	Opposite design	Vertical design	Lateral design	Token design
No. of experiments	5	6	8	5
General design	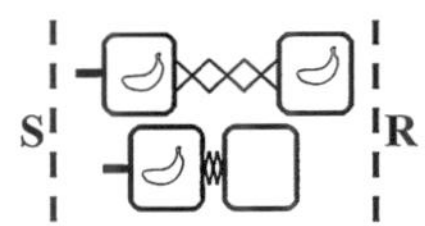	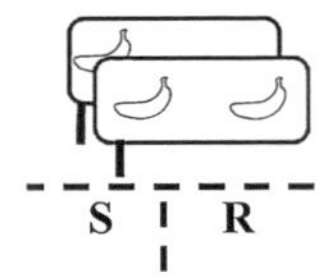	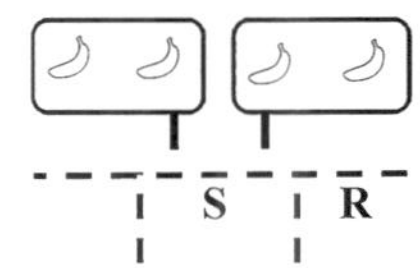	
Difficulty of self-regard pretest	Hard	Easy	Easy	Easy
Difficulty of within-subject counterbalancing	Easy: moving the food	Easy: moving the food	Hard: moving the animal	Hard: re-learning the symbols
% running self-regard pretest	0%	66.7%	75%	20%
% counterbalancing within-subject	100%	100%	12.5%	0%

The 'vertical' design seems the most desirable to conduct self-regard pretests and within-subject counterbalancing. In the general design images S stands for subject and R for recipient.

between test sessions given the small size of the species being tested (i.e., the experiment did not have to be designed around the constraints of the available housing as in ape studies). Overall the vertical design seems to be the most desirable: it accommodates the needs of a self-regard pretest and within-subject counterbalancing, plus it controls for local enhancement.

A lack of self-regard pretest and inadequate counterbalancing are only part of the challenges to the overall validity of the standard prosocial choice task. For example, Burkart & Rueth (2013) found that even 2–5-year-old human children failed to show prosocial preference when choosing between 1/1 and 1/0, but the same subjects were prosocial when choosing between 0/1 and 0/0. It is important to note that their subjects all received self-regard pretest and within-subject counterbalancing in a vertical design task (see House et al., 2012 for a similar result in children with an opposite design). As suggested by the authors, making a choice between 1/1 and 1/0 seems too taxing for the subjects because they had to encode and consider payoffs of four food dishes at once. That means any prosocial choice task using 1/1 and 1/0 might create false negative regarding prosocial preferences as it did in human children. Interestingly, an effective approach to reduce the cognitive load is to use 0/1 and 0/0 as payoff options (Burkart & Rueth, 2013). This essentially turns the standard prosocial choice task into an instrumental helping task (see also Cronin et al., 2010).

The question of ape prosocial and other regarding preferences is central to understanding the evolution of human sharing and cooperation. However, while a variety of tasks have consistently revealed helping and sharing in both bonobos and chimpanzees (Warneken et al., 2007; Hare & Kwetuenda, 2010; Melis et al., 2011; Yamamoto et al., 2012; Tan & Hare, 2013, 2014), studies using the standard prosocial choice task with apes have largely produced negative results (Silk et al., 2005; Jensen et al., 2006; Brosnan et al., 2009; Yamamoto & Tanaka, 2010; current study). In interpreting this literature some have weighted the negative findings in the prosocial choice task heavier than the positive results in other experimental tasks (Silk & House, 2011; Jensen, 2012). However, while positive results using other paradigms have included pretests in which subjects successfully demonstrated an understanding of the apparatus (e.g., Warneken et al., 2007; Tan & Hare, 2013) most prosocial choice tasks have either failed to show subjects understand the task or omitted the pretest all together (see Table 1). In addition, while quantitative comparisons between chimpanzees and human infants have shown highly similar results in instrumental helping tasks (Warneken et al., 2007), even human children show inconsistent results in standard prosocial choice tasks (Brownell et al., 2009; House et al., 2012; Burkart & Rueth, 2013). Our results and literature review suggest at least three constraints intrinsic to the standard prosocial choice paradigm: a lack of self-regard pretest, inadequate counterbalancing and unnecessary cognitive load. Taken together, it is difficult to conclude that standard prosocial choice tasks provide a valid measure of prosociality or other-regard. While a species failure to solve a problem or show a preference can often be interpreted as a meaningful difference between human and non-human apes (e.g., Hare & Tomasello, 2004; Melis et al., 2009; Wobber et al., 2014) that is currently not the case for the standard prosocial choice task. The failure of bonobos, chimpanzees and children in this task and the strength of evidence from other paradigms raise serious questions about the validity of this approach as a vehicle to powerfully compare human and non-human prosociality. Future prosocial choice studies using a vertical task design with 0/1 and 0/0 payoff options may allow researchers to circumvent the method constraints we outline, but essentially this would only replicate the decision subjects face in previously published instrumental tasks where there are already clear positive results across a range of species.

Acknowledgements

This work was supported by the European Research Council (No. 233297) and National Science Foundation (BCS-27552; BCS-25172) to B.H. We are grateful to Claudine André, Dominique Morel, Dhanani Valery, Fanny Mehl, Delphin Bilua and the staff of Lola ya Bonobo for their support. We thank Melissa Klein for assistance in coding. We are very thankful to two anonymous reviewers for their extensive and instructive comments that have largely improved the manuscript.

References

Brosnan, S.F., Silk, J.B., Henrich, J., Mareno, M.C., Lambeth, S.P. & Schapiro, S.J. (2009). Chimpanzees (*Pan troglodytes*) do not develop contingent reciprocity in an experimental task. — Anim. Cogn. 12: 587-597.

Brownell, C., Svetlova, M. & Nichols, S. (2009). To share or not to share: when do toddlers respond to another's needs? — Infancy 14: 117-130.

Burkart, J.M. & Rueth, K. (2013). Preschool children fail primate prosocial game because of attentional task demands. — PLoS One 8: e68440.

Burkart, J.M., Fehr, E., Efferson, C. & van Schaik, C.P. (2007). Other-regarding preferences in a non-human primate: common marmosets provision food altruistically. — Proc. Natl. Acad. Sci. USA 104: 19762-19766.

Clay, Z. & de Waal, F.B.M. (2013a). Bonobos respond to distress in others: consolation across the age spectrum. — PLoS One 8: e55206.

Clay, Z. & de Waal, F.B.M. (2013b). Development of socio-emotional competence in bonobos. — Proc. Natl. Acad. Sci. USA 110: 18121-18126.

Cronin, K. (2012). Prosocial behaviour in animals: the influence of social relationships, communication and rewards. — Anim. Behav. 84: 1085-1093.

Cronin, K., Schroeder, K.K.E., Rothwell, E.S., Silk, J.B. & Snowdon, C.T. (2009). Cooperatively breeding cottontop tamarins (*Saguinus oedipus*) do not donate rewards to their long-term mates. — J. Comp. Psychol. 123: 231-241.

Cronin, K., Schroeder, K.K.E. & Snowdon, C.T. (2010). Prosocial behaviour emerges independent of reciprocity in cottontop tamarins. — Proc. Roy. Soc. Lond. B: Biol. Sci. 277: 3845-3851.

de Waal, F.B.M., Leimgruber, K. & Greenberg, A.R. (2008). Giving is self-rewarding for monkeys. — Proc. Natl. Acad. Sci. USA 105: 13685-13689.

Di Lascio, F., Nyffeler, F., Bshary, R. & Bugnyar, T. (2013). Ravens (*Corvus corax*) are indifferent to the gains of conspecific recipients or human partners in experimental tasks. — Anim. Cogn. 16: 35-43.

Fletcher, G.E. (2008). Attending to the outcome of others: disadvantageous inequity aversion in male capuchin monkeys (*Cebus apella*). — Am. J. Primatol. 70: 901-905.

Furuichi, T. (2011). Female contributions to the peaceful nature of bonobo society. — Evol. Anthropol. 20: 131-142.

Gold, K. (2001). Group formation in captive bonobos: sex as a bonding strategy. — In: Proceedings of The Apes: Challenges for the 21st Century, Brookfield Zoo, Brookfield, IL. Chicago Zoological Society, Chicago, IL, p. 90-93.

Greenberg, J.R., Hamann, K., Warneken, F. & Tomasello, M. (2010). Chimpanzee helping in collaborative and noncollaborative contexts. — Anim. Behav. 80: 873-880.

Hare, B. & Kwetuenda, S. (2010). Bonobos voluntarily share their own food with others. — Curr. Biol. 20: R230-R231.

Hare, B. & Tan, J. (2011). How much of our cooperative behavior is human? — In: The primate mind: built to connect with other minds (de Waal, F.B.M. & Ferrari, P.F., eds). Harvard University Press, Cambridge, MA, p. 175-193.

Hare, B. & Tomasello, M. (2004). Chimpanzees are more skillful in competitive than in cooperative cognitive tasks. — Anim. Behav. 68: 571-581.

Hare, B., Melis, A.P., Woods, V., Hastings, S. & Wrangham, R.W. (2007). Tolerance allows bonobos to outperform chimpanzees on a cooperative task. — Curr. Biol. 17: 619-623.

Hare, B., Wobber, V. & Wrangham, R.W. (2012). The self-domestication hypothesis: evolution of bonobo psychology is due to selection against aggression. — Anim. Behav. 83: 573-585.

Horner, V., Carter, J.D., Suchak, M. & de Waal, F.B.M. (2011). Spontaneous prosocial choice by chimpanzees. — Proc. Natl. Acad. Sci. USA 108: 13847-13851.

House, B., Henrich, J., Brosnan, S.F. & Silk, J.B. (2012). The ontogeny of human prosociality: behavioral experiments with children aged 3 to 8. — Evol. Hum. Behav. 33: 291-308.

Idani, G. (1991). Social relationships between immigrant and resident bonobo (*Pan paniscus*) females at Wamba. — Folia Primatol. 57: 83-95.

Jaeggi, A.V., Stevens, J.M.G. & van Schaik, C.P. (2010a). Tolerant food sharing and reciprocity is precluded by despotism among bonobos but not chimpanzees. — Am. J. Primatol. 143: 41-51.

Jaeggi, A.V., Burkart, J.M. & van Schaik, C.P. (2010b). On the psychology of cooperation in humans and other primates: combining the natural history and experimental evidence of prosociality. — Phil. Trans. Roy. Soc. B 365: 2723-2735.

Jensen, K. (2012). Social regard: evolving a psychology of cooperation. — In: The evolution of primate societies (Mitani, J.C., Call, J., Kappeler, P.M., Palombit, R.A. & Silk, J.B., eds). The University of Chicago Press, Chicago, IL, p. 565-584.

Jensen, K., Hare, B., Call, J. & Tomasello, M. (2006). What's in it for me? Self-regard precludes altruism and spite in chimpanzees. — Proc. Roy. Soc. Lond. B: Biol. Sci. 273: 1013-1021.

Lakshminarayanan, V.R. & Santos, L.R. (2008). Capuchin monkeys are sensitive to others' welfare. — Curr. Biol. 18: R999-R1000.

Liebal, K., Vaish, A., Haun, D. & Tomasello, M. (2014). Does sympathy motivate prosocial behaviour in great apes? — PLoS One 9: e84299.

Massen, J.J.M., van den Berg, L.M., Spruijt, B.M. & Sterck, E.H.M. (2010). Generous leaders and selfish underdogs: pro-sociality in despotic macaques. — PLoS One 5: e9734.

Massen, J.J.M., Luyten, I.J.A.F., Spruijt, B.M. & Sterck, E.H.M. (2011). Benefiting friends or dominants: prosocial choices mainly depend on rank position in long-tailed macaques (*Macaca fascicularis*). — Primates 52: 237-247.

Melis, A.P., Hare, B. & Tomasello, M. (2009). Chimpanzees coordinate in a negotiation game. — Evol. Hum. Behav. 30: 381-392.

Melis, A.P., Warneken, F. & Hare, B. (2010). Collaboration and helping in chimpanzees. — In: The mind of the chimpanzee (Lonsdorf, E.V., Ross, S. & Matsuzawa, T., eds). The University of Chicago Press, Chicago, IL, p. 265-281.

Melis, A.P., Warneken, F., Jensen, K., Schneider, A.C., Call, J. & Tomasello, M. (2011). Chimpanzees help conspecifics obtain food and non-food items. — Proc. Roy. Soc. Lond. B: Biol. Sci. 278: 1405-1413.

Palagi, E. & Norscia, I. (2013). Bonobos protect and console friends and kin. — PLoS One 8: e79290.

Pfalzer, S. & Ehret, G. (1995). Social integration of a bonobo mother and her dependent daughter into an unfamiliar group. — Primates 36: 349-360.

Silk, J.B. & House, B.R. (2011). Evolutionary foundations of human prosocial sentiments. — Proc. Natl. Acad. Sci. USA 108: 10910-10917.

Silk, J.B. & House, B.R. (2012). The phylogeny and ontogeny of prosocial behavior. — In: The Oxford handbook of comparative evolutionary psychology (Vonk, J. & Shackelford, T.K., eds). Oxford University Press, Oxford, p. 381-398.

Silk, J.B., Brosnan, S.F., Vonk, J., Henrich, J., Povinelli, D.J., Richardson, A.S. & Shapiro, S.J. (2005). Chimpanzees are indifferent to the welfare of unrelated group members. — Nature 437: 1357-1359.

Skerry, A.E., Sheskin, M. & Santos, L.R. (2011). Capuchin monkeys are not prosocial in an instrumental helping task. — Anim. Cogn. 14: 647-654.

Stevens, J.R. (2010). Donor payoffs and other-regarding preferences in cotton-top tamarins (*Saguinus oedipus*). — Anim. Cogn. 13: 663-670.

Suchak, M. & de Waal, F.B.M. (2012). Monkeys benefit from reciprocity without the cognitive burden. — Proc. Natl. Acad. Sci. USA 109: 15191-15196.

Takimoto, A., Kuroshima, H. & Fujita, K. (2010). Capuchin monkeys (*Cebus apella*) are sensitive to others' reward: an experimental analysis of food-choice for conspecifics. — Anim. Cogn. 13: 249-261.

Tan, J. & Hare, B. (2013). Bonobos share with strangers. — PLoS One 8: e51922.

Tan, J. & Hare, B. (2014). Bonobos are xenophilic. — Am. J. Phys. Anthropol. 153: 251.

Vonk, J., Brosnan, S.F., Silk, J.B., Henrich, J., Richardson, A.S., Lambeth, S.P., Schapiro, S.J. & Povinelli, D.J. (2008). Chimpanzees do not take advantage of very low cost opportunities to deliver food to unrelated group members. — Anim. Behav. 75: 1757-1770.

Warneken, F. & Tomasello, M. (2009). Varieties of altruism in children and chimpanzees. — Trends Cogn. Sci. 13: 397-402.

Warneken, F., Hare, B., Melis, A.P., Hanus, D. & Tomasello, M. (2007). Spontaneous altruism by chimpanzees and young children. — PLoS Biol. 5: e184.

Wobber, V. & Hare, B. (2009). Testing the social dog hypothesis: are dogs also more skilled than chimpanzees in non-communicative social tasks? — Behav. Proc. 81: 423-428.

Wobber, V. & Hare, B. (2011). Psychological health of orphan bonobos and chimpanzees in African sanctuaries. — PLoS One 6: e17147.

Wobber, V., Wrangham, R.W. & Hare, B. (2010a). Bonobos exhibit delayed development of social behavior and cognition relative to chimpanzees. — Curr. Biol. 20: 226-230.

Wobber, V., Hare, B., Maboto, J., Lipson, S., Wrangham, R.W. & Ellison, P.T. (2010b). Differential changes in steroid hormones before competition in bonobos and chimpanzees. — Proc. Natl. Acad. Sci. USA 107: 12457-12462.

Wobber, V., Herrmann, E., Hare, B., Wrangham, R.W. & Tomasello, M. (2014). Differences in the early cognitive development of children and great apes. — Dev. Psychobiol. 56: 547-573.

Yamamoto, S. & Tanaka, M. (2010). The influence of kin relationship and reciprocal context on chimpanzees' other-regarding preferences. — Anim. Behav. 79: 595-602.

Yamamoto, S., Humle, T. & Tanaka, M. (2012). Chimpanzees' flexible targeted helping based on an understanding of conspecifics' goals. — Proc. Natl. Acad. Sci. USA 109: 3588-3592.

[When citing this chapter, refer to Behaviour 152 (2015) 545–562]

Experimental evidence that grooming and play are social currency in bonobos and chimpanzees

Kara Schroepfer-Walker [a,*], **Victoria Wobber** [b] **and Brian Hare** [a,c]

[a] Department of Evolutionary Anthropology, Duke University, Durham, NC, USA
[b] Department of Psychology, Harvard University, Cambridge, MA, USA
[c] Center for Cognitive Neuroscience, Duke University, Durham, NC, USA
[*]Corresponding author's e-mail address: ks163@duke.edu

Accepted 4 January 2015; published online 29 January 2015

Abstract

While natural observations show apes use grooming and play as social currency, no experimental manipulations have been carried out to measure the effects of these behaviours on relationship formation in apes. While previous experiments have demonstrated apes quickly learn the identity of individuals who will provide food in a variety of cooperative and non-cooperative situations, no experiment has ever examined how grooming and play might shape the preferences of apes for different individuals. We gave a group bonobos ($N = 25$) and chimpanzees ($N = 30$) a choice between an unfamiliar human who had recently groomed or played with them and one who had not. Both species showed a preference for the unfamiliar human that had interacted with them over the one who did not. The effect was largely driven by the males of both species while interacting with females showed little effect on their preferences for unfamiliar humans. Subjects showed this preference even though they only had social interactions with one of the unfamiliar humans for a few minutes before each trial and their choices were not rewarded with food differentially. Our results support the long held idea that grooming and play act as a form of social currency in great apes (and likely many other species) that can rapidly shape social relationships, particularly between unfamiliar individuals.

Keywords

bonobo, chimpanzee, groom, play, social bonds, social currency.

1. Introduction

Apes, like all primates, rely on social relationships to survive and reproduce. Evidence from several anthropoid species, including chimpanzees, shows that strong individual relationships, both with kin and non-kin, provide adaptive benefits and in many cases are correlated with reproductive success (Schülke & Ostner, 2008; Silk et al., 2009, 2010; Gilby et al., 2013). Apes can manipulate social relationships, including through the use of coalitions and alliances to increase rank within the dominance hierarchy (e.g., de Waal, 1982; Goodall, 1986; Surbeck et al., 2011). Social manipulation used to increase rank presumably benefits the individual as rank correlates with measures of fitness (Kano, 1996; Pusey et al., 1997; Gerloff et al., 1999; Boesch et al., 2006; Wroblewski et al., 2009).

Given the importance of social relationships for apes, observational work has been conducted to understand how these relationships are established and maintained. A number of studies in captivity and the wild have shown how bonobos and chimpanzees use grooming to form and maintain social bonds (de Waal, 1982; Kano, 1992). There is evidence for reciprocal grooming in both species (Watts, 2002; Stevens et al., 2006). In chimpanzees there is evidence that male chimpanzees groom their alliance and hunting partners more frequently than non-alliance and hunting partners (Nishida & Hosaka, 1996; Watts, 2000, 2002; Mitani & Watts, 2001). Though bonds between chimpanzees are strongest between male dyads (Gilby & Wrangham, 2008; Mitani, 2009), strong bonds can also form in male–female and female–female dyads (Gilby & Wrangham, 2008; Langergraber et al., 2009; Lehmann & Boesch, 2009). Meanwhile bonobo mothers and sons seem to preferentially travel together and groom most frequently in the wild (Kano, 1992; Furuichi, 1997; Hohmann et al., 1999; Surbeck et al., 2011). Female bonobos form coalitions to compete against males but strong bonds between males have not been observed (Parish, 1996; Hohmann et al., 1999). Play has also been observed to be another way apes can form and maintain bonds and is important even in adulthood (Goodall, 1986; Palagi et al., 2004; Palagi & Paoli, 2007; Nishida, 2012). Both species tend to play most frequently with kin and allies (Goodall, 1986; Palagi et al., 2004; Nishida, 2012).

Experiments have also been conducted to examine the cognitive abilities in apes that might play a role in relationship formation and maintenance (Tan & Hare, 2013). Bonobos and chimpanzees are both skilled at solving instrumental tasks through cooperation with conspecifics (Melis et al., 2006; Hare et al., 2007). Chimpanzees recruit help when cooperation is necessary and can quickly determine which of two partners is most skillful (Melis et al., 2006). Chimpanzees are able to maintain cooperation when they encounter a conflict of interest through non-verbal negotiation although skills at reciprocity are inconsistent (Melis et al., 2008, 2009; Brosnan et al., 2009). Bonobos retain a more juvenile level of individual tolerance into adulthood that facilitates cooperation in obtaining food, something not seen in chimpanzees (Hare et al., 2007). However, this leads to difficulty in inhibiting previously learned social associations (Wobber et al., 2010a, b). Bonobos also prefer to share food with strange bonobos over groupmates — a preference that may facilitate the extension of social networks in a way not observed in other apes (Hare & Kwetuenda, 2010; Tan & Hare, 2013). A number of experiments have also studied the social preferences of apes using human experimenters. Bonobos and chimpanzees show a preference for a human that was trying to share food with another human over one who was trying to steal the food during a triadic interaction (Russell et al., 2008; Subiaul et al., 2008; Herrmann et al., 2013a). Chimpanzees also demonstrated an almost immediate reversal of preference from a previously stingy to currently generous human in a reversal learning paradigm in which it took them dozens of trials to demonstrate a reversal when a non-social cue like colour was used (Wobber & Hare, 2009). Taken together experimental studies support the idea of bonobos and chimpanzees as flexible cooperators that monitor social relationships closely and rapidly change their preferences — even potentially in interactions with humans.

While observational studies have demonstrated the potential role of grooming and play in social relationships and experiments have shown how flexible bonobos and chimpanzees are in forming and maintaining cooperative relationships, no study has ever experimentally examined grooming and play as a currency in establishing social relationships in apes, where the amount of grooming or other affiliative behavior received leads to a change in preference toward one social partner over the other. While a strong role of grooming and play has long been suspected based on observational work, all previous experimental studies require apes to show social preferences or solve social problems for food rewards (although see Maclean & Hare, 2013). The Social Currency hypothesis suggests that both grooming and play are valued in social interactions and can be used to establish or shift social preferences depending on the amount of play or grooming that occurs between individuals. The central prediction being that an individual can improve their social relationship with another group member by grooming or playing with them. Therefore, an experimental manipulation of grooming and play should show a shift in preference toward those individuals who play and groom subjects the most. If confirmed a secondary question then becomes uncovering the mechanism that might drive such exchanges (e.g., biological markets, Noë & Hammerstein, 1995).

Given the rapid shift in preference for cooperative partners observed in some experiments (Melis et al., 2006; Wobber & Hare, 2009; Herrmann et al., 2013b), it may be apes alter their preferences extremely rapidly based on a relatively short social exchange. In this experiment we will test the social currency hypothesis by manipulating which of two social partners each subject interacts with before choosing which partner they prefer to receive food from. If grooming and play act as social currency, subjects should shift their preference toward an individual that recently played or groomed them. In testing the social currency hypothesis it is also important to examine both *Pan* species since they have very different response to strangers related to establishing new relationships (Tan & Hare, 2013), they handle social stress related to relationship maintenance differently (Wobber et al., 2010a, b), they exhibit temperament differences (Herrmann et al., 2011) and they differ in social cognitive abilities (Herrmann et al., 2010; Wobber et al., 2010a, b). In this context, we predicted bonobos would show greater short-term shifts due to their increased tolerance in general, and especially towards strangers. To control for differences in the two species general preference for interacting with conspecific strangers (i.e., chimpanzees are xenophobic while bonobos are xenophilic), human experimenters were used. When both species have been tested in the same context, they show the same strong preferences to interact with humans over playing alone and neither species has a xenophobic response to humans (Herrmann et al., 2011; Maclean & Hare, 2013).

2. Methods

30 chimpanzees and 24 bonobos housed at the Tchimpounga Centre for Chimpanzee Rehabilitation (Pointe Noire, Republic of Congo) and Lola ya Bonobo (Kinshasa, Democratic Republic of Congo) participated in this experiment. Most subjects arrived at the sanctuary as orphans and have been raised in mixed sex social groups with access to large outdoor forested enclosures where they engage in species-typical behavior including grooming and playing (for details see Wobber & Hare, 2011). Subjects interact with human caretakers on a daily basis when they return to their night dormitories at sunset. Chimpanzees ranged in age from 8–23 years and bonobos from

6–23 years. All had limited exposure to the human experimenters prior to the experimental conditions. A between-subjects design was utilized with 15 chimpanzees and 13 bonobos completing the groom condition and 15 chimpanzees and 12 bonobos completing the play condition (Table 1). Six subjects that were unable to complete the baseline session were dropped from all analyses.

Table 1.
List of subjects.

Chimpanzee				Bonobo			
Subject	Sex	Age	Condition	Subject	Sex	Age	Condition
Elykia	M	22	P	Kikwit	M	14	P
Jo	M	22	P	Lomami	M	12	P
Jay	M	21	P	Illebo	M	10	P
Yoko	M	14	P	Bandaka	M	10	P
Tabonga	M	12	P	Eleke	M	8	P
Chimpie	M	12	P	Yolo	M	8	P
Tiki	M	10	P	Kisantu	F	14	P
Kimenga	M	7	P	Bandundu	F	14	P
Pembele	F	18	P	Likasi	F	10	P
Low-Low	F	18	P	Muanda	F	8	P
Diba	F	15	P	Kinshasa	F	6	P
Fanitouek	F	11	P	Sake	F	6	P
Vitika	F	10	P	Makali	M	25	G
Lounama	F	10	P	Api	M	11	G
Marcelle	F	8	P	Boende	M	11	G
Tomy	M	21	G	Bili	M	10	G
Jacob	M	19	G	Maniema	M	9	G
Tamishi	M	18	G	Kasongo	M	9	G
Tchibanga	M	13	G	Chibombo	M	6	G
Wolo	M	13	G	Isiro	F	14	G
Kefan	M	11	G	Kalena	F	13	G
Petit Prince	M	10	G	Salonga	F	13	G
Lufumbu	M	9	G	Katako	F	7	G
Mayebo	F	22	G	Lukuru	F	6	G
Ramses	F	15	G	Masisi	F	6	G
Ouband	F	11	G				
Oumine	F	11	G				
Ulemvuka	F	10	G				
Makou	F	9	G				
Mvouti	F	8	G				

M, male; F, female; P, play; G, groom.

2.1. Test procedure

The test consisted of two phases and was conducted over 2–3 test sessions. All subjects began with a baseline preference test on day one followed by a test session that was divided across two more days. For a few subjects, the first test session occurred 30 min after the baseline session, rather than the following day, due to management constraints. Care was taken to assure an equal representation of species, sex and condition across the different testing schedules. Three human experimenters (E1, E2 and E3) took part in this experiment. E1 and E2 took part in all preference tests and following the baseline preference test, one was designated as the actor. E3 centred the subject but otherwise did not interact with them in any way. E1 and E2 were unfamiliar individuals though the subjects did have limited experience with them in different capacities. It was not possible to control for gender and race between E1 and E2 (see discussion) due to experimenter availability in the two sanctuaries. Five human experimenters served as E1 and E2 throughout the experiment. KSW was an experimenter for each subject and was known to the apes through minimal exposure during three weeks of study at the sanctuaries a year prior to the current experiment. Upon arrival to complete the current set of experiments, KSW did not interact with the apes prior to testing. At Tchimpounga two caretakers also served as E1 or E2. The primary experimenter was a female caretaker who worked exclusively with the juvenile group, located in a geographically separate area from the sub-adult and adult animals. This caretaker had very limited exposure to the sub-adult and adult individuals. Four older juveniles were tested at Tchimpounga and for these subjects and four additional adults a male caretaker who had a limited role with each group performed the role of second experimenter. At Lola ya Bonobo the second male experimenter was held constant and was only known to the apes through a three-week observational study he had conducted a month prior to the experiment. In this capacity he did not ever have physical contact with the animals and could only observe them from a distance of 10 or more meters. It is important to note that, overall, subjects did not have an a priori preference for one experimenter over the other. KSW served as the actor in 16/30 and 12/24 instances for the chimpanzees and bonobos, respectively.

2.1.1. Baseline

To assess any pre-existing preferences between E1 and E2 apes were given a baseline preference test. Subjects were brought one at a time into a testing room in their night dormitory and were allowed to acclimate to the space. Preference test trials began when E1 and E2 simultaneously gave the subject a slice of banana through the mesh that allowed human–ape interaction, in the centre of the room, before stepping away from the mesh. E3 then re-centered the ape using a banana slice. E1 and E2 stepped forward to the mesh and kneeled, 2 m apart, each holding half a banana in their outstretched hand as E3 stepped away from the testing room. Subjects were allowed to choose to beg from either E1 or E2. Regardless of choice, the subject never received the half banana. This procedure was then repeated for a total of eight trials.

2.1.2. Test session

The test session unfolded identically to the baseline preference with the addition of an interaction period during each trial. The actor became the experimenter (E1 or E2) that was least preferred in the baseline session. If the ape showed no preference then the actor was chosen randomly. Trials again began with E1 and E2 giving the subject a banana slice in the centre of the room. The non-actor then remained within arm's reach of the mesh and within a meter of the actor, facing the subject, throughout the interaction period to control for effects of proximity. Crucially, while the non-actor maintained proximity, s/he did not interact with the subject. Simultaneously, the actor began a 3-min interaction period (see below) which varied by condition. Following the interaction period, E3 centred the subject with a small piece of banana and E1 and E2 positioned themselves on either side of the room, counterbalanced by trial, holding half a banana. The subject was allowed to make a choice but was not provided with the banana. Each subject completed 8 test trials.

2.1.3. Interaction period

2.1.3.1. Groom condition. The actor sat in front of the mesh and engaged in grooming with the subject by sifting through the hair on the subject's body parts that were within reach of the mesh while making the grooming lipsmack vocalizations. Grooming was not reciprocal and if the subject attempted to groom the actor then the actor shifted positions to widen the space between subject and experimenter. If the subject left the mesh the actor made verbal attempts to call the subject back. Time away from the mesh was coded and included in analysis.

2.1.3.2. Play condition. The actor engaged the subject in high energy play that varied depending on individual preferences. Play could involve ticking, chase and poking and generally included all three. The actor alternated between a cheerful voice and their best attempt to mimic ape laughter vocalizations. If the subject left the mesh the actor made verbal attempts to call the subject back. Time away from the mesh was included in the analysis.

2.2. Coding and analysis

In both baseline and test trials choices were coded live by KSW and 30% of trials were later confirmed through reliability coding using an observer blind to the conditions and hypotheses and Cohen's kappa was 0.958. In the baseline and test trials choice was coded when the subject's fingers crossed the mesh in front of the experimenter. To control for motivation, time spent engaged was coded for each subject. Subjects were free to terminate an interaction with the actor by moving away from the mesh that allows human–ape interaction. Participants were considered engaged if they remained within arm's reach of the mesh barrier. For analysis, we used Poisson regression because the data consisted of counts (number of times the subject picked the actor). Baseline and test observations form repeated measures on the same individual, which are correlated. Generalized estimating equations (GEEs) were used to account for the dependent structure of the data. Inference focused on the treatment variable which had levels baseline and test condition (groom or play). Species, sex of the subject and sex of the human experimenters were included in the model as main effects as well as an interaction terms, condition by species and condition by sex of the subject. These analyses used the geeglm package (Yan, 2002; Yan & Fine, 2004; Højsgaard et al., 2006) in the R environment for statistical computing version 3.1.0 (R Core Team, 2014). An additional model, focusing on males only, was created to test for effects of period (first four vs. last four). Differences in motivation were compared using an independent samples t-test performed in JMP (JMP Pro 10, SAS Institute, Cary, NC, USA). Finally to see if motivation affected choice we ran an ordinary least squares regression on time spent unengaged and the change in preference between the baseline and test conditions, also in JMP.

3. Results

Bonobos increased their preference for the actor in the play condition from 35% in the baseline to 52% in the test session (Figure 1) and in the groom condition from 39% in the baseline condition to 57% in the test condition (Figure 2). Chimpanzees increased their preference for the actor in the play condition from 39% in the baseline to 52% (Figure 1) in the test trials and increased their preference for the actor in the groom condition from 32% in the baseline condition to 47% in the test condition (Figure 2). Species, condition and sex were included in the GEE model as well as interaction effects of species by condition and sex by condition (Table 2). There was no effect of species or experimenter sex but both condition and sex contributed to the model. The effect of sex was entirely driven by males (Figure 1). Males had a coefficient of 0.53 (SE $= 0.53$, $p < 0.001$) in the groom condition and 0.62 (SE $= 0.08$, $p < 0.001$) in the play condition. Females had a coefficient of 0.11 (SE $= 0.09$, $p = 0.25$) in the groom condition and 0.01 (SE $= 0.17$, $p = 0.96$) in the play condition. A second model (Table 3) that only included males looked at period effect (first four vs. last four). In this model, both groom and play were significant in both periods but the effect size diminished in the last four of the groom condition, indicating subjects were shifting their preference away from the actor in the final half of the session.

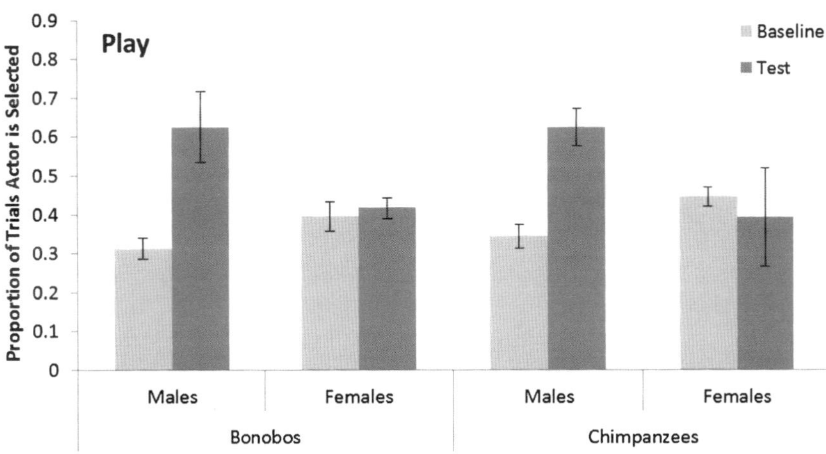

Figure 1. The y-axis represents the proportion of trials the experimenter who groomed the subject or 'actor' was chosen over an experimenter who did not in the baseline (light grey) and test conditions (dark grey) for both species separated by sex in the play condition. Error bars represent standard error.

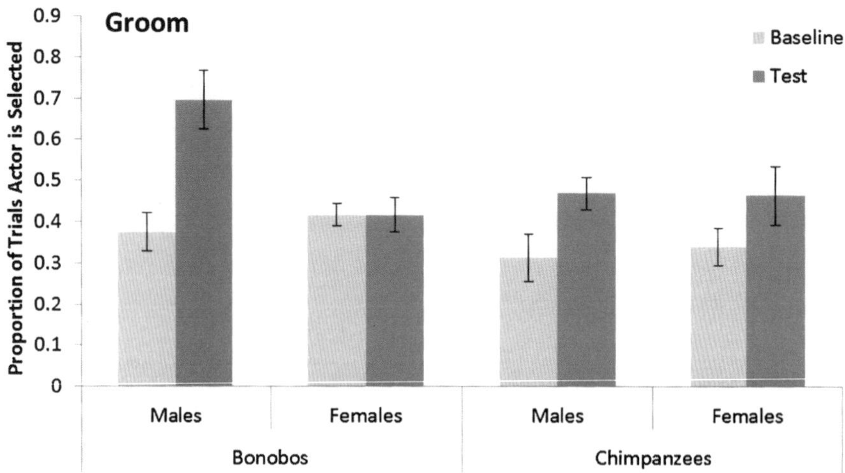

Figure 2. The y-axis represents the proportion of trials the experimenter who groomed the subject or 'actor' was chosen over an experimenter who did not in the baseline (light grey) and test conditions (dark grey) for both species separated by sex in the groom condition. Error bars represent standard error.

Chimpanzees were equally engaged in both conditions ($t = -0.85$, $p = 0.80$), spending 94 and 91% of the time engaged in the play and groom conditions, respectively (i.e., remaining in proximity of the human experimenter). Bonobos were more engaged in the play condition ($t = -2.47$, $p = 0.026$), spending 92% of their time engaged compared to 75% in the groom condition. The level of engagement did not affect the change in pref-

Table 2.
Results of the GEE model.

	Coefficient	p
Species		0.26
Species × Condition		0.05
Experimenter sex		0.20
Males		
Groom	0.53	<0.0001
Play	0.63	<0.0001
Females		
Groom	0.11	0.25
Play	0.01	0.96

Table 3.
Results of the supplementary GEE model, restricted to males but including periods (first four/last four trials).

	Coefficient	p
Period 1		
Groom	0.73	<0.001
Play	0.55	<0.001
Period 2		
Groom	0.31	0.01
Play	0.68	<0.001

erence between the baseline and test conditions (play: $R^2 = 0.05$, $p = 0.29$; groom: $R^2 = 0.06$, $p = 0.20$).

4. Discussion

Chimpanzees and bonobos rapidly changed their preference for a novel experimenter following an affiliative interaction, providing support for the Social Currency hypothesis. Although play and grooming behaviour likely arouse different emotional states in an individual (Rosati & Hare, 2012), apes used both grooming and play interactions equally as a currency to establish social relationships with human experimenters. Bonobos and chimpanzees did not differ in their response to human affiliative behaviors as social currency, despite differences in temperament (Hermann et al., 2011), frequency of cooperative behaviours (Kano, 1992; Muller & Mitani, 2005) and social cognitive abilities (Hermann et al., 2010; Wobber et al., 2010a, b).

The change in preference was strong in both species in both conditions but was primarily driven by males. Female chimpanzees and bonobos did not use play or grooming as a social currency in this particular paradigm. We did not initially predict such a dichotomy, and in fact, would have predicted the sex difference to differ between species due to differential bonding patterns between the sexes (Hohmann et al., 1999; Gilby & Wrangham, 2008). The observed pattern may be related to the shared socioecology between species where females disperse at adolescence (Pusey, 1980; Kano, 1992). In such a system males remain in their natal community for life and gain benefits from forming long-term relationships with kin and non-kin (Hohmann et al., 1999; Mitani, 2009; Surbeck et al., 2011). As males mature, the ability to use affiliative behaviour as a social currency to quickly ascertain the reputation of a partner should hasten the climb up the social ladder given the importance of allies in attaining rank (Nishida & Hosaka, 1996; Surbeck et al., 2011). Males are also likely to benefit from rapidly establishing a rapport with sexually receptive females (Idani, 1991; Kahlenberg et al., 2008). Dispersing females face different social challenges. Increasing evidence shows that females of both species do form differential bonds with both sexes and these relationships are likely to be adaptive (Gilby & Wrangham, 2008; Langergraber et al., 2009; Lehmann & Boesch, 2009). Moreover, upon immigration females must establish relationships with unfamiliar individuals and should benefit from being able to form reputations based on social currency. Therefore, it is unlikely that females do not use affiliative behaviour as a currency in forming new social interactions. Rather, it may be that females require more time to assess the value of a relationship and our short experimental interactions here are insufficient to see a change in preference. These results are more difficult to interpret in female bonobos. Unlike female chimpanzees who are hostile towards immigrants and less social than males, bonds between bonobo females are generally strong but not long-lasting (Parish, 1996; Hohmann et al., 1999), immigrants seek out specific high ranking females when entering a new community (Idani, 1991) and in experiments both male and female bonobos will pay a cost to have a social interaction with a stranger (Tan & Hare, 2013). These lines of evidence suggest they would show sensitivity to short-term affiliative interactions. Future research comparing these species using conspecific partners or longer periods of interaction may still reveal the expected species difference.

Subjects' rapid shift in preference for a human that either played or groomed with them can best be attributed to the social value of the experimenter's affiliative behaviour. Subjects were not differentially rewarded with food for their choices, making it difficult to explain a shift in preferences during the experiment based on anything but the interaction. Both species interacted with the experimenter similarly, choosing to maintain proximity throughout the interaction period and made choices on every trial. Both experimenters maintained proximity to the subject's room throughout each interaction period. Therefore the shift is not due to a lack of opportunity to interact equally with each experimenter. Although we were not able to hold constant the gender of the experimenters, this did not affect the outcome, primarily because the identity of the actor interacting socially with the subject was determined by selecting the least preferred experimenter for each subject in a baseline. While we observe apes quickly shift preferences based on affiliative interactions, this experiment does not speak to the precise cognitive mechanism involved. Future research will be needed to differentiate between mechanisms such as calculated reciprocity or physiological bonding, increased familiarity or trust through physical contact.

Regardless of mechanism the ability to quickly establish relationships using affiliative behaviors as social currency can have lifelong reproductive consequences for bonobos and chimpanzees. This is especially true for immigrating females who must establish a foraging territory and navigate a new social environment before reproducing (Idani, 1991; Pusey & Schroepfer-Walker, 2013). Females who can integrate more rapidly into their transfer community are expected to have an advantage in early reproduction. Males should also benefit from prioritizing the establishment of certain relationships (Mitani, 2009). As new males mature and move up the hierarchy they can become essential allies for older males and established males should then compete for their attention. They should also be attuned to establishing relationships with females as affiliative relationships may be important for reproductive success. Moreover, cooperation among bonobos and chimpanzees is often predicated on a previously established relationship. The ability to use both grooming and play as a social currency to guide partner choice should lead to stable and strong relationships among individuals.

Here we have established that chimpanzees and bonobos can use play and grooming as a social currency to form relationships. However, with further exploration, we expect to find differences in the speed, the developmental trajectory, the importance of different affiliative behaviours and the point at which social currency is overcome by the accumulation of social debt. We predict that the ability to use affiliative behaviour to form relationships and assess reputation is likely to develop over the juvenile period and should be especially important during the adolescent period when females transfer to their adult communities and males must enter the established hierarchy. Though we did not assess developmental changes, informal observation suggests juveniles and adolescents (10 years and younger) of both species showed larger shifts of preference than adults in the groom condition. Reciprocity is thought to be important for maintaining cooperative interactions in chimpanzees and bonobos. However, despite observational studies that note apes can exchange various commodities, including grooming, meat and coalitionary support (e.g., de Waal, 1997; Koyama et al., 2006; Mitani, 2006; Gomes & Boesch, 2009), experimental evidence is mixed and shows only weak support for contingent reciprocity (Melis et al., 2008; Brosnan et al., 2009). Melis et al. (2008) argue that short-term exchanges may be an inappropriate medium for assessing reciprocity in species that cooperate over long time periods. The current experiment was not designed to specifically address reciprocity but may be helpful in understanding the phenomenon because this experiment assesses interactions between strangers, rather than between individuals with a relationship history. Apes may be willing to pay a higher price in an initial encounter with a stranger to gain information about that individual, even sacrificing food to learn about a new competitor or ally through a social interaction (Hare & Kwetuenda, 2010; Tan & Hare, 2013). However, once the interaction begins, it may behove an individual to watch their accumulation of 'debt'. In this study we found limited evidence that chimpanzees may be attentive to the amount of grooming they receive with no reciprocation across a new interaction. Chimpanzees were less likely to beg from the actor in the latter half of the session (first 4 trials 55%, last 4 trials 38%). This phenomenon was not observed in the play condition (if anything, subjects continued to increase their preference across trials) and suggests grooming may be perceived differently from play in initial short-term interactions. Further work, addressing reciprocity among strangers, should be undertaken to clarify the extent to which apes are capable of contingent reciprocity.

Acknowledgements

We thank R. Atencia and D. Cox for permission to conduct experiments at the Tchimpounga Centre for Chimpanzee Rehabilitation and C. Andre for permission at Lola ya Bonobo. We thank D. Bilua and all caretakers at both sanctuaries for help in conducting the experiments, C. Mobley and A. Goulab for help with coding and S. Suchindran for help with statistics. Finally, we thank S. Yamamoto and two anonymous reviewers for their helpful comments. The studies had approval from the Institutional Animal Care and Use Committee of Duke University (protocol number A078-08-03) and strictly adhered to the legal requirements of the countries in which they were conducted. The chimpanzee research was carried out with permission from

Tchimpounga Chimpanzee Sanctuary in Pointe Noire, Republic of Congo and the Ministry of Scientific Research and Technological Innovation in Republic of Congo (permit: 009/MRS/DGRST/DMAST). The bonobo research was carried out with permission from Lola ya Bonobo Sanctuary in Kinshasa, Democratic Republic of Congo and the Ministry of Research and the Ministry of Environment in the Democratic Republic of Congo (permit: MIN.RS/SG/004/2009). This work was supported in part by National Science Foundation grants NSF-BCS-08-27552-02 and NSF-BCS-10-25172 to B.H.

References

Boesch, C., Kohou, G., Nene, H. & Vigilant, L. (2006). Male competition and paternity in wild chimpanzees of the Tai forest. — Am. J. Phys. Anthropol. 130: 103-115.

Brosnan, S.F., Silk, J.B., Henrich, J., Mareno, M.C., Lambeth, S.P. & Schapiro, S.J. (2009). Chimpanzees (*Pan troglodytes*) do not develop contingent reciprocity in an experimental task. — Anim. Cogn. 12: 587-597.

de Waal, F.B.M. (1982). Chimpanzee politics. — The Johns Hopkins University Press, Baltimore, MD.

de Waal, F.B.M. (1997). The chimpanzee's service economy: food for grooming. — Evol. Hum. Behav. 18: 375-386.

Furuichi, T. (1997). Agonistic interactions and matrifocal dominance rank of wild bonobos (*Pan paniscus*) at Wamba. — Int. J. Primatol. 18: 855-875.

Gerloff, U., Hartung, B., Hohmann, G. & Tautz, D. (1999). Intracommunity relationship, dispersal pattern and paternity success in a wild living community of bonobos (*Pan paniscus*) determined from DNA analysis of faecal samples. — Proc. Roy. Soc. Lond. B: Biol. Sci. 266: 1189-1195.

Gilby, I.C. & Wrangham, R.W. (2008). Association patterns among wild chimpanzees (*Pan troglodytes schweinfurthii*) reflect sex differences in cooperation. — Behav. Ecol. Sociobiol. 62: 1831-1842.

Gilby, I.C., Brent, L.J.N., Wroblewski, E.E., Rudicell, R.S., Hahn, B.H., Goodall, J. & Pusey, A.E. (2013). Fitness benefits of coalitionary aggression in male chimpanzees. — Behav. Ecol. Sociobiol. 67: 373-381.

Gomes, C.M. & Boesch, C. (2009). Wild chimpanzees exchange meat for sex on a long-term basis. — PLoS One 4: e5116.

Goodall, J. (1986). The chimpanzees of Gombe. — Belknap Press, Cambridge, MA.

Hare, B. & Kwetuenda, S. (2010). Bonobos voluntarily share their own food with others. — Curr. Biol. 20: R230-R231.

Hare, B., Melis, A.P., Woods, V., Hastings, S., Wrangham, R. & Carolina, N. (2007). Tolerance allows bonobos to outperform chimpanzees on a cooperative task. — Curr. Biol. 17: 619-623.

Herrmann, E., Hare, B., Call, J. & Tomasello, M. (2010). Differences in the cognitive skills of bonobos and chimpanzees. — PLoS One 5: e12438.

Herrmann, E., Hare, B., Cissewski, J. & Tomasello, M. (2011). A comparison of temperament in nonhuman apes and human infants. — Dev. Sci. 14: 1393-1405.

Herrmann, E., Keupp, S., Hare, B., Vaish, A. & Tomasello, M. (2013a). Direct and indirect reputation formation in nonhuman great apes and human children. — J. Comp. Psychol. 127: 32.

Herrmann, E., Keupp, S., Hare, B., Vaish, A. & Tomasello, M. (2013b). Direct and indirect reputation formation in nonhuman great apes (*Pan paniscus*, *Pan troglodytes*, *Gorilla gorilla*, *Pongo pygmaeus*) and human children (*Homo sapiens*). — J. Comp. Psychol. 127: 63-75.

Hohmann, G., Gerloff, U., Tautz, D. & Fruth, B. (1999). Social bonds and genetic ties: kinship, association and affiliation in a community of bonobos (*Pan paniscus*). — Behaviour 136: 1219-1235.

Højsgaard, S., Halekoh, U. & Yan, J. (2006). The R package geepack for generalized estimating equations. — J. Stat. Softw. 15: 1-11.

Idani, G. (1991). Social relationships between immigrant and resident bonobo (*Pan paniscus*) females at Wamba. — Folia Primatol. 57: 83-95.

Kahlenberg, S.M., Emery Thompson, M., Muller, M.N. & Wrangham, R.W. (2008). Immigration costs for female chimpanzees and male protection as an immigrant counterstrategy to intrasexual aggression. — Anim. Behav. 76: 1497-1509.

Kano, T. (1992). The last ape: pygmy chimpanzee behavior and ecology. — Stanford University Press, Stanford, CA.

Kano, T. (1996). Male rank order and copulation rate in a unit-group of bonobos at Wamba, Zaire. — In: Great ape societies (McGrew, T., Marchant, W. & Nishida, L., eds). Cambridge University Press, Cambridge, p. 135-143.

Koyama, N.F., Caws, C. & Aureli, F. (2006). Interchange of grooming and agonistic support in chimpanzees. — Int. J. Primatol. 27: 1293-1309.

Langergraber, K., Mitani, J. & Vigilant, L. (2009). Kinship and social bonds in female chimpanzees (*Pan troglodytes*). — Am. J. Primatol. 71: 840-851.

Lehmann, J. & Boesch, C. (2009). Sociality of the dispersing sex: the nature of social bonds in West African female chimpanzees, *Pan troglodytes*. — Anim. Behav. 77: 377-387.

Maclean, E. & Hare, B. (2013). Spontaneous triadic engagement in bonobos (*Pan paniscus*) and chimpanzees (*Pan troglodytes*). — J. Comp. Psychol. 127: 245-255.

Melis, A.P., Hare, B. & Tomasello, M. (2006). Chimpanzees recruit the best collaborators. — Science 311: 1297-1300.

Melis, A.P., Hare, B. & Tomasello, M. (2008). Do chimpanzees reciprocate received favours? — Anim. Behav. 76: 951-962.

Melis, A.P., Hare, B. & Tomasello, M. (2009). Chimpanzees coordinate in a negotiation game. — Evol. Hum. Behav. 30: 381-392.

Mitani, J.C. (2006). Reciprocal exchange in chimpanzees and other primates. — In: Cooperation in primates and humans (Kappeler, P. & van Schaik, C.P., eds). Springer, Berlin, p. 107-119.

Mitani, J.C. (2009). Male chimpanzees form enduring and equitable social bonds. — Anim. Behav. 77: 633-640.

Mitani, J.C. & Watts, D.P. (2001). Why do chimpanzees hunt and share meat? — Anim. Behav. 61: 915-924.

Muller, M.N. & Mitani, J.C. (2005). Conflict and cooperation in wild chimpanzees. — Adv. Stud. Behav. 35: 275-331.

Nishida, T. (2012). Chimpanzees of the lakeshore. — Cambridge University Press, Cambridge.

Nishida, T. & Hosaka, K. (1996). Coalition strategies among adult male chimpanzees of the Mahale Mountains, Tanzania. — In: Great ape societies (McGrew, T., Marchant, W. & Nishida, L., eds). Cambridge University Press, Cambridge, p. 114-134.

Noë, R. & Hammerstein, P. (1995). Biological markets. — Trends Ecol. Evol. 10: 336-339.

Palagi, E. & Paoli, T. (2007). Play in adult bonobos (*Pan paniscus*): modality and potential meaning. — Am. J. Phys. Anthropol. 134: 219-225.

Palagi, E., Cordoni, G. & Tarli, S.M.B. (2004). Immediate and delayed benefits of play behaviour: new evidence from chimpanzees (*Pan troglodytes*). — Ethology 962: 949-962.

Parish, A.R. (1996). Female relationships in bonobos (*Pan paniscus*): evidence for bonding, cooperation, and female dominance in a male-philopatric species. — Hum. Nat. 7: 61-96.

Pusey, A.E. (1980). Inbreeding avoidance in chimpanzees. — Anim. Behav. 28: 543-552.

Pusey, A.E. & Schroepfer-Walker, K. (2013). Female competition in chimpanzees. — Phil. Trans. Roy. Soc. Lond. 368: 20130077.

Pusey, A.E., Williams, J. & Goodall, J. (1997). The influence of dominance rank on the reproductive success of female chimpanzees. — Science 277: 828-831.

R Core Team (2014). R: a language and environment for statistical computing. — R Foundation for Statistical Computing, Vienna.

Rosati, A.G. & Hare, B. (2012). Decision making across social contexts: competition increases preferences for risk in chimpanzees and bonobos. — Anim. Behav. 84: 869-879.

Russell, Y.I., Call, J. & Dunbar, R.I.M. (2008). Image scoring in great apes. — Behav. Proc. 78: 108-111.

Schülke, O. & Ostner, J. (2008). Male reproductive skew, paternal relatedness, and female social relationships. — Am. J. Primatol. 70: 695-698.

Silk, J.B., Beehner, J.C., Bergman, T.J., Crockford, C., Engh, A.L., Moscovice, L.R. & Cheney, D.L. (2009). The benefits of social capital: close social bonds among female baboons enhance offspring survival. — Proc. Roy. Soc. Lond. B: Biol. Sci. 276: 3099-3104.

Silk, J.B., Beehner, J.C., Bergman, T.J., Crockford, C., Engh, A.L., Moscovice, L.R. & Cheney, D.L. (2010). Strong and consistent social bonds enhance the longevity of female baboons. — Curr. Biol. 20: 1359-1361.

Stevens, J.M.G., Vervaecke, H., De Vries, H. & Van Elsacker, L. (2006). Social structures in *Pan paniscus*: testing the female bonding hypothesis. — Primates 47: 210-217.

Subiaul, F., Vonk, J., Okamoto-Barth, S. & Barth, J. (2008). Do chimpanzees learn reputation by observation? Evidence from direct and indirect experience with generous and selfish strangers. — Anim. Cogn. 11: 611-623.

Surbeck, M., Mundry, R. & Hohmann, G. (2011). Mothers matter! Maternal support, dominance status and mating success in male bonobos (*Pan paniscus*). — Proc. Roy. Soc. Lond. B: Biol. Sci. 278: 590-598.

Tan, J. & Hare, B. (2013). Bonobos share with strangers. — PLoS One 8: e51922.

Watts, D.P. (2000). Grooming between male chimpanzees at Ngogo, Kibale National Park, Uganda. 1. Partner number and diversity and reciprocity. — Int. J. Primatol. 21: 189-210.

Watts, D.P. (2002). Reciprocity and interchange in the social relationships of wild male chimpanzees. — Behaviour 139: 343-370.

Wobber, V. & Hare, B. (2009). Testing the social dog hypothesis: are dogs also more skilled than chimpanzees in non-communicative social tasks? — Behav. Proc. 81: 423-428.

Wobber, V. & Hare, B. (2011). Psychological health of orphan bonobos and chimpanzees in African sanctuaries. — PLoS One 6: e17147.

Wobber, V., Hare, B., Maboto, J., Lipson, S., Wrangham, R. & Ellison, P.T. (2010). Differential changes in steroid hormones before competition in bonobos and chimpanzees. — Proc. Natl. Acad. Sci. USA 107: 12457-12462.

Wobber, V., Wrangham, R. & Hare, B. (2010). Bonobos exhibit delayed development of social behavior and cognition relative to chimpanzees. — Curr. Biol. 20: 226-230.

Wroblewski, E.E., Murray, C.M., Keele, B.F., Schumacher-Stankey, J.C., Hahn, B.H. & Pusey, A.E. (2009). Male dominance rank and reproductive success in chimpanzees, *Pan troglodytes schweinfurthii*. — Anim. Behav. 77: 873-885.

Yan, J. (2002). Geepack: yet another package for generalized estimating equations. — R News 2: 12-14.

Yan, J. & Fine, J.P. (2004). Estimating equations for association structures. — Stat. Med. 23: 859-880.

Index